Mathematical Modeling, Design, and Optimization of Complex Engineering Systems

The text provides a solid foundation in mathematical, modeling, and optimization techniques, ensuring that readers develop a deep understanding of the core concepts before delving into more complex topics. It discusses analytical techniques used to solve structural performance, potential flow analysis, fluid flow control in MEMS, a two-fluid nonlinear model, and unsteady blood flow.

Features

- Discusses fundamental equations and governing principles of heat transfer and fluid dynamics
- Covers topics such as analytical methods, numerical techniques, turbulence modeling, multiphase flow dynamics, circular economy, Industry 4.0, food security, different inventory models, and transportation problems
- Presents a detailed exploration of various computational methods, including finite difference, finite element, and spectral element methods
- Addresses the integrated approach needed for solving complex problems involving heat and fluid flow for a wide range of applications, including industrial, agricultural, and environmental systems
- Showcases how to use powerful computer tools and mathematical methods to analyze and improve real-world systems, such as building performance, materials property enhancement, wastewater treatment, genetically modified crops, and climate models

It is primarily written for senior undergraduates, graduate students, and academic researchers in manufacturing engineering, industrial engineering, engineering mathematics, mathematics, production engineering, industrial engineering, and environmental engineering.

Advances in Manufacturing, Design and Computational Intelligence Techniques

Series Editor-Ashwani Kumar—Professor and Head Mechanical Engineering, at Technical Education Department, Uttar Pradesh, Kanpur, India

The book series editor is inviting edited reference, and textbook proposals for submission to the book series. The main objective of this book series is to provide researchers a platform to present state-of-the-art innovations and research related to advanced materials applications, cutting-edge manufacturing techniques, innovative design, and computational intelligence methods used for solving nonlinear problems of engineering. The series includes a comprehensive range of topics and its application in engineering areas such as additive manufacturing, nanomanufacturing, micromachining, biodegradable composites, material synthesis and processing, energy materials, polymers and soft matter, nonlinear dynamics, dynamics of complex systems, MEMS, green and sustainable technologies, vibration control, AI in power station, analog-digital hybrid modulation, advancement in inverter technology, adaptive piezoelectric energy harvesting circuit, contactless energy transfer system, energy efficient motors, bioinformatics, computer aided inspection planning, hybrid electrical vehicle, autonomous vehicle, object identification, machine intelligence, deep learning, control-robotics-automation, knowledge based simulation, biomedical imaging, image processing, and visualization. This book series compiles all aspects of manufacturing, design, and computational intelligence techniques from fundamental principles to current advanced concepts.

Hybrid Metal Additive Manufacturing: Technology and Applications
Edited by Parnika Shrivastava, Anil Dhanola, and Kishor Kumar Gajrani

Thermal Energy Systems: Design, Computational Techniques, and Applications
Edited by Ashwani Kumar, Varun Pratap Singh, Chandan Swaroop Meena, Nitesh Dutt

Heat Exchanger Technologies for Sustainable Renewable Energy Systems
Mukesh Kumar Awasthi, Ashwani Kumar, Nitesh Dutt, Sivasakthivel Thangavel

Green Hydrogen Production: Storage, Transportation and Applications
Ashwani Kumar, Sivasakthivel Thangavel and Gaurav Dwivedi

Mathematical Modeling, Design, and Optimization of Complex Engineering Systems
Mukesh Kumar Awasthi, Ashwani Kumar, and Manoj Gupta

https://www.routledge.com/Advances-in-Manufacturing-Design-and-Computational-Intelligence-Techniques/book-series/CRCAIMDCIT?publishedFilter=alltitles&pd=published,forthcoming&pg=1&pp=12&so=pub&view=list?publishedFilter=alltitles&pd=published,forthcoming&pg=1&pp=12&so=pub&view=list

Mathematical Modeling, Design, and Optimization of Complex Engineering Systems

Edited by

Mukesh Kumar Awasthi

Ashwani Kumar

Manoj Gupta

CRC Press
Taylor & Francis Group
Boca Raton London New York

CRC Press is an imprint of the
Taylor & Francis Group, an **informa** business

First edition published 2026
by CRC Press
2385 NW Executive Center Drive, Suite 320, Boca Raton FL 33431

and by CRC Press
4 Park Square, Milton Park, Abingdon, Oxon, OX14 4RN

CRC Press is an imprint of Taylor & Francis Group, LLC

ISBN: 978-1-041-07767-1 (hbk)
ISBN: 978-1-041-07777-0 (pbk)
ISBN: 978-1-003-64215-2 (ebk)

DOI: 10.1201/9781003642152

Typeset in Sabon
by SPi Technologies India Pvt Ltd (Straive)

Contents

Aim and scope

The book "*Mathematical Modeling, Design, and Optimization of Complex Engineering Systems*" aims to provide an in-depth exploration of the principles, methodologies, and applications of mathematical modeling and optimization across three critical engineering disciplines: Mechanical Engineering, Civil Engineering, and Electrical Engineering. The book is designed to serve as a comprehensive resource for engineers, researchers, and students who are involved in the design and analysis of complex engineering systems, offering a structured approach to understanding and applying mathematical models to real-world engineering challenges.

This book begins with foundational concepts in mathematical modeling and optimization. The initial chapters introduce the essential tools and techniques required for modeling complex systems, along with advanced optimization algorithms. Chapter 1 provides a broad overview of the role of mathematical modeling in engineering, highlighting its importance across various fields. Chapter 2 explores into Future of Artificial Intelligence.

The book then focuses on application-specific chapters, starting with Chapter 3–4, which discusses computational modeling in structural engineering. Focusing on the design and optimization of column design and analysis of human hair composites. Chapter 5 explores Green synthesis of Gold Nanoparticles and their applications. The subsequent section of this work (Chapters 6 through 9) explores into the multifaceted world of fluid flow, exploring its diverse types, fundamental characteristics, and significant applications across various scales and contexts. Each chapter focuses on a specific aspect of fluid dynamics, employing a range of analytical, numerical, and review-based methodologies.

Chapter 6 investigates the phenomenon of Rayleigh-Taylor instability within a nanofluid layer using the principles of potential flow analysis. By examining the interplay of buoyancy and interfacial tension at the nanoscale, this study aims to provide insights into the stability characteristics of these advanced fluids, which have significant implications for heat transfer enhancement and other nanotechnology applications. Chapter 7 presents a comprehensive review of the critical role of materials integration in the development of pressure sensor mechanisms for precise fluid flow control in

Microelectromechanical Systems (MEMS) microfluidic devices. Chapter 8 focuses on the complex dynamics of pulsatile blood flow in the presence of arterial stenosis (narrowing). Employing a two-fluid nonlinear model, the study utilizes mathematical and computational techniques to analyze the intricate interactions between blood components and the arterial wall. Building upon the study of blood flow in stenosed arteries, Chapter 9 presents a numerical simulation of unsteady blood flow through an artery with multiple stenoses. This investigation utilizes computational fluid dynamics (CFD) to model the complex flow patterns, pressure variations, and shear stress distributions that arise in such intricate vascular geometries.

This section, comprising Chapters 10 through 13, highlights into contemporary strategies and technological advancements aimed at enhancing efficiency, sustainability, and automation within industrial and commercial sectors. This Chapter 10 investigates the potential for integrating the lean principles of Just-in-Time (JIT) manufacturing with the tenets of the Circular Economy. It explores how minimizing waste, optimizing resource utilization, and promoting closed-loop material flows – core concepts of the Circular Economy – can be effectively implemented within a JIT framework. This Chapter 11 focuses on the critical aspect of energy efficiency in commercial utilities through power factor enhancement. It details the development and application of mathematical models and simulations to analyze and optimize power factor. This Chapter 12 presents a comprehensive feasibility study examining the potential of deploying Industry 4.0 technologies in Material Handling Systems (MHS) specifically within the pipe manufacturing industry. It will likely assess the technical, economic, and operational viability of integrating advanced automation, data analytics, and connectivity into the material flow processes.

This section Chapter 13–15 provides a multifaceted exploration of contemporary scientific challenges, ranging from the practical application of separation technologies for environmental protection to the systemic analysis of a critical agricultural issue and the comparative evaluation of essential analytical methodologies. Chapter 13 focuses on environmental engineering and remediation. This chapter presents an in-depth investigation into the effectiveness of fixed-bed column systems for eliminating specific pollutants from simulated industrial or domestic wastewater. Chapter 14 shifts the focus to the intersection of biotechnology, agriculture, and global food systems. This chapter adopts a macro-level analytical approach, employing bibliometric techniques to systematically analyze the existing body of research on genetically modified (GM) crops and their impact on food security. Finally, Chapter 15, explore into the realm of analytical chemistry and environmental monitoring. This chapter likely presents a detailed comparative study of two distinct analytical techniques – Thin Layer Chromatography (TLC) and Atomic Absorption Spectroscopy (AAS) – for the identification and quantification of metallic toxicants. By examining the strengths and limitations of each method in terms of sensitivity, selectivity, cost-effectiveness, and ease of use, this chapter aims to provide valuable insights for researchers and practitioners involved in environmental analysis, food safety, and toxicology.

Spanning Chapters 16–23, this section offers a comprehensive exploration of key concepts ranging from logistical optimization to human behavior and technological innovation. It begins by examining quantitative methods for optimizing resource flow, specifically through inventory models (Chapter 16–17) and the Transportation Problem (Chapter 18–19). The focus then broadens to encompass understanding market dynamics via an analysis of buying behavior of consumers (Chapter 20). This is followed by an exploration of emerging technologies for sustainability, exemplified by the Piezometric Energy Harvester (Chapter 21). The concluding chapters pivot to the realm of modern organizational leadership, investigating the application of Designing Optimization Frameworks for ICT-Enabled e-Leadership Strategies and the Models for Analyzing the critical Impact of Leadership and Followership Values on Organizational Outcomes (Chapters 22 and 23). This interdisciplinary approach highlights the multifaceted nature of contemporary organizational challenges and opportunities.

This book "*Mathematical Modeling, Design, and Optimization of Complex Engineering Systems*" serves as a valuable reference for experienced engineers while also offering a comprehensive guide for students and researchers aiming to deepen their understanding of mathematical modeling and optimization in engineering. By seamlessly combining theoretical concepts with practical applications and case studies, the book bridges the gap between academia and industry, equipping readers with the tools and knowledge to tackle the complex challenges of modern engineering. Each chapter is thoughtfully designed to include a series of problems and exercises that reinforce the mathematical and computational concepts discussed, allowing readers to test their understanding and apply what they have learned. Solutions and detailed explanations are provided to facilitate self-study and independent learning. This integration of theoretical rigor with practical insights and simulation techniques makes the book an indispensable resource for anyone involved in the study or application of mathematical modeling and computational dynamics. It provides a comprehensive and up-to-date reference that supports the development of advanced skills and knowledge in these critical areas of applied mathematics, engineering, and science.

Editors
Mukesh Kumar Awasthi
Department of Mathematics, Babasaheb Bhimarao Ambedkar University, Lucknow, India

Ashwani Kumar
Department of Mechanical Engineering, Technical Education Department Uttar Pradesh, Kanpur, India

Manoj Gupta
Department of Mechanical Engineering, National University of Singapore, Singapore

Preface

In the rapidly evolving field of engineering, the ability to effectively model, design, and optimize complex systems is paramount. Engineers today face unprecedented challenges, from the development of sustainable energy solutions to the design of resilient infrastructure and the creation of advanced materials. To address these challenges, a deep understanding of mathematical modeling and optimization techniques is essential. This book, *Mathematical Modeling, Design, and Optimization of Complex Engineering Systems*, is crafted to serve as a comprehensive guide, providing insights and practical tools for engineers, researchers, and students engaged in the design and analysis of complex systems across three key engineering disciplines: Mechanical Engineering, Civil Engineering, and Electrical Engineering and Applied Sciences.

The book is structured to first establish a solid foundation in mathematical modeling and optimization. The opening chapters lay the groundwork by introducing the core principles and methodologies that underpin these disciplines. *Chapters 1 and 2* provide a foundational overview of mathematical modeling in engineering and explore the future of artificial intelligence. *Chapters 3 and 4* explore computational modeling in structural engineering, specifically focusing on column design and the analysis of human hair composites. *Chapter 5* investigates the green synthesis of gold nanoparticles and their applications. The subsequent section (*Chapters 6–9*) focuses on fluid dynamics. *Chapter 6* examines Rayleigh-Taylor instability in nanofluids using potential flow analysis. *Chapter 7* reviews materials integration in MEMS microfluidic pressure sensors for fluid flow control. *Chapters 8 and 9* explore pulsatile blood flow in arteries with single and multiple stenoses, respectively, employing nonlinear models and CFD. *Chapters 10–12* highlight industrial and commercial optimization strategies. *Chapter 10* investigates integrating lean JIT manufacturing with the Circular Economy. *Chapter 11* focuses on power factor enhancement for energy efficiency in commercial utilities. *Chapter 12* presents a feasibility study on Industry 4.0 technologies in material handling within pipe manufacturing.

Chapters 13–15 explore diverse scientific challenges. *Chapter 13* examines fixed-bed column systems for wastewater treatment. *Chapter 14* analyzes research on genetically modified crops and food security using bibliometrics. *Chapter 15* compares Thin Layer Chromatography and Atomic Absorption Spectroscopy for metallic toxicant analysis.

The final section (*Chapters 16–23*) covers logistical optimization, consumer behavior, and organizational leadership. *Chapters 16–19* address inventory models and the Transportation Problem. *Chapter 20* analyzes buying behavior. *Chapter 21* explores the Piezometric Energy Harvester. *Chapters 22 and 23* investigate designing optimization frameworks for ICT-Enabled e-Leadership Strategies and the impact of leadership and followership values on organizational outcomes.

This book is intended to bridge the gap between theory and practice, providing readers with both the theoretical tools and practical examples needed to tackle the complex challenges of modern engineering. Whether you are an engineer seeking to enhance your design and optimization skills, a researcher exploring new frontiers in mathematical modeling, or a student eager to learn about the latest advances in engineering, this book will serve as a valuable resource. The integration of case studies throughout the book ensures that the concepts and techniques discussed are grounded in real-world applications, making this book not only informative but also practical.

We hope that this book will inspire readers to explore new ideas, develop innovative solutions, and contribute to the advancement of engineering as a discipline. The challenges we face in the coming years are significant, but with the right tools and knowledge, we are confident that engineers will continue to lead the way in solving these challenges and building a better future.

Editors

Mukesh Kumar Awasthi
Department of Mathematics
Babasaheb Bhimarao Ambedkar University
Lucknow, India

Ashwani Kumar
Department of Mechanical Engineering
Technical Education Department
Uttar Pradesh, Kanpur, India

Manoj Gupta
Department of Mechanical Engineering
National University of Singapore
Singapore

About the editors

Mukesh Kumar Awasthi has done his Ph.D. on the topic "Viscous Correction for the Potential Flow Analysis of Capillary and Kelvin-Helmholtz instability." He is working as an Assistant Professor in the Department of Mathematics at Babasaheb Bhimrao Ambedkar University, Lucknow. Dr. Awasthi specializes in the mathematical modeling of flow problems. He has taught courses in Fluid Mechanics, Discrete Mathematics, Partial Differential Equations, Abstract Algebra, Mathematical Methods, and Measure Theory to postgraduate students. He has acquired excellent knowledge in the mathematical modeling of flow problems, and he can solve these problems analytically as well as numerically. He has a good grasp of subjects like viscous potential flow, electro-hydrodynamics, magneto-hydrodynamics, heat, and mass transfer. He has excellent communication skills and leadership qualities. He is self-motivated and responds to suggestions in a more convincing manner. He has qualified for the National Eligibility Test (NET) conducted at all-India level in the year 2008 by the Council of Scientific and Industrial Research (CSIR) and got Junior Research Fellowship (JRF) and Senior Research Fellowship (SRF) for doing research. He has published **125-plus** research publications (journal articles/books/book chapters/conference articles) in many national and international books, journals, and conferences. Also, he has published **ten** books. He has attended many symposia, workshops, and conferences in mathematics as well as fluid mechanics. He has won the "Research Awards" consecutively four times from 2013 to 2016 by the University of Petroleum and Energy Studies, Dehradun, India. He has also received the start-up research fund for his project "Nonlinear study of the interface in multilayer fluid system" from UGC, New Delhi. He is also listed in the **top 2% of influential researchers in the world, prepared** by Stanford University based on Scopus data in 2022. His ORCID ID is *0000-0002-6706-5226*, Google Scholar web link is ***https://scholar.google.co.in/citations?user=Dj3ktGAAAAAJ*** and Research Gate web link is ***https://www.researchgate.net/profile/Mukesh-Awasthi-2*.**

Ashwani Kumar holds a Ph.D. in Mechanical Engineering and has a diverse and impressive career spanning over 15 years. He currently holds the position of Professor and Head of Department Mechanical Engineering (Gazetted Officer Class 1) at the Technical Education Department, Uttar Pradesh (Government of Uttar Pradesh) in Kanpur, India. Prior to his current role, Dr. Kumar served as Senior Lecturer and Lecturer in the same organization and Assistant Professor in the Department of Mechanical Engineering at Graphic Era University, Dehradun (ranked 52nd by NIRF India) from July 2010 to November 2013. Beyond his academic contributions, he possesses extensive leadership and administrative experience. He has held numerous key positions within the education sector, including Coordinator for AICTE-Extension of Approval, Nodal Officer for PMKVY-TI Scheme (Government of India), Internal Coordinator—CDTP Scheme (Government of Uttar Pradesh), Industry Academia Relation Officer, Assistant Centre Superintendent (ACS)—Institute Examination Cell, Zonal Officer for Joint Entrance Examination (JEE-Diploma), and Sector Magistrate for State Assembly and General Elections. He is a passionate researcher and a prominent figure in the field of academic publishing. His dedication to knowledge dissemination is evident in his role as Series Editor for eight distinguished book series published by reputed publishers. These series encompass a wide range of topics, titled as Advances in Manufacturing, Design, and Computational Intelligence Techniques/Renewable and Sustainable Energy Developments/Smart Innovations and Technological Advancements in Mechanical and Materials Engineering/Solar Thermal Energy Systems: Advancements in Engineering, Ergonomics, and Sustainable Development/Artificial Intelligence and Machine Learning for Intelligent Engineering Systems/Computational Intelligence and Biomedical Engineering/Advances in Green Technologies for Sustainable Energy Solutions/Advances in Materials, Manufacturing, and Computational Intelligence Techniques. His editorial expertise extends beyond book series. He serves as Editor-in-Chief for the International Journal of Materials, Manufacturing, and Sustainable Technologies (IJMMST, *ISSN: 2583-6625*), and Editor of the International Journal of Energy Resources Applications (IJERA, *ISSN: 2583-6617*). He has also been a guest editor for a special issue of Buildings titled Sustainable Buildings, Resilient Cities, and Infrastructure Systems (*ISSN: 2075-5309*, I.F. 3.1) and holds editorial board positions on eight international journals and acts as review board member of 20 prestigious (Indexed in SCI/SCIE/Scopus) international journals with high impact factors. His dedication to scholarship is further demonstrated by his prolific authorship of over 200+ articles in prestigious journals, book chapters, and conference proceedings. Additionally, he has co-authored or edited over 40+ books in Mechanical Engineering, Materials Science, and Renewable Energy Engineering. His achievements have been recognized with prestigious awards, including a Best Teacher award for excellence in academics and research. He has also successfully guided 15 students through their BTech, MTech., and Ph.D. theses and currently serves as an external doctoral committee member at SRM University, New Delhi. He is actively involved in cutting-edge research, focusing on areas such as Artificial Intelligence and Machine Learning in Mechanical Engineering, Smart Materials and Manufacturing Techniques, Thermal Energy Storage, Building Efficiency, Renewable Energy Harvesting, Sustainable Transportation, and Heavy Vehicle Dynamics. He holds four patents and frequently participates in international conferences as an invited speaker, session chair, and member of advisory boards, review boards, and program committees. His ORCID ID is 0000-0003-4099-935X and Google Scholar web link is https://scholar.google.com/citations?hl=en&user=KOILpEkAAAAJ.

Manoj Gupta is a distinguished materials scientist and engineer with a remarkable track record of innovation and academic leadership. Formerly the Head of the Materials Division within the Mechanical Engineering Department and the Director designate of the Materials Science and Engineering Initiative at the National University of Singapore (NUS), his contributions have significantly advanced the field. Dr. Gupta earned his Ph.D. from the University of California, Irvine, United States, in 1992, followed by postdoctoral research at the University of Alberta, Canada, also in 1992. His exceptional research impact is consistently recognized globally. He is ranked among the top 0.6% of researchers worldwide according to Stanford's list, within the top 1% of scientists globally by The Universal Scientific Education and Research Network (achieving this recognition as early as August 2017), and within the top 2.5% of scientists as per ResearchGate metrics. His innovative spirit is evident in the development of two significant materials processing techniques. His extensive scholarly output includes over 600 peer-reviewed journal publications, demonstrating his significant contributions to scientific literature. He also holds two US patents and one Trade Secret, highlighting the practical and industrial relevance of his research. His current academic influence is underscored by an h-index of 90, an i10 index of 471, and over 30,000 citations. Furthermore, he has co-authored eight influential books published by leading academic publishers, further disseminating his expertise. Beyond his research and publications, he is actively engaged with the global scientific community, serving as Editor-in-Chief or Editor for 12 international peer-reviewed journals. A recipient of numerous awards, he fosters international collaborations through active visits and engagements as a visiting researcher, professor, and chair professor in countries, including Japan, France, Saudi Arabia, Qatar, China, the United States, and India. His Google Scholar web link is https://scholar.google.com.sg/citations?user=PGzgTLoAAAAJ&hl=en.

Contributors

Bhawna Agrawal
Department of Mathematics
Rabindranath Tagore University
Bhopal, India;
Department of Physical Science
Rabindranath Tagore University
Bhopal, India

K. Viswanath Allamraju
Department of Mechanical Engineering
Institute of Aeronautical Engineering
Hyderabad, (Telangana) India

Mukesh Kumar Awasthi
Department of Mathematics
Babasaheb Bhimarao Ambedkar University
Lucknow

Mehul Barai
Dr. D. Y. Patil Vidyapeeth (Deemed to be University)
Global Business School and Research Centre
Tathawade, Pune

Jitendra Bhaiswar
Department of Physics
Nagpur Institute of Technology
Nagpur, India

Abhishek Bhandari
Department of Mechanical Engineering
Rabindranath Tagore University
Bhopal, India

Bhaskar Chatterjee
Research Scholar
Rabindranath Tagore University
Bhopal, India

Mona P. Dange
Department of Mathematics
Rabindranath Tagore University
Bhopal, India

Prashant Dange
Rabindranath Tagore University
Bhopal, India

Sandip Das
Department of Mechanical Engineering
Rabindranath Tagore University
Bhopal, India

Nitin Dheeman
COER University Roorkee
Uttarakhand, India

Nitesh Dutt
Department of Mechanical Engineering
COER University
Roorkee, Uttarakhand, India

Ravindra Gandhe
Department of Mechanical Engineering
Government Polytechnic
Nagpur, India

Krishnakant Gautam
Department of Electrical and Electronics Engineering
Rabindranath Tagore University
Bhopal, India

Joeeta Mukherjee Ghosh
Mechanical Engineering
Dr. C.V. Raman University
Kargi road, Kota, Bilaspur (C.G)

Buddharatna Godboley
Department of Civil Engineering
St.John College of Engineering and Management
Palghar, India

Yatika Gori
Department of Mechanical Engineering
Graphic Era deemed to be University
Dehradun, India

Ankit Goyal
Technocrat Institute of Technology Science
Bhopal, India

Mayank Gupta
Department of Civil Engineering
Sagar Institute of Research Technology
Bhopal, India

Mohit Kumar Gupta
COER University Roorkee
Uttarakhand, India

Neeraj Kumar Gupta
Department of Electrical and Electronics Engineering
Rabindranath Tagore University
Bhopal, India

Jaicky H. Gurnani
School of Civil and Environmental Sciences
JSPM University Pune
Pune, India

Pallavi Jain
Department of Physical Science
Faculty of Science
Rabindranath Tagore University
Bhopal, India

Meghna Jumde
Department of Chemistry
Nagpur Institute of Technology
Nagpur, India

Meghna H. Jumde
Department of Chemistry
Nagpur Institute of Technology
Nagpur, India

Amit Kharwade
Department of Civil Engineering
Nagpur Institute of Technology
Nagpur, India

T. Ravi Kiran
Department of Mechanical Engineering
Rabindranath Tagore University
Bhopal, India

Hrishikesh Kokate
Dr. D. Y. Patil Vidyapeeth (Deemed to be University)
Global Business School and Research Centre
Tathawade, Pune

Ashwani Kumar
Professor and Head Mechanical Engineering
Technical Education Department
Uttar Pradesh, Kanpur, India
and
Department of Mechanical Engineering
Graphic Era deemed to be University
Dehradun, India

Sanjit Kumar
Department of Mathematics
Lakshmi Narain College of Technology & Science
Bhopal, India

D. Lingamaia
Department of Electronics & Communication Engineering
Bharat Institute of Engineering & Technology
Hyderabad, India

Aishwarya Maharshi
Dr. D. Y. Patil Vidyapeeth (Deemed to be University)
Global Business School and Research Centre
Tathawade, Pune

Utpal Majee
Research Schoolar
Rabindranath Tagore University
Bhopal, India

Akram Masood
COER University Roorkee
Uttarakhand, India

Bheemrao Mazumdar
Department of Electrical and Electronics Engineering
Rabindranath Tagore University
Bhopal, India

Balkrishna Namdeo
Faculty of Agriculture
Rabindranath Tagore University
Raisen, India

Ravendra Prasad Namdeo
Department of Mathematics
Technocrats Institute of Technology
Bhopal, India

Amar Nath
COER University Roorkee
Uttarakhand, India

Tejkaran Narolia
Department of Mechanical Engineering
Rabindranath Tagore University
Raisen, Bhopal Madhya Pradesh, India

Prateek Nigam
Department of Electrical and Electronics Engineering
Rabindranath Tagore University
Bhopal, India

Anjali Pathak
Department of Mechanical Engineering
Rabindranath Tagore University
Raisen, Bhopal, India
Department of Mechanical Engineering Govt. Polytechnic College
Pachore, Rajgarh, India

Neha Singh Raghuvanshi
Department of Applied Sciences, ABES Engineering College
Ghaziabad, Uttar Pradesh, India

Sanjay Rastogi
School of Commerce, Finance and Accountancy Christ (Deemed to be University)

Surendar Reddy
Department of Electronics & Communication Engineering
Bharat Institute of Engineering & Technology
Hyderabad, India

Amitabh Roy
Institute of Management Sciences
University of Lucknow
Lucknow

Md. Sadullah
Department of Electronics & Communication Engineering
Bharat Institute of Engineering & Technology
Hyderabad, India

Abid Salati
Dr. D. Y. Patil Vidyapeeth (Deemed to be University)
Global Business School and Research Centre
Tathawade, Pune

Ankur Saxena
Department of Electronics & Communication Engineering
Bharat Institute of Engineering & Technology
Hyderabad, India

Barkha Sharma
Department of Applied Science
Sagar Institute of Research and Technology
Bhopal, India

Gaurav Sharma
Department of Applied Science
Sagar Institute of Research and Technology
Bhopal, India

Gaurav Sharma
Department of Mathematics
Sagar Institute of Research and Technology
India

Neelu Sharma
Department of Engineering and Science
Sri Aurobindo Institute of Technology
Indore, India

Arti Shrivastava
Department of Mathematics
Technocrats Institute of Technology and Science
Bhopal, India

S.P. Singh
Technical Education Department Uttar Pradesh
Kanpur, India

Satendar Singh
Sharda School of Business Studies

Satyabrata Singha
Department of Electronics & Communication Engineering
Bharat Institute of Engineering & Technology
Hyderabad, India

Vikas Solanki
Department of Electrical Engineering
Government Polytechnic College
Mandsaur

Ajay Soni
IES College of Technology
Bhopal, India

Amrita Soni
SAGE University
Indore, India

Dinesh Kumar Soni
Department of Mechanical Engineering
Rabindranath Tagore University
Bhopal, India

Prachi Tadge
Department of Physical Science
Faculty of Science
Rabindranath Tagore University
Bhopal, India

Kishor Thakre
Department of Electrical and Electronics Engineering
Rabindranath Tagore University
Bhopal, India

Nitesh Thikare
Department of Civil Engineering
Priyadarshini Bhagwati College of Engineering
Nagpur, India

Shweta Tidake
Department of Chemistry
Nagpur Institute of Technology
Nagpur, India

Akhilesh Tiwari
School of Business and Management
Christ (Deemed to be University)

Madhuranjan Vatsa
Akhildev IPR and Research Services
Greater Noida, India

H. D. Verma
Faculty of Agriculture
Rabindranath Tagore University
Raisen, India

Dhananjay Yadav
Department of Mathematical & Physical Sciences
University of Nizwa, Oman

Shalini Yadav
Department of Civil Engineering
Rabindranath Tagore University
Bhopal, India

Acknowledgement

The completion of "*Mathematical Modeling, Design, and Optimization of Complex Engineering Systems*" signifies the successful conclusion of a deeply collaborative endeavor, one that has been enriched by the dedication and expertise of numerous individuals and insightful reviewers. The editors extend their sincere gratitude to all who generously shared their knowledge, encouragement, and valuable time throughout this process. At the heart of this endeavor are the contributing authors. Their profound knowledge and steadfast commitment to dissecting the intricacies of mathematical modeling, design methodologies, and optimization techniques for complex engineering systems have been fundamental to the creation of this volume. Each chapter reflects their rigorous research, analytical prowess, and innovative strategies for tackling multifaceted engineering challenges. Their collective passion and scholarly dedication have transformed the initial concept into a tangible resource, offering both advanced theoretical understanding and practical insights for the field. Their contributions are deeply appreciated and form the core strength of this book.

Editors are immensely indebted to our esteemed academic and professional colleagues, whose intellectual generosity and constructive critiques have been invaluable in refining the quality and depth of this work. We specifically acknowledge the crucial role of the diligent peer reviewers, whose meticulous evaluations and insightful recommendations have been instrumental in ensuring the scholarly rigor and accuracy of each chapter.

"*Mathematical Modeling, Design, and Optimization of Complex Engineering Systems*" stands as a testament to shared effort, a synthesis of contributions from diverse individuals united by a common objective: the advancement of mathematical modeling, design, and optimization for complex engineering systems. It is our hope that this work will serve as a valuable resource for researchers, practitioners, and students seeking to understand and implement sophisticated mathematical and computational tools in addressing intricate engineering problems. We aspire to contribute meaningfully to the ongoing evolution of design and optimization methodologies and to inspire further innovation in this critical domain, ultimately fostering the development of more efficient, robust, and sustainable engineered systems."

Editors

Mukesh Kumar Awasthi
Department of Mathematics,
Babasaheb Bhimarao Ambedkar University
Lucknow, India

Ashwani Kumar
Department of Mechanical Engineering,
Technical Education Department
Uttar Pradesh, Kanpur, India

Manoj Gupta
Department of Mechanical Engineering,
National University of Singapore
Singapore

Chapter 1

Mathematical modeling and optimization techniques

Applications in engineering, economics, and data science

Neha Singh Raghuvanshi, Yatika Gori, Shalini Yadav, Ashwani Kumar, and Mukesh Kumar Awasthi

1.1 INTRODUCTION

The purpose of this chapter is to provide readers with a well-rounded understanding of mathematical modeling and optimization techniques, which are fundamental to informed decision-making in diverse scientific and industrial fields [1]. In an age defined by complexity and data abundance, these tools have become indispensable for analyzing systems, forecasting outcomes, and identifying optimal strategies. Whether in engineering, economics, healthcare, logistics, or data science, mathematical models serve as the backbone for representing reality in abstract yet practical forms. Coupled with optimization techniques, they enable us to derive the most efficient, cost-effective, or risk-minimized solutions in scenarios often constrained by limited resources and competing objectives [2]. The evolution of mathematical modeling is both ancient and fascinating. Rooted in early human efforts to quantify and predict natural phenomena, modeling has long been a tool of philosophers and scientists alike. From Archimedes' work on buoyancy to Newton's laws of motion, mathematics was used to model the physical world with remarkable precision. The 20th century marked a significant leap with the formalization of operations research during World War II, where mathematical models were applied to military logistics, resource allocation, and strategic planning. These efforts laid the groundwork for the modern, algorithm-driven techniques that we now take for granted. Mathematical modeling grew beyond physics and engineering into economics, biology, and the social sciences, evolving into a central language for interdisciplinary research [3].

The real revolution, however, began with the advent of digital computation. With the development of high-speed computers in the mid-20th century, models could be simulated, solved, and analyzed with unprecedented accuracy and scale. This paradigm shift allowed for nonlinear, dynamic, and stochastic models to be explored, which had been previously too complex to tackle analytically [4]. Today, digital computation enables not only more complex modeling but also real-time optimization, large-scale simulations, and integration with machine learning and artificial intelligence. Fields like predictive analytics, smart manufacturing, and precision agriculture owe

DOI: 10.1201/9781003642152-1

their advancement to the synergy between mathematical modeling and computational power.

Optimization, in this context, plays a crucial role in real-world decision-making. It's not enough to simply model a system; we must also determine the best way to operate within it. From minimizing production costs to maximizing investment returns, optimization transforms models into actionable strategies [5]. It helps in balancing trade-offs—be it cost versus quality, speed versus accuracy, or risk versus reward. In domains like healthcare, this can mean optimal scheduling of surgeries or equitable vaccine distribution. In environmental science, it may involve reducing emissions without stalling development. This chapter is structured to offer a foundational yet forward-looking view. It begins by defining core concepts in mathematical modeling and optimization, then delves into classical and modern techniques, software tools, and real-world applications through case studies. Contributions are drawn from engineering, economics, data science, and systems biology to illustrate the versatility of these tools. The goal is to highlight not only how mathematical modeling and optimization have shaped the past but also how they are essential to navigating the future (Figure 1.1).

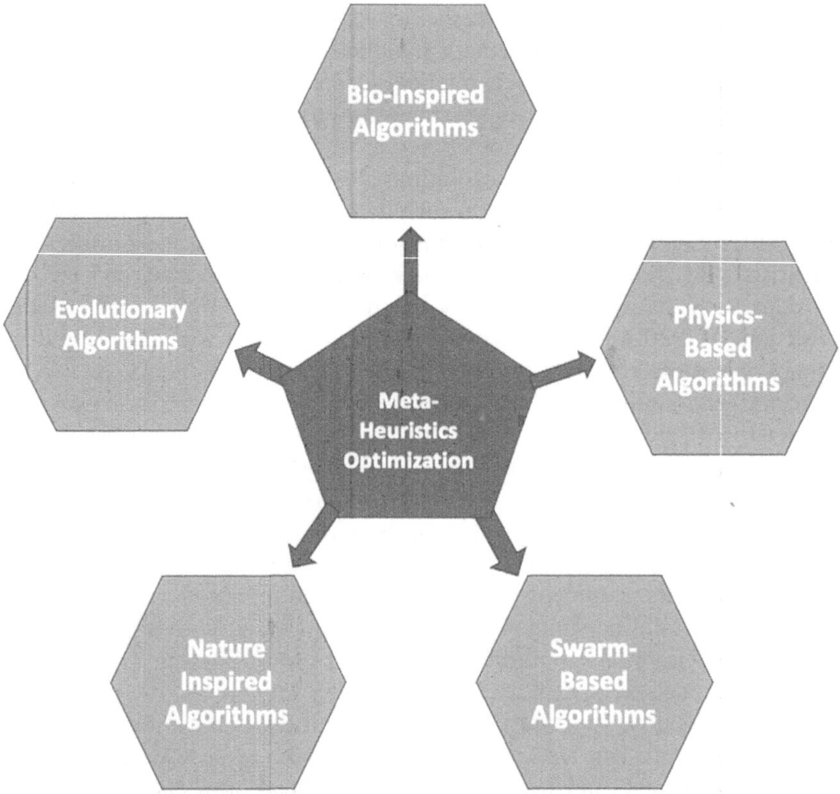

Figure 1.1 Different optimization techniques.

1.2 FUNDAMENTAL CONCEPTS

At the core of any scientific inquiry or engineering design lies the need to understand, represent, and predict the behavior of systems. Mathematical modeling serves this very purpose—translating real-world phenomena into mathematical language that can be analyzed, simulated, and optimized [6]. A **mathematical model** is essentially a simplified representation of a system, process, or relationship using mathematical expressions such as equations, inequalities, functions, or algorithms. These models serve as blueprints that help predict outcomes, test scenarios, or derive insights before any physical or economic actions are taken.

1.2.1 Definition and types of mathematical models

Mathematical models can be broadly categorized based on their structure and function. The first major classification is between **deterministic** and **stochastic models**. In **deterministic models**, the outcomes are precisely determined through known relationships without any randomness involved [7]. For instance, a physics-based model calculating the trajectory of a projectile using Newtonian laws assumes perfect knowledge of initial conditions and does not account for randomness or uncertainty [8]. These models are useful when the environment is well understood, controlled, or predictable. On the other hand, **stochastic models** incorporate elements of randomness and uncertainty. These are particularly useful when dealing with systems influenced by unpredictable factors, such as consumer behavior, financial markets, or weather patterns [9]. A classic example is the Monte Carlo simulation, where probability distributions are used to model uncertain inputs, generating a range of possible outcomes rather than a single prediction.

Another way to classify mathematical models is as **static** or **dynamic**. **Static models** represent systems at a single point in time or assume time-invariant relationships [10]. They are often simpler and used for systems in equilibrium or for snapshot analysis. For example, a linear regression model predicting sales based on advertising spend at one point in time is a static model. **Dynamic models**, in contrast, account for changes over time and typically involve differential or difference equations. They are crucial in modeling systems that evolve population growth, economic cycles, or chemical reactions [11].

1.2.2 Role of assumptions and constraints

No mathematical model is ever a perfect mirror of reality. Models are inherently **simplifications**—and this is by design. To make problems tractable, modelers introduce **assumptions** that reduce complexity. For instance, assuming linear relationships, constant temperature, or frictionless surfaces is common in early-stage models. These assumptions, while not always accurate, are valuable starting points for analysis.

Constraints play an equally crucial role, especially in optimization contexts. These are the limitations or boundaries within which a solution must lie. Constraints could be physical (like capacity limits), economic (budget constraints), temporal (deadlines), or policy-driven (regulations). Effective modeling must strike a balance—too many constraints can overcomplicate the problem, while too few may make it irrelevant or overly idealized. Clear articulation of assumptions and constraints ensures the model's scope is realistic and its results are interpretable.

1.2.3 Model validation and accuracy

Once a model is built, it must be tested to ensure that it performs reliably—a process known as **model validation**. Validation involves comparing model outputs with real-world data or known outcomes to assess their accuracy and relevance. If a model predicts economic growth, but consistently underestimates inflation rates when compared to historical data, it may require recalibration or rethinking of underlying assumptions. Validation techniques vary depending on the model's type and domain. In data-driven models, techniques like cross-validation and residual analysis are common. In simulation-based models, sensitivity analysis is used to assess how changes in input parameters affect outcomes. Regardless of the method, the goal remains the same: ensure that the model does not just fit past data, but also generalizes well to future or unseen scenarios. Together, these foundational concepts form the bedrock upon which more sophisticated modeling and optimization frameworks are built. Understanding the types of models, the assumptions they rest upon, and the need for validation equips researchers and practitioners with the insight necessary to build robust, reliable, and actionable tools.

1.3 CORE OPTIMIZATION TECHNIQUES

1.3.1 Deterministic versus stochastic models

Deterministic models yield consistent results given fixed inputs; stochastic models include random variables, offering multiple possible outcomes. Deterministic is ideal for controlled systems; stochastic is suited to uncertain environments like markets (Table 1.1).

1.3.2 Static versus dynamic models

Static models analyze systems at one moment; dynamic models show change over time. Static is a snapshot; dynamic adds a timeline—crucial for time-sensitive systems like epidemics or finance (Table 1.2).

Table 1.1 Sales forecast (in units)

Week	Ad spend ($)	Deterministic forecast	Stochastic forecast (range)
1	1000	500	450–550
2	1500	750	700–800
3	2000	1000	950–1100

Table 1.2 Population growth

Year	Static estimate	Dynamic estimate (logistic growth)
2020	1000	1000
2021		1150
2022	1000	1300
2023	1000	1425

1.3.3 Role of assumptions and constraints

Assumptions simplify reality (e.g., "linear demand"); constraints define limits (e.g., budget). Both influence model realism. Ignoring constraints can lead to impractical solutions, while too many may oversimplify (Figure 1.3).

1.3.4 Model validation and accuracy

Validation checks how well model predictions match actual outcomes. A good model should generalize, not just fit past data. Accuracy is assessed using error metrics like MAE or RMSE (Figure 1.2 and Table 1.4).

1.4 COMPUTATIONAL TOOLS AND SOLVERS

In the contemporary landscape of mathematical modeling and optimization, **computational tools** and **solvers** serve as the workhorses that bring theory into practical execution. Whether it's minimizing cost functions in logistics, simulating engineering systems, or predicting economic trends, these tools

Table 1.3 Product mix under budget constraint

Product	Profit/units ($)	Cost/unit ($)	Max units (Budget = $1000)
A	20	10	100
B	30	25	40

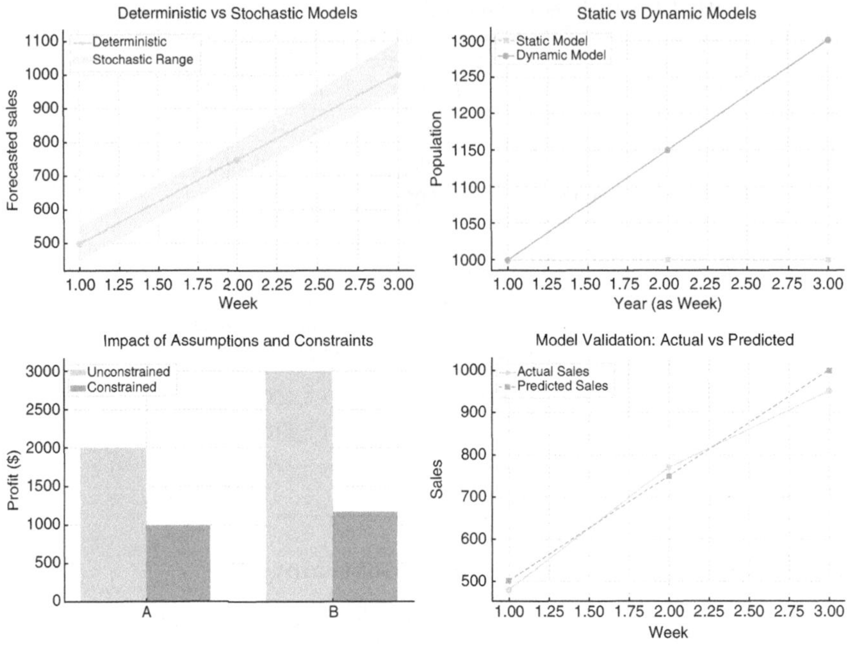

Figure 1.2 Comparison of core optimization techniques.

Table 1.4 Predicted versus actual sales

Week	*Actual sales*	*Predicted sales*	*Error (actual – predicted)*
1	480	500	−20
2	770	750	+20
3	950	1000	−50

offer the processing power, algorithmic flexibility, and precision necessary for real-world problem-solving.

1.4.1 Common software packages

1. **MATLAB:** Used extensively in engineering and applied sciences, MATLAB shines in matrix computations, algorithm development, and data visualization. Its Optimization Toolbox supports both linear and nonlinear problems. Its Simulink environment is ideal for modeling dynamic systems such as control mechanisms and mechanical simulations.
2. **Python (SciPy, Pyomo):** Python is an open-source powerhouse with libraries like SciPy for numerical optimization and Pyomo for algebraic modeling of optimization problems. It's favored for its versatility

Table 1.5 Comparison of solvers

Software	*Speed*	*Accuracy*	*Scalability*	*Area of specialization*
MATLAB	Moderate	High	Moderate	Engineering simulations, control systems
Python	High	High	High	Data science, supply chain, ML
R	Moderate	High	Moderate	Statistics, economics
GAMS	High	Very High	Very High	Energy, macroeconomic modeling
LINGO	Moderate	High	Moderate	Manufacturing, logistics

and ease of integrating with machine learning libraries like TensorFlow and scikit-learn. It's widely used in data science, finance, and operations research.

3. **R:** Originally designed for statistics, R has grown into a potent environment for optimization in econometrics and data analysis. Packages like optim() and ROI (R Optimization Infrastructure) are utilized in modeling regression-based optimization, portfolio selection, and forecasting.
4. **General Algebraic Modeling System (GAMS):** GAMS is tailored for large-scale optimization and economic modeling. Its strength lies in handling complex linear, nonlinear, and mixed-integer programming problems. It's commonly used in energy sector modeling, transportation planning, and resource allocation.
5. **LINGO:** LINGO offers an intuitive interface for formulating and solving optimization models. Its ease of use makes it popular in academic environments and industries needing quick deployment of optimization solutions, especially in logistics and production planning (Table 1.5).

1.4.2 Role of artificial intelligence and machine learning

AI and ML are rapidly transforming optimization by introducing intelligent adaptability to otherwise rigid models.

1.4.2.1 Surrogate modeling

This involves creating a simpler model (or "surrogate") that approximates the behavior of a complex system, significantly speeding up computation. Common in engineering design optimization, surrogate models like kriging or radial basis functions can replicate outcomes of computationally expensive simulations.

1.4.2.2 Reinforcement learning in optimization

Reinforcement learning (RL), a type of machine learning where an agent learns by trial and error, is now being applied to dynamic optimization problems—think autonomous navigation, adaptive inventory management, or smart energy grids. RL optimizes actions over time under uncertainty, surpassing traditional techniques in environments that evolve unpredictably.

The integration of **powerful computational tools** and **intelligent solvers** has revolutionized how mathematical models are executed. By automating complex problem-solving and enhancing predictive accuracy, they serve as indispensable assets in engineering, economics, logistics, and AI-driven domains. Each tool has its sweet spot, and selecting the right one depends on the problem's size, complexity, and domain requirements.

1.5 APPLICATIONS ACROSS DISCIPLINES

Mathematical modeling and optimization techniques have become foundational across various disciplines, solving complex, high-stakes problems with precision and foresight. These methods transcend theoretical mathematics, bringing tangible improvements to engineering, economics, healthcare, logistics, and environmental systems. Each discipline applies modeling uniquely, but all aim for the same outcome: optimal performance, minimal waste, and informed decision-making (Figure 1.3).

1.5.1 Engineering

In engineering, optimization is the backbone of innovation and efficiency. It ensures systems perform at peak levels while minimizing cost, material use, and risk. From designing lightweight bridges to fine-tuning robotics,

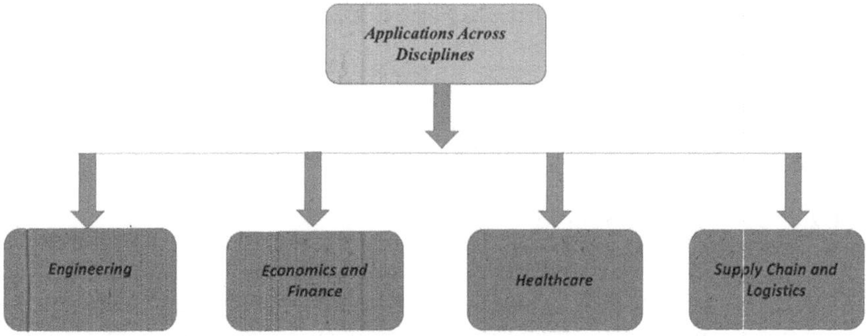

Figure 1.3 Applications of mathematical modeling and optimization across disciplines.

engineers rely on optimization to solve complex, real-world problems with mathematical precision and reliability.

1. **Structural Design:** Civil and mechanical engineers use mathematical models to design structures that withstand stress, minimize material usage, and comply with safety regulations. Optimization helps determine the best dimensions, materials, and load distributions. For instance, finite element analysis (FEA) is used in conjunction with optimization algorithms to create cost-efficient, durable bridges and buildings.
2. **Control Systems:** In fields like robotics and automation, control systems must respond accurately and in real time. Mathematical models optimize feedback loops to maintain system stability and performance. Techniques like proportional-integral-derivative (PID) tuning or model predictive control (MPC) are rooted in optimization strategies.

1.5.2 Economics and finance

In the world of numbers and markets, modeling informs decisions that affect economies and investors.

1. **Portfolio Optimization:** Using models like the Markowitz Mean-Variance framework, investors can construct portfolios that balance risk and return. Algorithms identify the optimal asset allocation given constraints like risk tolerance or market volatility.
2. **Cost-Benefit Analysis:** Optimization helps assess whether an investment or policy is worth pursuing by modeling projected costs and comparing them with expected benefits. Governments and businesses use this to make funding or policy decisions, especially when resources are limited.

1.5.3 Healthcare

The healthcare sector benefits enormously from modeling and optimization, especially under tight resource constraints.

1. **Resource Allocation:** Hospitals use optimization to allocate beds, schedule surgeries, or assign staff to shifts in a way that maximizes efficiency while maintaining care quality. During crises like pandemics, models help prioritize resources like ventilators or vaccines.
2. **Epidemic Modeling:** Epidemiologists apply mathematical models to simulate disease spread (e.g., SIR models) and evaluate containment strategies. Optimization refines intervention strategies like lockdowns or vaccination rollouts to minimize infections and economic impact.

1.5.4 Supply chain and logistics

This field thrives on efficiency, and modeling makes global logistics smart and responsive.

1. **Route Optimization:** Companies use models like the Traveling Salesman Problem (TSP) and Vehicle Routing Problem (VRP) to minimize delivery times and fuel consumption. These models are central to logistics giants like FedEx and Amazon.
2. **Inventory Control:** Mathematical models forecast demand, optimize order quantities, and reduce holding costs. Just-in-time (JIT) systems are underpinned by continuous optimization to ensure smooth operations without surplus.

1.5.5 Environmental systems

As sustainability takes center stage, modeling and optimization help manage our planet's future.

1. **Climate Modeling:** Climate scientists use simulation models to predict future climate scenarios. Optimization refines these models for greater accuracy, aiding in international policy planning.
2. **Energy Consumption Reduction:** Energy systems engineers apply optimization to reduce power consumption in industries and homes. Smart grids, load balancing, and renewable integration all benefit from predictive, optimized control strategies.

1.6 CASE STUDIES

CASE STUDY 1 Optimizing Renewable Energy Systems using Genetic Algorithms (Figure 1.4)

Problem Definition: A hybrid renewable energy system (solar + wind) needs to be optimized for a remote village. The objective is to minimize the cost of energy (COE) while ensuring a reliable energy supply.

Modeling Approach: A mathematical model was developed to simulate hourly energy generation from solar and wind based on historical weather data, energy demand, and storage requirements. The model includes energy balance equations, cost functions, and reliability constraints.

Optimization Method Used: Genetic algorithm (GA) was applied due to its ability to handle nonlinear, multi-objective problems with numerous local minima.

Outcomes: GA yielded an optimal configuration with 12 solar panels, 3 wind turbines, and a 25 kWh battery bank, reducing the COE from $0.42/kWh to $0.26/kWh with 98% reliability (Table 1.6).

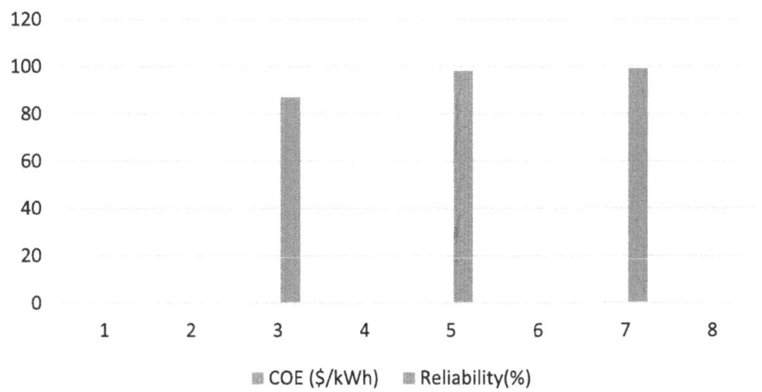

Figure 1.4 Comparing configurations versus COE and reliability.

Table 1.6 Sample data table

Configuration	*Solar panels*	*Wind turbines*	*Battery capacity (kWh)*	*COE ($/kWh)*	*Reliability (%)*
Initial setup	8	2	15	0.42	87
GA-optimized setup	12	3	25	0.26	98
Overprovisioned setup	15	5	40	0.39	99

CASE STUDY 2 Mathematical Model for COVID-19 Spread and Vaccine Distribution

Problem Definition: The government aims to control COVID-19 spread while maximizing vaccine utilization efficiency across five cities with varied population densities and infection rates.

Modeling Approach: A compartmental Susceptible-Exposed-Infectious-Recovered (SEIR) model was used to simulate disease spread. Constraints included daily vaccine availability, transportation delays, and city-specific infection rates.

Optimization Method Used: Linear programming (LP) was used to optimize daily vaccine allocation, minimizing infection rate over a 30-day period.

Outcomes: Strategic vaccine distribution helped reduce the infection rate by 43% and accelerated immunity development, especially in high-density areas (Figure 1.5 and Table 1.7).

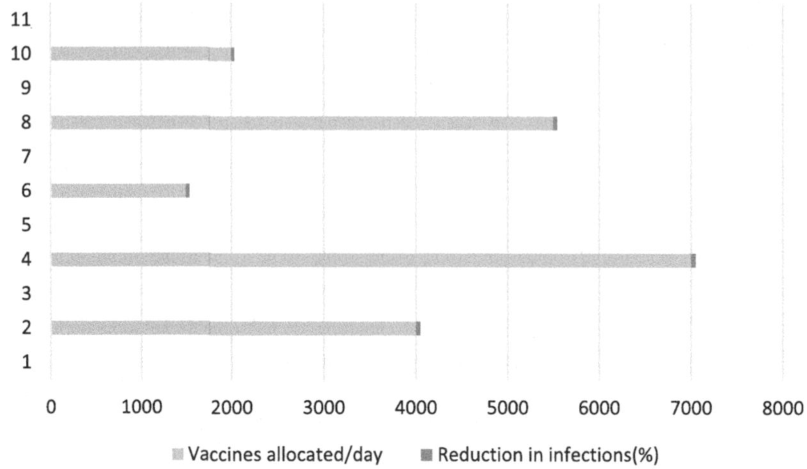

Figure 1.5 Comparing vaccine allocation versus infection reduction per city.

Table 1.7 Sample data table

City	*Population*	*Initial infected*	*Vaccines allocated/day*	*Reduction in infections (%)*
A	500000	8000	4000	47
B	1000000	15000	7000	51
C	250000	3000	1500	39
D	750000	10000	5500	44
E	600000	5000	2000	35

CASE STUDY 3 Cost Optimization in Large-Scale Manufacturing Plants

Problem Definition: A manufacturing company operating three plants seeks to minimize operational costs while meeting production demand and resource constraints.

Modeling Approach: A mixed-integer linear programming (MILP) model was developed. Decision variables included the number of workers, machine run-time, and raw material procurement. Constraints captured production limits, labor availability, and machine maintenance schedules.

Optimization Method Used: LINGO optimization software was used to solve the MILP for cost minimization.

Outcomes: Total operational cost dropped from $1.8 million to $1.3 million/month. Machine utilization improved by 17%, and material wastage reduced by 12% (Table 1.8).

Table 1.8 Sample data table

Plant	*Pre-optimization cost ($K)*	*Post-optimization cost ($K)*	*Machine utilization (%)*	*Material wastage (%)*
1	600	420	85→95	10→8
2	700	510	78→92	14→11
3	500	370	80→93	13→10

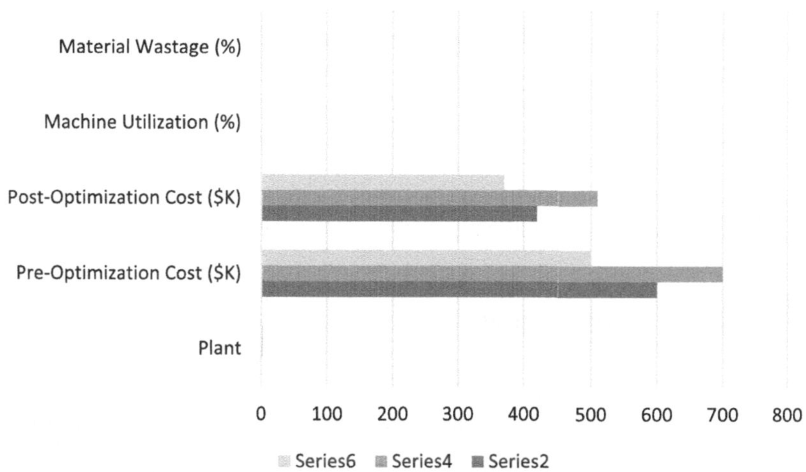

Figure 1.6 Cost reduction and performance improvements across plants.

These real-world case studies demonstrate the versatility and power of mathematical modeling and optimization in solving critical, large-scale challenges (Figure 1.6). Whether it's reducing energy costs, managing pandemics, or enhancing industrial efficiency, tailored models combined with the right solvers produce data-driven, measurable improvements that can be visualized, scaled, and replicated across sectors.

1.7 CHALLENGES AND LIMITATIONS

1. **Computational Complexity:** Optimization models often require solving large, nonlinear, or combinatorial problems, which can be computationally intensive. As variables and constraints increase, the problem's dimensionality grows exponentially, leading to longer processing times and the risk of non-convergence. For instance, solving a VRP for 100 trucks and 1000 locations can take hours or even days using traditional solvers, making real-time decisions difficult.

2. **Model Overfitting/Underfitting:** Striking the right balance in model complexity is crucial. Overfitting occurs when a model captures noise instead of underlying patterns, leading to poor generalization. Underfitting, on the other hand, produces overly simplistic models that miss important dynamics. In healthcare diagnostics, an overfitted model may predict perfectly on training data but fail with new patient records, making it unreliable for deployment.
3. **Real-Time Application Constraints:** Real-world systems, such as autonomous vehicles or stock trading algorithms, require optimization decisions in milliseconds. Models that take minutes to compute aren't viable here. Therefore, the challenge lies in reducing latency without sacrificing accuracy or robustness.
4. **Data Availability and Uncertainty:** Reliable data is the backbone of accurate modeling. In many fields—like climate science or pandemic modeling—data can be incomplete, outdated, or noisy. This uncertainty hampers model reliability and increases the risk of suboptimal or even harmful decisions.

1.8 FUTURE DIRECTIONS

1. **Integration with Big Data Analytics:** With the exponential rise of data from sensors, IoT, social media, and transactional systems, optimization is being revolutionized. Combining mathematical models with big data tools like Hadoop or Spark enables near real-time decision-making in complex environments, such as predictive maintenance in manufacturing or personalized marketing in e-commerce.
2. **Hybrid Optimization Techniques:** The future lies in combining multiple methods—like GAs with gradient descent or particle swarm optimization with simulated annealing—to harness the strengths of each. These hybrids improve convergence speed, solution diversity, and robustness, especially in highly nonlinear or stochastic systems.
3. **Quantum Computing in Optimization:** Quantum optimization algorithms, such as the Quantum Approximate Optimization Algorithm (QAOA), promise exponential speedups for solving NP-hard problems. While still in its infancy, quantum computing has shown early potential in solving complex logistics, finance, and cryptography optimization tasks much faster than classical machines.
4. **Ethics in Decision-Making Algorithms:** As optimization algorithms increasingly influence critical areas—like loan approvals, hiring, or law enforcement—ethical considerations become paramount. Bias in data or model design can lead to unfair or discriminatory outcomes. Future research must incorporate fairness constraints, explainability, and accountability into model development to ensure ethical decision-making.

1.9 CONCLUSION

Mathematical modeling and optimization are powerful tools driving innovation across disciplines. By blending theoretical rigor with computational advances, they solve real-world problems efficiently. As challenges evolve, embracing emerging technologies and ethical frameworks will ensure these techniques remain relevant, impactful, and responsible in shaping our future.

REFERENCES

[1] Rao, S. S. *Engineering Optimisation: Theory and Practice*; John Wiley & Sons: Hoboken, NJ, 2009. b) Bertsimas, D.; Tsitsiklis, J. N. *Introduction to Linear Optimization*; Athena Scientific: Belmont, MA, 1997. c) Hillier, F. S.; Lieberman, G. J. *Introduction to Operations Research*, 10th ed.; McGraw-Hill Education: New York, NY, 2014.

[2] Arora, J. S. *Introduction to Optimum Design*, 4th ed.; Academic Press: Cambridge, MA, 2017.

[3] Bressan, A. *Lecture Notes on Mathematical Modelling in the Life Sciences*; American Mathematical Society: Providence, RI, 2010.

[4] Murray, J. D. *Mathematical Biology: I. An Introduction*, 3rd ed.; Springer: New York, 2002.

[5] Bazaraa, M. S.; Jarvis, J. J.; Sherali, H. D. *Linear Programming and Network Flows*, 4th ed.; Wiley: Hoboken, NJ, 2009.

[6] Giordano, F. R.; Fox, W. P.; Horton, S. B. *A First Course in Mathematical Modeling*, 5th ed.; Cengage Learning: Boston, MA, 2013.

[7] Rardin, R. L. *Optimization in Operations Research*; Prentice Hall: Upper Saddle River, NJ, 1998.

[8] Halliday, D.; Resnick, R.; Walker, J. *Fundamentals of Physics*, 10th ed.; Wiley: Hoboken, NJ, 2014.

[9] Hull, J. C. *Options, Futures, and Other Derivatives*, 10th ed.; Pearson: Boston, MA, 2018.

[10] Taha, H. A. *Operations Research: An Introduction*, 10th ed.; Pearson: Boston, MA, 2017.

[11] Murray, J. D. *Mathematical Biology I: An Introduction*, 3rd ed.; Springer: New York, 2002.

Chapter 2

The future of Artificial Intelligence

Innovations, challenges, and ethical considerations toward 2030 and beyond

Nitin Dheeman, Amar Nath, Mohit Kumar Gupta, and Akram Masood

2.1 INTRODUCTION

The frontline in this transformative revolution is artificial intelligence (AI), which is fundamentally going to change nearly everything within society. From simple, rule-based systems through to the neural networks driving the new applications of AI, the speed with which the technology has moved forward has been breathtaking. By 2030, it is believed AI will make leaps in the unlocking of quantum computing, it will introduce the concept of artificial general intelligence (AGI), and neuromorphic computers will be developed. However, this creates new challenges for social, ethical, and regulatory purposes (Venkatasubramanian, 2018; Zhang, 2024). Envisioning AI in the next decade cannot be avoided, since it holds both the potential that can help advance the world, together with the risks associated with growth. It will have huge impacts on society, far in sectors, governance, and individual levels. It will alter the way humanity will face other challenges like climate change, public health, and inequality in the future, but simultaneously reshape the global workforce, too. This is according to Kulkov et al. (2023) and Sawers et al. (2021). Consequently, it makes the urgency of dealing with ethical and social implications created due to the deployment of AI an important need. Algorithmic bias, data privacy, accountability, and transparency would be some of the crucial issues to consider while achieving responsible development in AI.

This chapter presents an ambitious undertaking to comprehensively outline the direction AI is expected to take by 2030, with three general themes, namely innovation, ethics, and impact (Ryan & Stahl 2020). The first part discusses these as expected technological breakthroughs in such areas as AGI and quantum AI, which promise to redefine the limits of machine intelligence. The second half revolves around the ethical issues this would entail, such as issues of fairness, transparency, and governance. The paper finally explores AI's general impact on society while taking into consideration its possibility to contribute to sustainable development or play its role in worsening or mitigating some problems that the world is facing at

 DOI: 10.1201/9781003642152-2

the moment, for instance, inequality and unemployment. The positive transformations AI can bring might disrupt the very same improvements; therefore, now is the time for proactive moves with regard to governance, regulation, and ethical oversight. This chapter aims to contribute by discussing current trends and future projections toward such a necessary ongoing dialogue for the responsible harnessing of AI toward shaping a better future for all.

Table 2.1 Key datasets on AI innovations, ethics, and societal impacts (2024)

Sr. No.	*Dataset name*	*Description*	*Focus areas*	*Timeframe*	*Link*
I.	AI Index Report 2024	Comprehensive insights into global AI trends, including technological advancements and ethics.	Innovations, Ethics, Societal Impact	Up to 2024	https://aiindex.stanford.edu/report/
II.	Global AI Ethics Guidelines	Compilation of AI ethics guidelines worldwide, comparative analysis on ethical considerations.	Ethics, Governance	Up to 2024	https://algorithmwatch.org/en/ai-ethics-guidelines-global-inventory/
III.	AI and Employment Dataset	Statistics on job displacement, creation, and transformation due to AI across various sectors.	Societal Impact, Workforce	Up to 2024	https://arxiv.org/abs/2206.11922
IV.	AI in Healthcare Innovations	Information on AI applications in healthcare and their impacts.	Innovations, Societal Impact, Ethics	Up to 2024	https://arxiv.org/abs/2405.19522
V.	AI and Climate Change Mitigation	Data on AI-driven solutions for mitigating climate change and their societal implications.	Innovations, Sustainability	Up to 2024	https://arxiv.org/abs/2008.02275

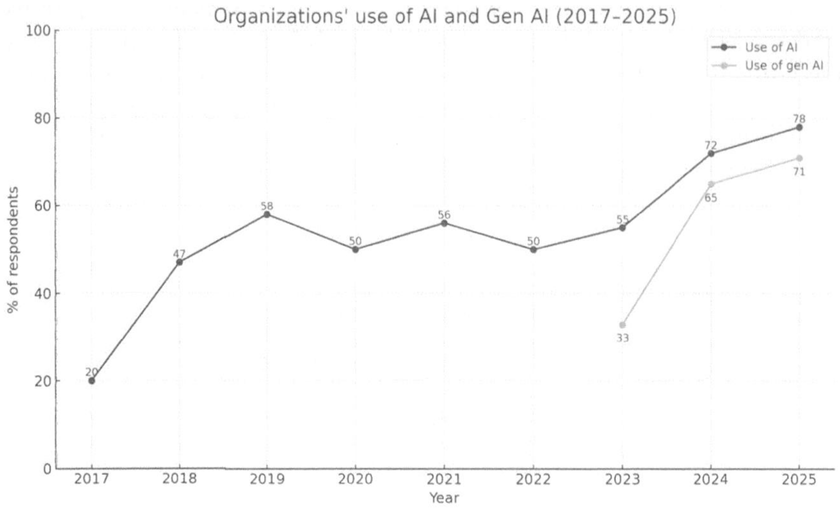

Figure 2.1 Trends in organizational use of AI and generative AI from 2017 to 2025.

2.2 ETHICAL CHALLENGES AND FRAMEWORKS

Certainly, these are the parts of an AI's development and have presented ongoing ethical dilemmas in the design and utilization of AI. These concerns include bias, accountability, transparency, and governance, since these factors define how people view the role of AI in building fair, reliable, and secure AI systems.

2.2.1 Bias in AI

One of the greatest ethical problems that affects AI would probably be bias found in the algorithms of an AI system. Bias could show from anywhere, but it is often found in the training data supposed to reproduce existing inequalities in society. Too many types of biases exist, such as gender bias, race bias, and socioeconomic bias, which result in unfair or discriminatory outcomes on applications of AI related to hiring, criminal justice, healthcare, and financial services (Ryan and Stahl, 2020). For example, AI-based hiring could be prone to gender and race biases if it draws upon historic data which relates to past discrimination practices. Even more, the widening of societal inequalities through AI raises critical ethical issues concerning fairness and equality. Therefore, the remedy lies in improving data collection practices, ways of detecting and correcting biases, and auditing and testing AI systems for fairness.

2.2.2 Accountability and responsibility

AI interfaces used in most organizations are of their boxed natures, as the trend keeps "black boxes," meaning that one cannot understand by what logic such a given decision has been made, not even by the developers themselves. Accountability gaps turn into a challenging task whenever there is no transparency given from an AI system, in case it has decided so, or its performance was not up to the mark. As is perhaps evident, establishing who is to blame in circumstances such as incorrect diagnosis by the healthcare AI in operation becomes somewhat complex. Therefore, for example, not having clear-cut structures of accountability undermines public confidence in AI while denying meaningful redress to harmed individuals. This is now increasingly addressed through the development of frameworks in AI accountability, such as embedded clear responsibility lies within the developing, deploying, and maintaining of AI systems (Sigfrids, 2022).

2.2.3 Transparency and trustworthiness

Another important ethical challenge in AI is that of transparency. Most AI systems, especially deep learning model-based ones, are so complex that users cannot understand how decisions are made. This lack of explainability raises concerns about fairness in that individuals may not understand or trust why certain decisions were made about them (Ryan and Stahl, 2020). For instance, if an AI-powered loan approval system rejects credit without showing a clear explanation, the situation could be perceived with suspicion and frustration. The push for more transparency and explainability in AI systems, to this end, would be important features that will make them ethical to deploy; transparency enhances accountability and at the same time provides users with an opportunity to appeal against decisions which they perceive to be unfair. Emerging efforts toward explainable AI, or XAI, aim to make the decision-making process of AI more accessible and understandable. However, challenges persist in balancing transparency with performance.

2.2.4 Governance and regulation

The rapid development in AI has raised a call for appropriate governance and regulatory frameworks so it is ensured that AI will be deployed with due care in a very responsible and ethical manner. Presently, AI governance is fragmentary, as different countries/regions decide upon their independent policies/guidelines. Nonetheless, on the deployment of AI unbiasedly and ethically, what becomes important is a global sense of cooperation or harmonization of standards (Shanker, NaN, n.d.). Another possibility is that international collaboration could play a major role in the establishment of common standards and guiding principles with regard to AI development, deployment, and regulation on issues such as data privacy, accountability, and the mitigation of bias, among others.

National regulatory bodies are expected to concentrate on tackling significant ethical concerns associated with AI, including enhancing transparency, guaranteeing fairness, safeguarding data privacy, and ensuring that software developers and organizations are accountable for the impacts of their systems (Spiro, NaN, n.d.). The evolution of this technology will have implications for maintaining flexibility and responsiveness in the context of regulations, ensuring that innovations keep up with the benefit of society. The establishment of ethical AI protocols for developers and users alike will be crucial in converting rather vague ethical principles into applicable standards for the different industries (Ryan and Stahl, 2020).

2.2.5 Balancing regulation and innovation

Balancing regulation with the need for innovation has long been one of the core challenges in AI ethics. If applied too stringently, regulations may smother emerging technologies and foreclose their potential benefits; if they are not careful enough, dangerous practices could run rampant without oversight. Reuel noted,

> To build the future for AI, encouraging innovation and keeping it straightforward and ethical would be the crux. As AI systems grow larger and begmore important, bridge regulations shall be made, which would enable encouraging responsible innovation along such lines while also addressing the concerns of bias, security, or accountability.

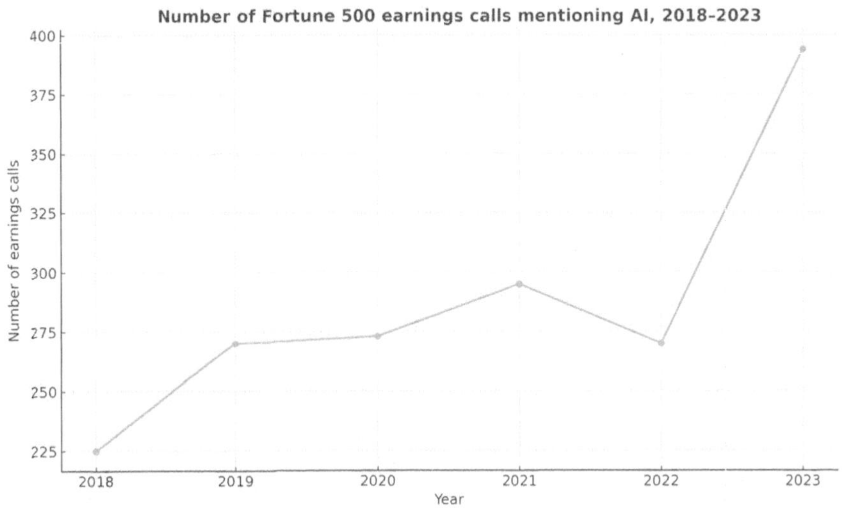

Figure 2.2 Number of Fortune 500 earning calls mentioning AI (2018-2023).

Source: Adapted from Quid, as cited in the 2024 AI Index Report.

Efficiency and ethics in AI are not mutually exclusive; the challenge lies in finding the balance between them and doing so at scale. As AI technologies continue to evolve and leave us behind, it is also important to speak up about and implement the regulatory frameworks that need to be in place, therefore making global cooperation and collaboration even more essential.

2.2.6 The role of international cooperation

As the AI explosion spreads across the globe, international cooperation will be essential for developing consistent ethical standards. Country-to-country regulatory differences divide up what moves across borders and in what way, and therefore can slow down AI deployment in companies operating in global industry intersections—healthcare, finance, and transport (Sigfrids, 2022). One will need to establish international treaties, agreements, and research collaborations to devise standards for the ethical use of AI in different parts of the world. Absent international cooperation, the ethical landscape for AI could be disjointed, with individual states applying varied standards that undercut global efforts to tackle issues such as bias, accountability, and transparency.

2.2.7 Ethical AI guidelines for developers and users

With AI tech still evolving at a remarkable pace, we'll need to create ethical AI guidelines not only for developers but also for users to promote responsible use of these systems. Guidelines should also address how these plans ensure fair treatment of the humans involved, minimize potential harm, enhance transparency of AI, and foster accountability. Such guidance would help developers who want to bring ethics to the design, development, and deployment of AI systems. Guidelines for users will communicate how they can draw on AI systems in ways consistent with ethical standards and aligned with protecting their rights (Ryan and Stahl, 2020). These new guidelines could yield trust in AI systems, as they would specify a common definition of ethical use in AI, which in turn would hasten the adoption of socially responsible AI.

2.3 SOCIETAL AND ECONOMIC IMPACT

AI could prove an enormous revolution across wide swathes of different spheres by presenting novel ways of solving world problems. Inevitably, such proliferation carries social and economic implications that must be managed with great sensitivity. Almost certainly, AI will be one of the most potent forces that will affect the shape of tomorrow—for the better but, most likely, also for the worse—within our society and economy.

2.3.1 AI in addressing global challenges

It is definitely more productive to address these issues—whether climate change, healthcare, or education—on a global scale with the assistance of AI. Especially, AI is valuable for processing large data sets, modeling different parts of environmental systems, and improving researchers' predictions about how climate will change and how humans might adapt. This includes, for instance, the creation of climate models through AI programs (Kulkov, 2023b). With AI tools today, renewable energy production is optimized, energy consumption is increased, and carbon footprints are reduced so that global climate objectives are maintained.

Today, like this, AI is transforming the best aspects of diagnostics, drug discovery, and personalized medicine in the healthcare industry. By providing more accurate diagnoses, speeding up drug development, and creating individualized treatment plans, AI is improving healthcare access and outcomes, especially in underserved populations (Sawers, 2021b). AI-driven advancements like telemedicine and predictive analytics are enhancing healthcare delivery, lowering costs, and tackling disparities in access to care. In the field of education, AI tools are personalizing learning experiences, catering to students' specific needs and learning styles. Additionally, AI systems are automating administrative tasks, which allows educators to focus on more vital teaching responsibilities (Satyam).

2.3.2 Risks of job displacement and economic inequality

Indeed, AI could positively affect the larger society, but it does carry a large amount of risk, primarily in the workplace. Statistics state that millions of jobs will be on the line through automation using AI technologies, primarily in the sectors of manufacturing, retail, and transportation (Srihith, 2022). It could threaten occupations in industries that depend on performing repetitive tasks, as it is much more efficient compared to human input in performing repetitive processes. The impact of such disruption could be more severe on the economic status of many, particularly if they lose jobs from lower-skilled occupations; these workers may find it harder than others to find alternative employment in an economy led by AI.

Displacement in the job comes from the uneven relation in the profits accruing from advances in AI; the number of people and sectors reaping the benefits from AI has increased, while others grapple with disturbance and loss of jobs. This rift would stretch the already widespread opening between winners and losers of AI investments, with its consequences amplifying over time as conditions become more crucial in perpetuating social and income inequalities. In extreme cases, the social differences caused may lead to political instability, social unrest, and high polarization among groups within the global society (Satyam, 2023). Part of prevention from such threats would be by having the governments and private sector both set up retraining and reskilling programs for the people who would otherwise have their jobs displaced by the innovations, or even in the future by some other innovations in AI-driven workplaces. Hence, holistic social safety nets that actualize reality, such as universal basic income or unemployment benefits, will cushion completely against any displacement through automation. In addition, policies should guarantee the growing economy inclusively and the general assurance that AI benefit allocation will be cast widely enough to not create any further entrenched inequality (Satyam, 2023). Also, put such policies in place to ensure that no group or society is left behind while the entire economy transforms by AI.

Benefits such as universal basic income or unemployment benefits will therefore complete the picture of safety nets these days: they really cover all bases against possible displacements through automation. Beyond that, such policies must also empower the economies to grow inclusively and ensure that, as a group, the benefits of AI will be shared as broadly as possible to prevent any existing inequalities from deepening (Satyam, 2023). Such policies, too, will guarantee that no group or society is left behind while the economy transforms in its entirety using AI.

The safety nets, for example, provisions for universal basic income or unemployment benefits, will provide almost absolute cushioning against the displacement caused by automation. In addition, such policies have to empower economies to grow in an inclusive approach and also assure the general society that the benefits of AI will be availed widely enough to avoid further entrenching current inequalities (Satyam, 2023). Such policies will ensure that no group or society will be cut during the entire process of economic transformation using AI. A certain kind of holistic social safety net, such as universal basic income or unemployment benefits, will cushion completely against any displacement through automation. The policies should be such that they could empower even the growing economy inclusively, as well as guarantee that benefits from AI will be cast broadly enough to cover any existing inequality beyond the deepened ones (Satyam, 2023). These are the policies, also, which will guarantee no group or society is left behind, as all would be transformed by AI. Very soon, in short, safety nets, like universal basic income or unemployment benefits, would completely cover all bases against potential displacements caused by automation. Beyond that, such policies must also empower the economies to grow inclusively and ensure that, as a group, the benefits of AI will be shared out as widely as possible to prevent any existing inequalities from deepening (Satyam, 2023). These policies will also ensure that none of the groups or societies gets left behind as the whole economy transforms with AI.

2.3.3 Transformation of AI-driven industries

It gives rise to new emerging industries that emerge as a product of data science, the advent of AI, and AI-based services. New job creation and improved economy: these AIs develop new industries. A competent workforce is required for AI, and the requirements for data scientists, machine learning engineers, and AI specialists are on a the rise (Satyam, 2023). A new economic ecosystem created by AI-dependent startups and companies opens numerous doors for investments with the invention. Partial reasons for economic transformation in existing industries are brought by AI. With the implementation of AI in operations, productivity is enhanced, and efficiency enables the sector to serve the customers appropriately, especially in the finance, retail, and manufacturing industries (Satyam, 2023). For instance, supply chains and inventory, as well as market forecasts, improve to cut costs as well as significant profit margins resulting from AI-driven automation. Though all these economic benefits of AI are visible, their realization is a function of business and governmental effectiveness in managing the transition.

2.3.4 Inequality and access to AI benefits

AI makes it quite huge in opportunities regarding the economics, but has some inequality in reaping the benefits: high-income countries and, mostly, tech companies are the primary beneficiaries of AI and advanced developing economies, as much as possible, in order for them to be unable to absorb or assimilate into these systems. This makes the issue all the more acute from the perspective of world inequality, as the richer countries and corporations will be able to monopolize the economic benefits of AI while the poorer regions are left behind. More specifically, targeted policy measures for inclusive growth in such areas transformed by the digital revolution would be required to include all people in what AI can deliver. Development is needed in infrastructure, education, and training programs of governments that seek to enable the gains of AI to reach every corner of the world. Such international cooperation will also play a very vital role in promoting global generation and sharing of AI technologies in a more equitable way; this ensures that poorer countries would not be excluded from the process of the revolution toward AI (Kulkov, 2023a).

2.3.5 Managing the economic transition

Ultimately, it will be determined by the management of governments, businesses, and societies during the transition to an AI-driven economy. It is the challenge for policymakers to strike a balance between the demand of the need for innovation and businesses in protecting workers as well as maintaining economic stability. Companies will naturally become important actors when they implement their innovations, but government policy will need to put in place policy instruments that support their actions with respect to their workers, reinforcing such interventions with investments in education, reskilling of workers, and job creation in emerging sectors. To achieve an inclusive and equitable economic transformation, it will be important to have a comprehensive strategy reflective of all workers, including those at risk of displacement from their previous employment. Businesses play an important role in this change, too. They must be ready to shoulder a part of the commitment, particularly aiding their employees in the areas of career development and wellness, and instituting programs for retraining and career transition for those that get affected by automation, among other things. Efforts of this nature should be predicated on a well-thought-out ethical AI strategy whose goal is to put the well-being of workers and consumers at heart rather than maximizing productivity alone in the development and deployment of AI technologies.

2.4 TECHNICAL MILESTONES BY 2030

AI will experience enormous progress by 2030, with major advancements in embedding AI with other impending technologies and further improvements in the scalability, tools, and simplicity of AI frameworks. These advancements will not only fuel innovation among various sectors but will also bring challenges that they must tackle, especially related to security, ethics, and social impact.

2.4.1 Integration of AI with other emerging technologies

The convergence of AI with other emerging technologies will be one of the major factors in unlocking revolutionary potential in any field. The most anticipated development is the integration of AI and the Internet of Things (IoT). Linking these two would open access to the development of intelligent interconnected systems that could empower smart homes, smart cities, and smart infrastructures with IoT-capable devices embedded with AI, allowing real-time monitoring and decision-making automation on an entirely unprecedented scale. For instance, smart cities could optimize traffic flow and energy distribution while even prediction and automation enhance public safety (Alahi, 2023). These will not be achievable without very effective data management and security frameworks to handle the enormous amounts of data produced by these connected devices. This will bring forth the emergence of self- and adaptive robots that will conduct complex tasks in a variety of environments. Such advanced robots will rely on high-performance machine learning algorithms to conduct performance, including navigating through unexplored terrains, performing precision tasks in healthcare or manufacturing, and functioning in dangerous situations where human presence would be ineffective or dangerous. AI will further build memory capabilities into these robots from experience to enhance adaptation and efficiency over time (Venkatasubramanian, 2018). In sectors such as healthcare, agriculture, and logistics, the operations enhanced by AI in robots would bring a revolution in operations in three ways: they would increase productivity, improve safety, and reduce cost.

It is expected that the integration of AI and blockchain technology will further enhance the security, transparency, and trustworthiness of data within AI structures. The decentralized and immutable nature of the blockchain is hopeful in assuaging fears about the quality of data integrity and accountability in the AI decision-making process. It is thought that the combination of AI computing power and the safe and transparent nature of blockchain would facilitate enterprise, easily verifiable data processing by AI systems, thus making them more trusted and less amenable to misuse. This synergy will be primarily appropriate for the case of supply chains, financial transactions, and the management of healthcare data (Dalkey, 1963).

2.4.2 Advancements in scalability, reliability, and accessibility

The AI systems will experience a growing level of performance improvement between now and 2030, accompanying technological integration. The most important impelling factor of this improvement will be in hardware and software innovations. The new generations of AI chips would be optimized for deep learning tasks, thus developing newer AI with more power and energy efficiency so that more massive datasets can be processed or performed on increasingly complex tasks. These hardware improvements cause the applications of AI to be cheaper in cost and easier to use for businesses and governments, thereby accelerating AI use all across sectors (Venkatasubramanian, 2018). Indeed, more reliable progressive AI algorithms will be responsible for further enhancement in the dependability of AI systems. Algorithms will be able to process data more effectively, thus reducing errors and facilitating a sense of reliability about decisions in contexts that are high stakes, for example, healthcare, autonomous driving, and financial services. Moreover, AI systems would be designed to be more transparent and thus more easily explainable and interpretable to alleviate many concerns regarding the "black box" nature of AI systems and hence help build public confidence in the systems (Venkatasubramanian, 2018).

There would also be more efforts in democratizing AI so that millions of users would have access to it instead of just a privileged few. By 2030, we envision such platforms to be user-friendly and accessible even to laypersons so that they can use AI for specific applications, such as business analytics, healthcare diagnostics, or personalized education. This would lead to tremendous innovation through democratization instead of confining the benefits to a few large corporations or small techno-savvy regions (Venkatasubramanian, 2018). It is also necessary to ensure equitable access to AI technologies and expertise to avoid unequal distribution of the benefits of AI. If this is not done, the technological divide may continue to widen and increase the already existing inequalities.

2.4.3 The convergence of AI and human intelligence

By 2030, convergence between AI and human intelligence is likely to be accelerated, albeit not yet achieving the long-term goals of AGI, which would embrace human cognitive capabilities across a broad range of tasks. By 2030, AI will rapidly transition into an increasingly central part in the conduct of human life—from decision-making in healthcare to personalization in education. This could bring with it a revolution in areas such as mental health, human-machine collaboration, or even creativity (Venkatasubramanian, 2018).

On the one hand, all these developments will be accompanied by enormous promises; on the other, they raise ethical and social questions about the nature of human work, autonomy, and decision-making involvement in that work. As the technological world increasingly starts to penetrate more profoundly into the lives of humans, questions will emerge about where the boundaries end or perhaps the middle ground between human and machine intelligence. Related, too, will be issues concerning the individual's identity and privacy, as well as those of other components within the system. Ethical challenges will emerge from the engagement of individuals through AI in possibly developing human capacity (such as cognitive enhancement and human collaboration with robots). For instance, augmentation of the human being, through AI, would offer greater physical or mental capabilities, but, again, this raises issues of equality, fairness, and access (Venkatasubramanian, 2018).

The growth of AI in the workplace will create a whole new impact on the human dimension of labor. There will be efforts to improve productivity without letting actual human workers become redundant; they must be left with no choice regarding employment and the displacement of jobs. A new context will have to be elaborated, within which the relationship between human and machine intelligence must be framed: one that honors human welfare, dignity, and equitable rights.

2.4.4 Ethical and societal challenges of AI integration

Larger issues of ethicality and sociality will also arise with AI technology being coupled with other emerging technologies. With decision-making increasingly automated in areas such as life, healthcare, finance, and criminal justice—and many other possible areas of decision-making—human judgment will gradually be supplanted by AI models and human biases. Such systems may lead to inequitable results by perpetuating discrimination and biases that already exist in training datasets.

There will certainly arise new privacy-related issues as the influence of AI spreads more into sectors like healthcare and security. While examining personal data in a manner not seen before, protecting the data of individuals and making this usage ethical will have to be top priorities. The organization needs to install a strong safeguard against any misuse of their personal information, as well as develop transparency in AI decision-making.

2.5 AI DISASTER PREPARATION IN 2030

AI is advancing and growing into wider realms of activities in society. Significant challenges will come along with the development as well as deployment of AI (Gautam and Thapaliya, 2024). Ranging from cybersecurity threats to failures in autonomous systems, AI-related disasters could have grave impacts on society at large. In fact, preparation for these impending risks needs the design of proactive strategies to avert and mitigate their negative consequences, as well as the safe and responsible deployment of AI technologies.

2.5.1 Assigned risk and disaster

The quick and succeeding pace of the improvement of AI technology continues to make these systems more vulnerable to different cyberattacks. As AI gradually penetrates vital infrastructures such as power systems, healthcare systems, and financial networks, it is expected that cyberattacks will indeed have disastrous impacts. For example, a successful attack on the AI-controlled infrastructure systems will most likely result in widespread disruption of primary services that may lead to extensive financial losses. Exploiting a weakness in AI systems would also lead to nefarious data manipulation and any change in the processes involved in decision-making, as it further incurs risks (Sebastian, 2023).

The second type of novel threat introduced by autonomous systems is the risk of malfunction. Autonomous automobiles, unmanned aircraft vehicles, and every form of self-driving armament depend on AI to operate safely and properly. However, AI systems will commit errors that cause accidents as well as make terrible decisions, and bring about unnecessary death. A failure of the decision-making algorithms of self-driving cars, for instance, could lead to road or traffic accidents or even condemning life to death (Gautam and Thapaliya, 2024). Similar to autonomous weapons, they could also cause unintended casualties because these systems might not be well managed and possibly misused during conflict or warfare. Thus, while more such technologies develop, there should also be sufficient guarantees about the high safety and accountability standards by which these systems operate. Yet another great danger is the malicious use of AI. Think of deepfakes, which can be quite realistic but are fake media that can then be used to twist public opinion, tarnish reputations, or spark depreciation (Al-Kfairy et al., 2024). AI-powered bots have also been widely broadcast into fact-misinforming social media networks, from elections to dragging the institutions from which they come. Since it stands to multiply these threats, it becomes necessary to seek mitigations against the misuse of AI to ensure sound socio-political stability and confidence.

2.5.2 Strategies for mitigating AI disasters

Among the various proactive strategies to mitigate the chances of AI disasters, the major one lies in the development of strong safety and security measures for AI. One primary classification among them is creating preventive measures against misuse of AI technology by malicious interference and provision of armored features of AI systems against penetrating cyberthreats or unauthorized access. There should also be safety nets incorporated into the architecture of AI designs to prevent a catastrophic error in decision-making. For instance, an autonomous car should have an emergency stop in case any of the principal part systems experience a failure (Gautam and Thapaliya, 2024).

Strengthening defense systems will be another addition to prevent AI systems from invading threats. The more AI is embraced in critical infrastructure, the more the need for specific cyber defense grows. This includes real-time detection and response mechanisms for managing cyber threats using AI models, as well as enabling AI systems and other devices on the network to have secure communications with the help of such real-time detection. This will require layering, which means adopting the traditional approach that will then incorporate AI solutions (Gautam and Thapaliya, 2024).

There shall be certain oversight mechanisms for autonomous systems, with the possibility of preventing accidents and misuse. Autonomous systems, for example, in the transportation, defense, and healthcare sectors, must be closely regulated and actively monitored under high-risk applications. Regulation should have the government and political will to develop accountability frameworks around these systems to enable tracing failures or abuses along a responsible party line (Gautam and Thapaliya, 2024). These standards should also include regulatory coverage for performance expectations that any AI system should have, among other important areas, that of safety, reliability, and transparency.

Another important milestone in securing responsible AI development is creating ethical guidelines on AI development and deployment. Such guidelines should direct the focus of AI's judgment toward transparency, fairness, and accountability without exposing AI to inflict harm on the most disadvantaged and further complicating existing inequalities in society (Gautam and Thapaliya, 2024). Ethical integration into the lifecycle of AI development would mitigate the risk of exposure while at the same time improving public welfare.

2.5.3 International cooperation and risk management frameworks

AI development and application, considering many of the risks involved, will need to be internationally cooperative so that it can be done safely and ethically. This can be made possible by collaborative efforts between countries, regulatory bodies, and international organizations in creating such global standards for AI safety, security, and ethics, and it will be highly important in preventing fragmentation of regulatory frameworks and ensuring responsible deployment of trans-border AI technologies (Gautam and Thapaliya, 2024).

Similarly, an important aspect of a well-designed AI risk management framework will be identifying and measuring as well and mitigating any form of possible risk. Moreover, that entailed a multi-dimensional approach, bringing developers, regulators, ethicists, and any other stakeholder to evaluate the possibility and impact of such AI disasters so that stakeholders would have a clear understanding of possible risks targeted by developing relevant initiatives ranging from prevention to disaster relief planning (Gautam and Thapaliya, 2024). Ultimately, continuous review and refinement of indexes need to take place as technology keeps evolving.

Enabling the primary feature of AI risk management would be to devise early-warning systems for AI-initiated calamities. Such systems would monitor real-time operations and performance of AI systems, sending early alerts concerning possible system failures or breaches of security incidents. For example, an AI surveillance system would track an abnormal behavior pattern in the navigation system of an autonomous vehicle and close down the system automatically just before an accident were to occur. Early warning will be a vital avenue through which to protect against disasters before they manifest and cause damage to human life and infrastructure (Gautam and Thapaliya, 2024).

2.6 THE ROLE OF AI IN SUSTAINABLE DEVELOPMENT GOALS (SDGs)

AI has enormous potential in transforming the Sustainable Development Goals (SDGs) by 2030. Its various applications can improve the living standards of people while conserving nature in areas such as health, education, agriculture, and environmental monitoring (Kulkov, 2023a; Tomaev, 2020). Giving support on many of the SDGs—from eradicating poverty to making access to clean water to promoting innovation and fighting climate change—broad support can be given by AI by better allocation of resources, enhancement of efficiencies, and assistance in sustainable practices (Kulkov, 2023a). It thus follows that developing and using AI in a way that makes it contribute to sustainable development is uneven and not equal in an exclusive manner; its contribution has to be fair in an inclusive way (Truby, 2020).

2.6.1 AI's impact on healthcare and education

AI is capable of making great leaps in expanding access to and boosting the efficiency of healthcare services, thus furthering SDG 3—the ensuring of good health and well-being. AI technologies are contributing significantly to closing healthcare service gaps between areas that are underserved. Diagnostic devices driven by AI, telemedicine, and predictive analytics will improve the delivery of healthcare and patient outcomes, reduce equality gaps, and bring healthcare services closer to the population around the world (Sawers, 2021a). It is the recognition of vast amounts of medical data that would inherently lead away from generalization to personalized medicine, new therapies, and prevention strategies, thus directly contributing to SDG 3 (Tomaev, 2020).

In education, AI can make learning more personalized and available (Arini, 2024). From here, teachers can use AI-enabled educational tools to personalize courses and tests to the ordinary student experience, making learning much better and thus fulfilling the ideals of inclusive and equitable quality education for all (SDG 4). By having teachers do less admin work, they can teach better and, thus, this will improve the effectiveness of the education system. Fountains of literacy, even in remote or marginalized areas with a lack of educational resources, have opened opportunities in lifelong learning and reduced the gap in global education (Tomaev, 2020).

2.6.2 AI in agriculture and environmental sustainability

Agriculture is being transformed by AI in terms of SDG 2 achievement, which targets Zero Hunger. Tools powered by AI are improving crop yields and balancing water, pesticide, and fertilizer use so that farm inputs are much cheaper without sacrificing agricultural productivity through sustainable agriculture (Efremova, 2019). Precision farming technologies use AI in real-time monitoring and decision support to achieve productivity gains with reduced environmental damage. AI has been instrumental in predicting food demand and optimizing supply chains directed toward minimizing food loss and waste, thus also contributing to more sustainable agricultural practices (Sawers, 2021a). This, therefore, means that AI has the role of ensuring that agriculture increases survival while simultaneously solving global hunger and the issues of sustainable food systems.

Similarly, in the environmental sector, AI is used to monitor and reduce environmental damage. AI technologies are used in the monitoring of deforestation, forecasting of natural disasters, and their optimization in water resources management (Gupta, 2022). Above all, for example, satellite data consolidated with an AI model can analyze patterns of deforestation so that governments and organizations can respond immediately to preserving the very basis for climate change and the loss of biodiversity (Kulkov, 2023a).

Also, predictive capabilities of AI are for natural disasters: floods, droughts, and wildfires. They mostly give early warnings that can save lives while minimizing economic losses after these calamities (Gupta, 2022).

2.6.3 Ethical considerations and equitable access to AI

While AI has great potential, it has also turned a lot of corners in providing solutions to the SDGs. This, however, has put forth the ethical dilemmas regarding implementation in sustainable development, in which fairness, transparency, and inclusion are needed for AI applications. Gradually, a heightening chance is that of aggravating already pre-existing inequalities, which would be so much worse with the unequal access and benefits of AI technologies (Truby, 2020). This might mean, for example, that the current rise in the concentration of expertise and infrastructure for AI in high-income countries would further widen the already existing gap for low-income countries, thereby worsening inequality globally. Also, if left unaddressed, the biases found in algorithms of AI can deepen or entrench inequalities regarding access to or the quality of health, education, and employment (Kulkov, 2023a).

Create vital frameworks and ethical guidance that include appropriate incorporation and equity so that AI can serve useful roles for sustainable development. The initial frames or guidelines of AI development and deployment should be cognizant of social and cultural contexts relevant to the application of technology and foresee unintended consequences (Truby, 2020). The issue of privacy and security of data is also important since the decisions made by AI would affect sensitive information. Besides, developing capacity building for AI across different continents, especially in developing countries, thematically broadens the opportunity for equitable access to AI benefits for SDG attainment (Kulkov, 2023a).

2.6.4 Collaboration for sustainable AI solutions

Indeed, collaborative work by researchers, policymakers, and practitioners will make sure AI gives sustainable development in a responsible manner (Kulkov, 2023a). A multi-stakeholder partnership can create conditions under which AI tools and solutions would be developed to help improve local challenges while working in line with global sustainability goals. In addition, such partnerships would facilitate knowledge and resource sharing and help scale up validated AI solutions from one region to the rest of the world. Third, governments and organizations should put in place regulations that will encourage an ethical environment for developing AI, but allow innovations.

2.7 THE VISION OF AI BEYOND 2030

The future of AI beyond 2030 holds mysterious yet tremendous possibilities in regard to both challenges and the tremendous prospects one may have (Venkatasubramanian, 2018). These possibilities mainly relate to AGI—an objective for which machines are developed to replicate the cognitive capabilities of humans at the highest levels. If such machines prove successful in stepping beyond AGI, then humans are going to live a totally different life; in comparison to previous generations, all new discoveries, technologies, and economic growth will be completely unparalleled (Gautam and Thapaliya, 2024). The advances seem to carry, however, profound risks stemming in particular from the uncontrolled self-improvement of AGI and its consequences, which might be unintended (Gautam and Thapaliya, 2024). It will be very important to install strong mechanisms for AI safety and security, so AGI would be a boon to mankind and not result in significant catastrophic risks (Gautam and Thapaliya, 2024). Just as the development of AGI raises other ethical questions, such as the fact that it might circumvent human control and decision-making, collaboration between various developing countries will need to be engaged for consideration (Gautam and Thapaliya, 2024).

2.7.1 The role of human-AI collaboration

As AI gets more sophisticated, the future will be determined by an evolving human-AI collaboration (Venkatasubramanian, 2018). Instead of substituting human labor, AI would complement human skills, making for a future that is characterized by a more effective collaboration between humans and machines. This change in the nature of work could revolutionize entire industries by having AI do repetitive and data-intensive tasks while humans focus on creative, strategic, and interpersonal roles (Venkatasubramanian, 2018). The incorporation of AI into daily life can open up new avenues for productivity and innovation in all sectors. Now, a little more concise: our increasingly penetrative association to blur the lines between human and machine work more or less transforms into fresh, flexible, and dynamic workplaces.

Possibly, the most exciting part of this future will be human augmentation by AI, which will, in the very near future, radically transform the possibilities. For instance, BCIs will allow humans to communicate with machines without even the physical presence of their body—but instead through thought alone, thus augmenting cognition and creating new advanced formats for communication and interaction (Venkatasubramanian, 2018). A future of advanced prostheses and AI-based health interventions may restore or enhance human physical abilities and allow some persons with disabilities to act as humans are sometimes thought capable of acting, but could not have managed to do so in the past (Venkatasubramanian, 2018). All these aspects of human augmentation indeed promise much toward the enrichment of quality of life, but they also open a Pandora's box of ethical dilemmas involving issues on equity, privacy, and the creation of new inequalities (Venkatasubramanian, 2018).

2.7.2 Ethical implications of AI and human enhancement

This article is devoted to the complex ethical concerns surrounding human enhancements via AI, tracing this subject from some of its possible historical roots. The more the AI systems are integrated within human life, even in terms of usage as a lifestyle prop, the more the boundaries of biological and AIs will blur. Questions of consent, privacy, and access will then take center stage. Who will be able to make use of such life-changing technologies? Most would answer that the vast majority would not be able to afford it and, thus, make the gap between different groups bigger. How would one society construct the rules that would outline the limits of AI in the lives of humans? These questions shall need vigorous thought and a global ethical framework that ensures that human augmentation technologies are developed and used in ways that are just, equitable, and beneficial to all (Venkatasubramanian, 2018).

The need for strong governance structures and international cooperation in advance of AGI and its implications in human augmentation also exists. The ethical guidelines would evolve as the technology does, to answer new challenges raised by new technologies (Venkatasubramanian, 2018). International collaboration will be very much needed to ensure that the AI and human enhancement technologies are developed in a manner that is not going to harm and beneficial for humanity as a whole.

2.7.3 Long-term impacts on human society and the environment

The long-term effects of AI on human society and the environment are still speculative, but we must start now to monitor and address them as they occur. As AI systems become more capable, they will play an increasingly central role in decision-making processes that affect critical aspects of life, from healthcare and governance to environmental management and social structures (Gautam and Thapaliya, 2024). The potential of using AI in countering the difficulties of globalized challenges such as climate change and energy distribution resources will be great; however, for these systems not to exacerbate the current prevailing environmental and social issues, a sustainable approach shall be applied to their design. Sustainable development, driven by AI, will be relentlessly measured against long-term ecological and social objectives (Gautam and Thapaliya, 2024). AI will be a driver for social systems restructuring. As it continues to take on more work, we may see some redistributions of labor and sometimes some different economic models. Future work may see fewer traditional jobs, with those jobs being replaced by AI-driven tasks that will require new forms of collaboration and education. Job displacement and the social implications of an AI-dominated workforce need better planning and preparation in order not to worsen inequality and social unrest (Venkatasubramanian, 2018).

2.8 CONCLUSIONS AND RECOMMENDATIONS

AI, a vast area of multidimensionality, would reach its horizon by the year 2030. This innovation can be viewed from the ethical aspect of its social impact, its development, and its implementation into human life. The application of AI brings with it substantial possibilities to face global challenges, increase economic activity, and improve people's welfare all over the world. Innovations like deep learning, quantum AI, AGI, and neuromorphic computing will rewrite the rules of industries and ordinary life with extreme impacts on all sectors from health to education and climate science through transportation. The risks brought by opportunities include biases and a lack of accountability as well as job displacement and misuse of power vested in them, and they call for proactive approach strategies. The future beyond the year 2030 is really bright when looked at from the perspective of AI, but it is not without its challenges. More and closer alliances between human and machine partners are in the advance of AGI, AI, the IoT, robotics, and blockchain stream toward the future. Still no less frightening are the ethical, societal, and environmental risks, which might create more. But all this must go hand in hand with global collaboration and well-founded ethical foundations as well as proper regulatory controls to make AI safe and grow the same uniformly across the globe.

2.9 RECOMMENDATIONS

2.9.1 Policy recommendations

It is about developing comprehensive national and international policies that will translate into navigating existing challenges and maximizing the potential of AI. These include the following:

1. Major ethical issues will be addressed, such as bias, responsibility, transparency, and data privacy (Patil, 2024; Sigfrids, 2022).
2. Promotes international cooperation for standardization to avoid fragmented regulation (Sigfrids, 2022).

3. Facilitated ethical development of AI through regulatory frameworks compelling transparency and fairness in the algorithms of AI (Sigfrids, 2022).
4. Sponsored research with AI safety and security against malicious uses, and minimize the chances of (Gautam and Thapaliya, 2024).

2.9.2 Research recommendations

There is an urgent necessity for interdisciplinary research between the technical as well as ethical aspects of AI development. • They relate to highlighting ethical issues created by AI, whereby it operates increasingly autonomously and integrates into societal decision-making (Gautam and Thapaliya, 2024). Provide funding support for developing technical mechanisms and protocols for ensuring the safety as well as managing the risks as AI systems become more powerful (Shanker, n.d., NaN). • Further study different human-AI partnership models to ensure that these advanced systems augment human capabilities rather than replace the key human roles (Venkatasubramanian, 2018). • Review long-term societal and environmental impact of AI, including issues of sustainability and social equity (Venkatasubramanian, 2018). • Involve cross-disciplinary expertise from computer science to ethics, social sciences, and policy-making in designing holistic solutions to the complex issues posed by AI (Al-Kfairy et al., 2024).

2.9.3 Call to action

AI will only be assisting mankind if cooperation with the stakeholders is achieved with it. Governments must create policies that promote innovation and at the same time mitigate the dangers posed by AI advancements so that benefits do not accrue while harms or disadvantages befall a part of that society (Sigfrids, 2022; Reuel, NaN, n.d.). • Industries should integrate ethical considerations throughout the AI development process, prioritizing transparency, fairness, and inclusivity (Shanker, NaN, n.d.). • It calls for the lines of exploration of AI while remaining convinced of safety, security, and ethical practices (Gautam and Thapaliya, 2024; Venkatasubramanian, 2018). Hence, dialoguing with government and industry, and researchers will contribute significantly toward ensuring that AI bears a positive mark on society and that it brings forward a collective understanding of what AI could do for a sustainable and equitable future (Kulkov, 2023a).

2.10 CONCLUSION

In summary, AI will transform human life in several ways by 2030 and beyond, potentially solving some of the world's greatest problems—for instance, climate change, the gaps in healthcare, and issues pertaining to educational access. But the rapid advance of AI has raised troubling ethical, social, and technical issues that must be addressed proactively. Global governance frameworks, interdisciplinary investigations, and ethical development of AI will guarantee the realization of this complex technology for the good of all humanity. By 2030, AI will no longer consist of sets of specialized tools, but modern life itself will integrate the technologies. Joint and fruitful policymaking would also entail practice toward continuous ethical deliberation for AI promises to be beneficial and justifiable. The future of AI rests within us, shoulder to shoulder, making responsible efforts at creating a world that becomes more equitable, sustainable, and genuinely prosperous.

Figure 2.1 and Figure 2.2 https://aiindex.stanford.edu/report/

REFERENCES

Alahi, A. (2023). The Intersection of AI and IoT: Future Trends and Challenges. *Journal of AI and Internet of Things.*

Al-Kfairy, M., Mustafa, D. G., Kshetri, N., Insiew, M., &Alfandi, O. (2024). Ethical Challenges and Solutions of Generative AI: An Interdisciplinary Perspective. *Informatics.* https://doi.org/10.3390/informatics11030058

Arini, S. (2024). AI in Education: Transforming Learning Experiences for All. *Journal of Educational Technology*, 14(2), 78–92.

Dalkey, D. (1963). Blockchain Technology for Secure and Transparent AI Systems. *Journal of Cryptography and AI Security.*

Efremova, M. (2019). The Role of AI in Sustainable Agriculture: Boosting Productivity and Reducing Waste. *Journal of AI and Sustainability*, 11(4), 67–83.

Gautam, A., & Thapaliya, S. (2024). A chronology of AI failures in safety and cybersecurity. None. https://doi.org/10.3126/nprcjmr.v1i6.71734

Gupta, R. (2022). AI for Environmental Sustainability: Monitoring, Predicting, and Protecting Nature. *Journal of Environmental AI*, 6(1), 35–50.

Kulkov, A. (2023a). AI for Sustainable Development: Optimizing Resources for Global Challenges. *International Journal of AI and Sustainable Development*, 15(2), 43–61.

Kulkov, I., *et al.* (2023). Artificial intelligence-driven sustainable development: Examining organizational, technical, and processing approaches to achieving global goals. *Sustainable Development.* https://doi.org/10.1002/sd.2773

Kulkov, S. (2023b). AI in Climate Change: Applications and Future Prospects. *Environmental Science & Technology.*

Patil, R. (2024). *Accountability in Artificial Intelligence Systems.* MIT Press.

Reuel, C. (NaN). (n.d.) *Balancing Regulation and Innovation in AI.* Wiley.

Ryan, M., & Stahl, B. C. (2020). *Artificial Intelligence Ethics Guidelines for Developers and Users: Clarifying their Content and Normative Implications.* Emerald Publishing Limited. https://doi.org/10.1108/jices-12-2019-0138

Satyam, A. (2023). AI's Economic Impact: Opportunities and Risks. *Business Economics Review*, 56(4), 421–437.

Sawers, J. (2021a). AI for Social Good: Addressing Inequalities in Healthcare and Education. *Nature Communications.*

Sawers, L. (2021b). AI and Healthcare: Improving Access and Efficiency. *Journal of Healthcare AI*, 12(3), 100–115.

Sawers, N., Bolster, N. M., & Bastawrous, A. (2021). The Contribution of Artificial Intelligence in Achieving the Sustainable Development Goals (SDGs): What Can Eye Health Learn from Commercial Industry and Early Lessons from the Application of Machine Learning in Eye Health Programmes. *Frontiers in Public Health*. https://doi.org/10.3389/fpubh.2021.752049

Sebastian, G. (2023). Do ChatGPT and other AI chatbots pose a cybersecurity risk? An exploratory study. *Social Science Research Network*. https://doi.org/10.2139/ssrn.4363843

Shanker, B., & Neyigapula, (NaN). (n.d.) Ethical considerations in AI development: Balancing autonomy and accountability. None. https://doi.org/10.18178/jaai.2024.2.1.138-148

Sigfrids, H. (2022). *AI Governance and Global Collaboration*. Routledge.

Spiro, G. (NaN). (n.d.) *AI Transparency: Why It Matters*. Oxford University Press.

Srihith, P. (2022). The Future of Work in an AI-Driven Economy. *Journal of Labor Economics*, 40(2), 253–285. https://doi.org/10.1086/719130

Tomaev, T. (2020). AI in the Global Race for Sustainable Development: Opportunities and Risks. *AI and Global Development*, 8(1), 55–71.

Truby, J. (2020). AI and Sustainable Development: Ethical Considerations and Policy Frameworks. *Sustainable Development Policy Journal*, 9(2), 120–133.

Venkatasubramanian, P. (2018). AI and the Future: Trends and Technologies. *AI and Robotics Journal*, 15(4), 342–367.

Zhang, Z. (2024). Advancements and Challenges in AI-Driven Language Technologies: From Natural Language Processing to Language Acquisition. *Applied and Computational Engineering*. https://doi.org/10.54254/2755-2721/57/20241325

Chapter 3

Finite element analysis for structural performance of flat slabs with column-adjacent openings of varying shapes

Jaicky H. Gurnani, Meghna Jumde, Amit Kharwade, Buddharatna Godboley, Nitesh Thikare, and Mayank Gupta

3.1 INTRODUCTION TO FLAT SLAB

The flat slab system is composed of a reinforced concrete panel that is directly supported by concrete columns, thereby eliminating the need for intermediate beams. The advantages associated with this construction method are being acknowledged more widely and at an accelerated pace. This approach, characterized as a two-way reinforced concrete slab, is broadly implemented in projects such as expansive supermarkets, retail centers, subterranean parking facilities, and bridge decks, among other applications (Guandalini & Muttoni, 2004). Furthermore, flat slabs are frequently employed in high-rise constructions for diverse functional requirements and in buildings designed to accommodate adaptable three-dimensional configurations over their service life. This technique represents a well-established and effective structural system in which columns support floor slabs directly, avoiding the use of beams—a methodology often preferred in the construction of in situ concrete frame buildings (Ahmed et al., 2018).

Architectural considerations necessitate the incorporation of apertures within the slab to accommodate elements such as staircases, elevator shafts, installations for gas, electrical, and water supply, and air conditioning systems. The integration of these apertures disrupts the natural load-transfer path of the structure, thereby introducing imbalances in both shear forces and bending moments (Al-Jaberi et al., 2021). One of the primary design challenges associated with flat slabs lies in ensuring adequate punching shear capacity. Given that punching shear failure exhibits a brittle and sudden nature, it is considered one of the most critical failure modes, with the potential to cause the collapse of the entire structure (Beutel, 2001).

Both the dimensions and the positioning of these openings play a significant role in affecting the bending moment distribution and the overall punching shear capacity of the system (Einpaul et al., 2015). As the size of an aperture increases (Eder et al., 2010), the quantity of reinforcement steel

DOI: 10.1201/9781003642152-3

required also escalates. Although the presence of an opening inherently raises safety concerns (El-Salakawy & Polak, 2005), the situation becomes most critical when an aperture is located immediately adjacent to a column (Birkle & Dilger, 2008). An opening positioned close to the column can substantially diminish the punching shear capacity of the flat slab (Bompa & Elghazouli, 2020). Considering that flat plate slabs are inherently susceptible to punching shear, the occurrence of an opening at the column face may impose notable constraints on the design of the system (Broms, 2000).

It is imperative that the effects of these apertures on the punching shear behavior within flat slab–column assemblies are examined meticulously (Choi & Song, 2007). For reasons of safety, punching shear failure must be rigorously evaluated, and a conservative, traditional design approach is recommended (Deng & Matsumoto, 2020). In particular, the influence of openings placed at the column face on the punching shear performance of the flat slab–column system requires careful and detailed investigation. This paper investigates the behavior of flat slabs with regard to varying aperture sizes, positions, combinations, and aspect ratios, ultimately providing a series of recommendations based on the analytical findings.

3.2 OBJECTIVES

The primary aim of this investigation is to evaluate how various parameters affect the performance of flat slab systems. In detail, the study is designed to:

- Examine the influence of varied opening dimensions on the structural behavior of flat slabs.
- Investigate the effect that the distance of an opening from the column face has on the slab's overall response.
- Analyze how differing geometric configurations of openings modify the performance characteristics of flat slabs.
- Assess the impact of the opening's aspect ratio (i.e., the length-to-width proportion) on the system's structural behavior.

3.3 ANALYTICAL MODELING

The finite element analysis of the flat slab system was performed using the CSI SAFE2016 software package, a tool well-regarded for its capabilities in structural concrete analysis (CSI SAFE). The investigation focused on an interior flat slab panel with overall dimensions of 6 m by 6 m, which incorporated both a drop panel and a column head as an integral design feature. The specified columns featured dimensions of 0.5 m by 0.5 m. A live load intensity of 5 kN/m^2 and an additional floor finish load of 2 kN/m^2

were applied uniformly across the slab panel. Openings, which were considered in both square and circular shapes, were designed to possess dimensions equal to those of the supporting columns. The concrete utilized throughout the study was of grade M30, while the reinforcement employed was of Fe500 grade. In addition, the drop panel was proportioned such that its size corresponded to one-third of a representative length (L/3), and the diameter of the column head was determined to be one-fourth of that length (L/4).

3.4 DESIGN INPUT

The design parameters for this investigation are outlined with precise dimensional and material specifications to support the analysis of the flat slab system. The study employs an internal slab panel with plan dimensions of 6 m by 6 m. Each supporting column is configured to measure 500 mm in both width and depth, ensuring a robust framework for load transfer. The overall structural slab is cast with a thickness of 200 mm, and the building is designed with a floor-to-floor height of 4.00 m, accommodating both functional and architectural needs.

Furthermore, an integral drop panel is incorporated into the design, possessing a plan dimension of 2000 mm by 2000 mm and a depth of 300 mm. In addition to this, the column head—or capital—is dimensioned at 1330 mm by 1330 mm with a thickness of 500 mm, serving as a critical transitional element between the columns and the slab. The structural components are constructed using M30 grade concrete, ensuring adequate compressive strength, while Fe500 grade reinforcement steel is employed to provide the necessary tensile capacity. These choices are aligned with standard industry practices and design codes, such as those detailed in IS 456:2000 (Bureau of Indian Standards) and supported by software-specific guidelines from CSI SAFE.

The slab is subjected to a combination of live and dead loads, specifically a live load of 5 kN/m^2 along with a floor finish load of 2 kN/m^2, ensuring that the design is both serviceable and durable under typical usage conditions. Regarding the architectural and structural openings integrated into the slab, two distinct configurations are adopted: a square opening measuring 500 mm by 500 mm and a circular opening defined by a diameter of 564.20 mm. These openings are included to accommodate various service conduits and other building utilities while simultaneously influencing the overall structural behavior of the flat slab system.

Each of these design inputs is critical for the subsequent detailed analysis and simulation, contributing to the robustness and accuracy of the finite element investigation performed using CSI SAFE2016.

3.4.1 Nonlinear analysis

CSI-SAFE v16 software is employed to execute nonlinear evaluations of the slab models investigated in this research. The integrated automated design feature is accessible for every model examined, whether or not it includes an aperture. Consequently, determining and contrasting essential performance metrics—namely, punching shear capacity and structural deflection—across the various scenarios is achieved with relative ease (CSI-SAFE).

3.4.2 Material properties

The investigation employs a concrete mix with a specified compressive strength of 30 N/mm^2, complemented by Fe500 grade reinforcement steel to ensure optimal structural performance (Bureau of Indian Standards, IS 456:2000).

3.4.3 Boundary condition

Since the focus of this study is the interior panel of the slab, it is assumed that the periphery of the slab is fully continuous, ensuring an unbroken load-transfer path along its edges (Bureau of Indian Standards, IS 456:2000).

3.4.4 Mesh size

For the computational analysis, a discretization strategy was implemented whereby the maximum element size was limited to 1000 mm.

3.4.5 Notation for the specimen

In this research, a systematic nomenclature is employed to categorize various flat slab specimens, facilitating clear identification and comparison of different structural configurations. The notation scheme is defined as follows:

- FS-D-C: Represents a flat slab system without a drop panel and without a column head.
- FS+D+C: Denotes a flat slab system incorporating both a drop panel and a column head.
- NO: Indicates the absence of any openings in the slab.
- SO: Refers to a square-shaped opening within the slab.
- CO: Denotes a circular-shaped opening within the slab.
- @0: Specifies that the opening is located directly at the face of the column.
- @0.2: Indicates that the opening is positioned 0.2 meters away from the face of the column.
- @C: Signifies that the opening is situated at the center of the slab panel.

For instance, the notation FS-D-C_SO@0.2 describes a flat slab configuration lacking both a drop panel and a column head, featuring a square opening located 0.2 meters from the column face. This systematic approach to labeling ensures precise communication of the specific structural configurations under investigation, thereby enhancing the clarity and reproducibility of the study's findings.

To facilitate a comprehensive understanding of the structural behavior of flat slabs incorporating openings, the analytical framework of this study is systematically organized into two primary groups. This bifurcation allows for a detailed examination of the influence exerted by various opening configurations on key performance indicators such as punching shear resistance and deflection characteristics.

3.5 RESEARCH FRAMEWORK

Group A encompasses flat slab models devoid of both drop panels and column heads. This group serves as a baseline to assess the structural implications of openings in the absence of additional reinforcement features.

Group B includes flat slab models equipped with both drop panels and column heads. The inclusion of these elements aims to evaluate their effectiveness in mitigating the adverse effects introduced by openings.

Each group is further subdivided based on the presence and geometry of openings:

- NO: Models without any openings, providing a control scenario for comparative analysis.
- SO: Models featuring square-shaped openings, allowing for the assessment of angular discontinuities in the slab.
- CO: Models incorporating circular-shaped openings, facilitating the evaluation of curved discontinuities.

The location of these openings is also varied to investigate positional effects:

- @0: Openings positioned directly at the face of the column, representing the most critical scenario due to proximity to high-stress regions.
- @0.2: Openings located 0.2 meters away from the column face, offering insight into the impact of slight positional shifts.
- @C: Openings situated at the center of the slab panel, enabling the study of effects in regions removed from direct column support.

This structured approach ensures a meticulous analysis of how different opening shapes and positions influence the structural integrity of flat slabs,

Table 3.1 Details of model specification

Group	*Case*	*Opening shape*	*Opening size*	*Opening location*
Group A	Case I	Without Opening	-	-
	Case II	Circular Opening	(0.5642 m ϕ)	Measurements were conducted along the column face, ranging from 0 to 3 meters, at consistent intervals of 0.2 meters.
	Case III	Square Opening	(0.5 x 0.5 m)	
Group B	Case I	Without Opening	-	-
	Case II	Circular Opening	(0.5642 m ϕ)	Measurements were conducted along the column face, ranging from 0 to 3 meters, at consistent intervals of 0.2 meters.
	Case III	Square Opening	(0.5 x 0.5 m)	

thereby providing valuable insights for design optimization and safety assessments.

The dimensional attributes and positional configurations of the openings incorporated in the test specimens are systematically presented in Table 3.1. This table delineates the specifics pertaining to each specimen category.

Group A pertains to the set of slab specimens designed without the inclusion of drop panels or column capitals, thus representing a structurally unenhanced scenario.

In contrast, Group B comprises specimens that integrate both drop panels and column heads, thereby reflecting an improved structural configuration aimed at resisting shear and deflection more effectively.

Each group is further segmented into three subcategories for a comprehensive evaluation. The initial subgroup within each category serves as a baseline or control model, characterized by the absence of any openings. The subsequent subgroups incorporate openings of different geometries—namely circular and square—strategically positioned at varying distances from the column face.

Figures 3.1–3.4 visually illustrate these configurations, aiding in the spatial understanding of the experimental setup. The diversity in opening shape and placement across these models is intended to provide a detailed insight into their influence on the mechanical behavior and performance characteristics of the flat slab system. This methodology facilitates a nuanced analysis of the structural response, particularly in relation to punching shear, deflection, and moment redistribution.

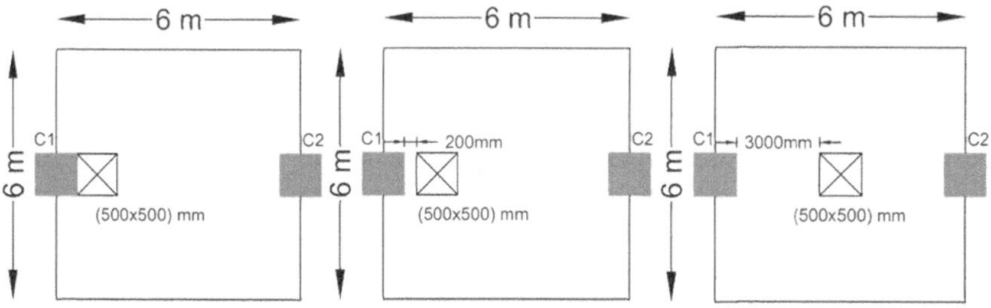

Figure 3.1 Flat slab without drop panel and column capital featuring a square opening.

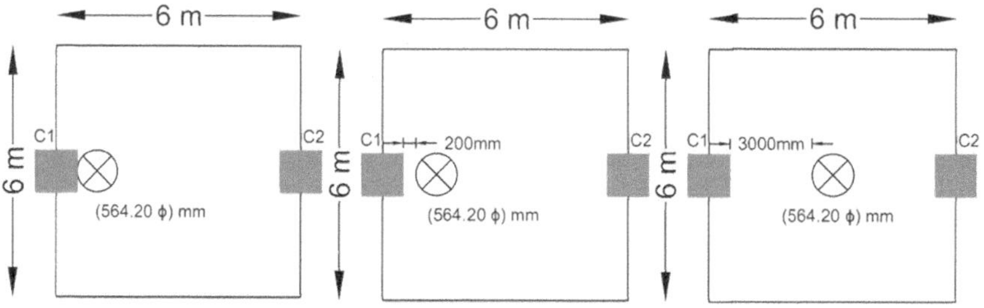

Figure 3.2 Flat slab without drop panel and column capital featuring a circular opening.

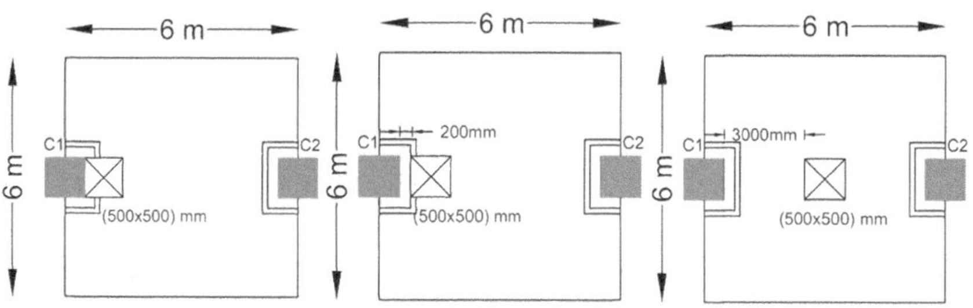

Figure 3.3 Flat slab with drop panel and column capital incorporating a square opening.

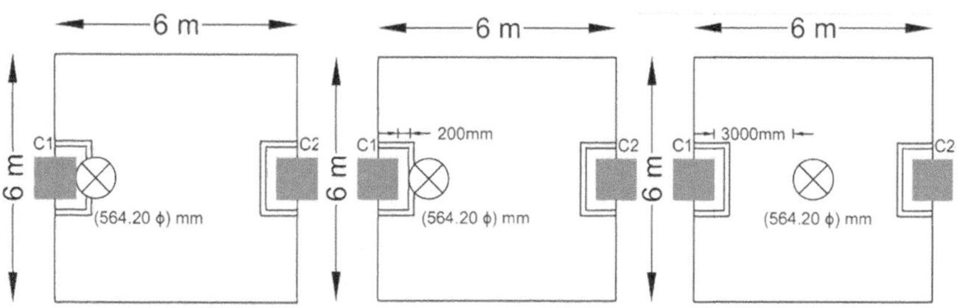

Figure 3.4 Flat slab with drop panel and column capital incorporating a circular opening.

3.6 ANALYTICAL RESULTS AND OBSERVATION

The structural analysis of flat slab systems was conducted using CSI SAFE software, encompassing two distinct configurations: one incorporating both drop panels and column heads, and another omitting these features. The outcomes of these analyses are systematically presented in graphical form across Figures 3.5–3.14, illustrating the variations in structural behavior under different configurations (Table 3.2).

The presence of openings adjacent to the column face significantly compromises the ultimate load-bearing capacity of flat slabs, primarily due to a reduction in the critical punching shear perimeter. This reduction is especially pronounced in specimens where openings are positioned directly at

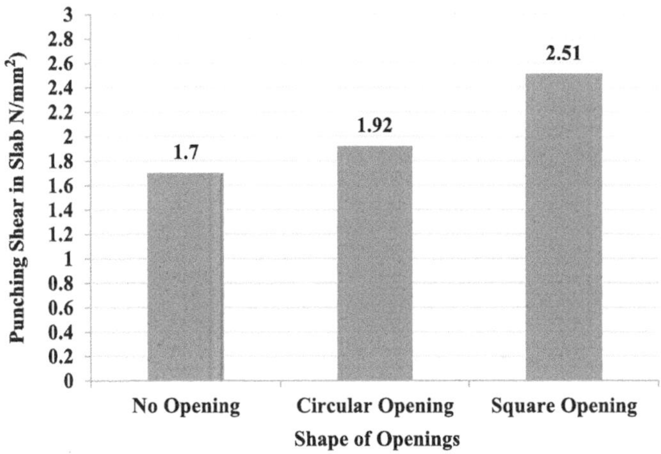

Figure 3.5 Peak two-way punching shear in the slab for Group A.

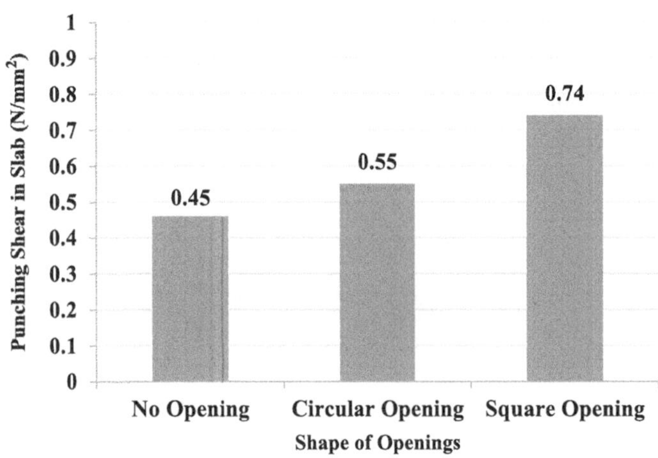

Figure 3.6 Peak two-way punching shear in the slab for Group B.

Table 3.2 Percentage enhancement in the punching shear capacity of the slab

Shape of opening	*Increase in punching shear relative to the no-opening condition*	*Increase in punching shear relative to the no-opening condition*
	Group A	**Group B**
Circular Opening	12.94%	22.22%
Square Opening	47.65%	64.44%

the column face, such as FS-D-C_SO@0 and FS+D+C_SO@0, which exhibit the highest punching shear stresses within their respective groups. The diminished perimeter around the column in these configurations leads to elevated stress concentrations, thereby increasing susceptibility to punching shear failure.

Analytical investigations reveal that relocating openings further from the column face mitigates their detrimental impact on punching shear capacity. As the distance between the opening and the column face increases, the critical shear perimeter is less compromised, resulting in a more uniform stress distribution and enhanced structural performance. This observation aligns with findings from experimental studies, which indicate that openings situated beyond a certain threshold distance—typically four times the slab thickness—exert negligible influence on punching shear strength.

In summary, the strategic placement of openings away from the column face is crucial in preserving the punching shear capacity of flat slabs. Design considerations must account for the size and location of openings to ensure structural integrity and prevent premature failure due to punching shear (Table 3.3).

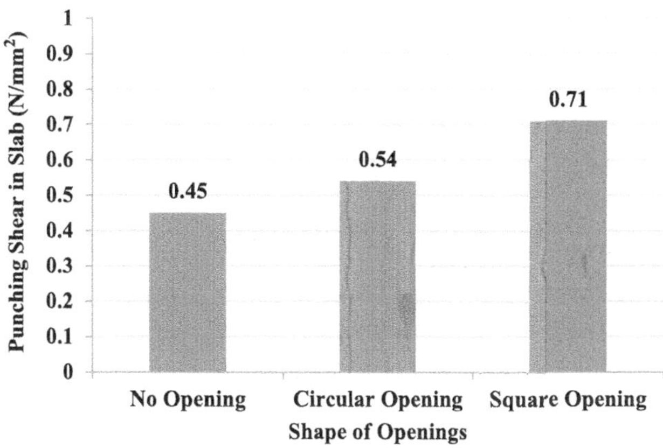

Figure 3.7 Peak two-way punching shear in the drop panel for Group B.

Table 3.3 Percentage augmentation in the punching shear strength at the drop region

Opening shape	*Percentage increase in punching shear at the drop region compared to the scenario with no openings for Group B*
Circular Opening	20%
Square Opening	57.78%

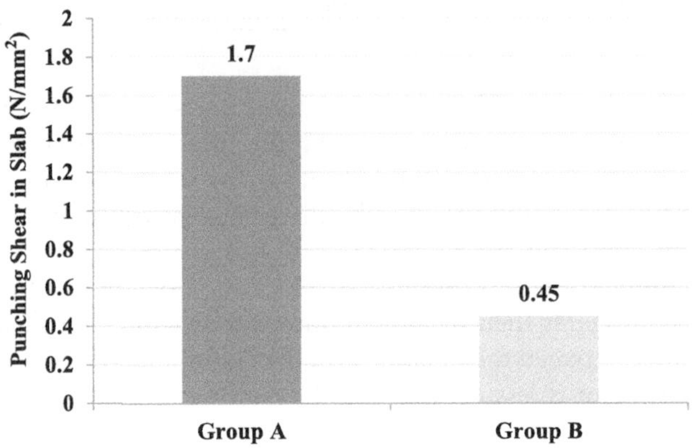

Figure 3.8 Peak two-way punching shear in the slab for Group A and Group B without any openings.

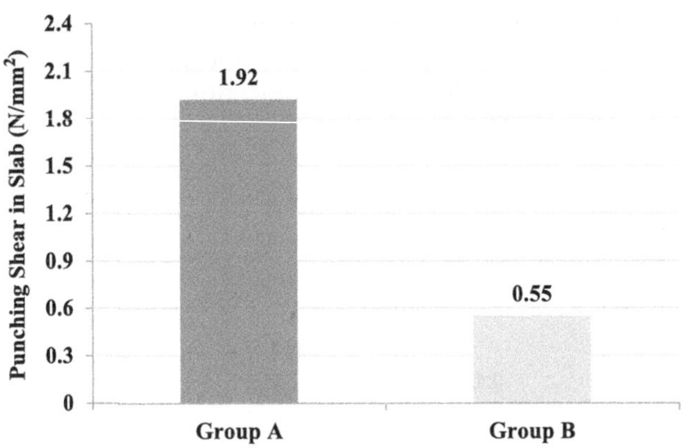

Figure 3.9 Peak two-way punching shear in the slab for Group A and Group B with a circular opening.

The presence of openings adjacent to the column face significantly compromises the ultimate load-bearing capacity of flat slabs, primarily due to a reduction in the critical punching shear perimeter. This reduction is especially pronounced in specimens where openings are positioned directly at

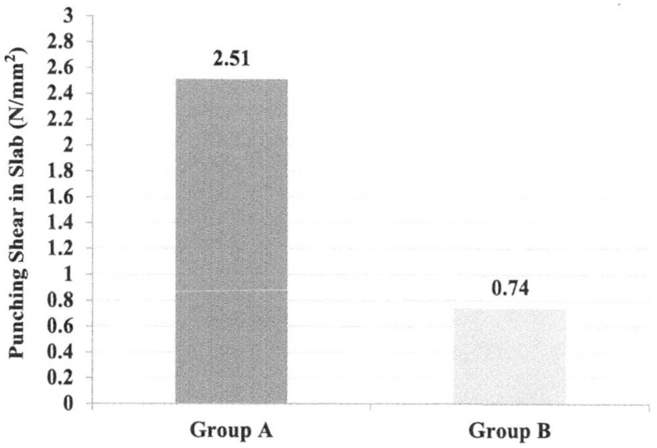

Figure 3.10 Peak two-way punching shear in the slab for Group A and Group B with a square opening.

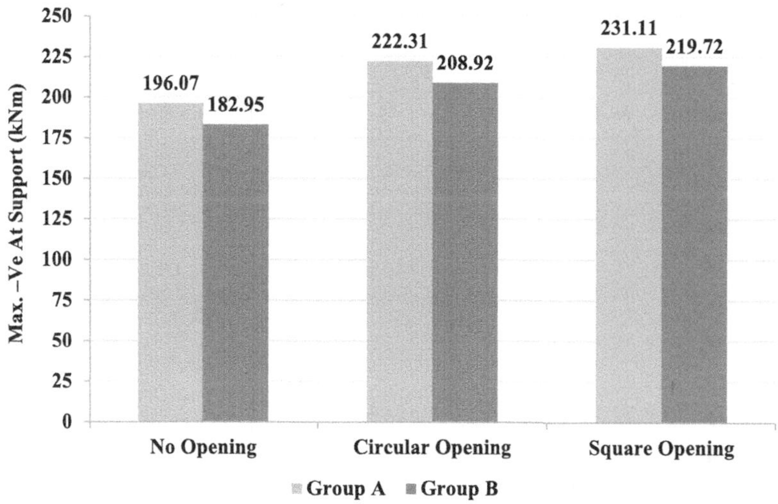

Figure 3.11 Maximum negative bending moment at the support (kN·m) within the column strip.

the column face, such as FS-D-C_SO@0 and FS+D+C_SO@0, which exhibit the highest punching shear stresses within their respective groups. The diminished perimeter around the column in these configurations leads to elevated stress concentrations, thereby increasing susceptibility to punching shear failure.

Analytical investigations reveal that relocating openings further from the column face mitigates their detrimental impact on punching shear capacity. As the distance between the opening and the column face increases, the

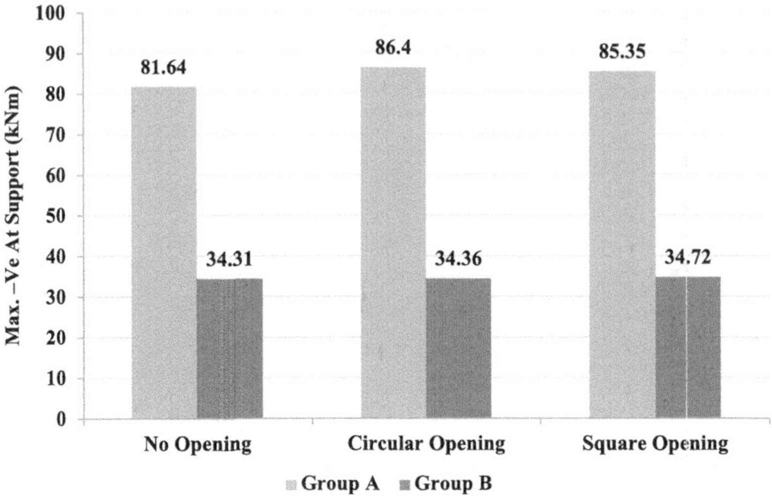

Figure 3.12 Maximum negative bending moment at the support (kN·m) within the middle strip.

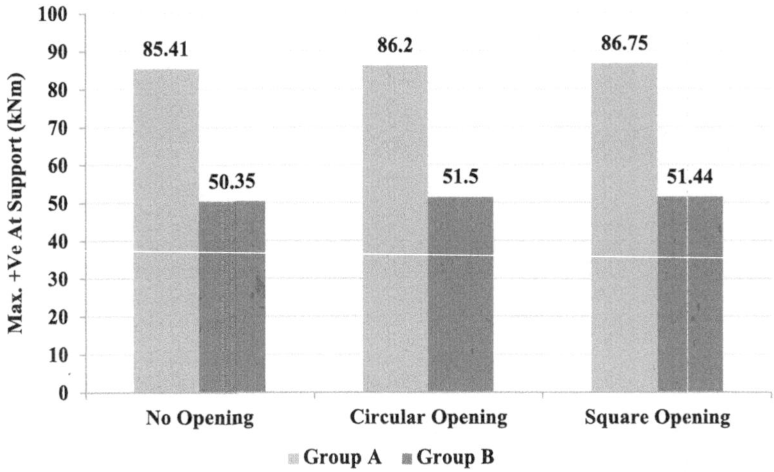

Figure 3.13 Peak positive bending moment at the support (kN·m) within the column strip.

critical shear perimeter is less compromised, resulting in a more uniform stress distribution and enhanced structural performance. This observation aligns with findings from experimental studies, which indicate that openings situated beyond a certain threshold distance—typically four times the slab thickness—exert negligible influence on punching shear strength.

In summary, the strategic placement of openings away from the column face is crucial in preserving the punching shear capacity of flat slabs. Design

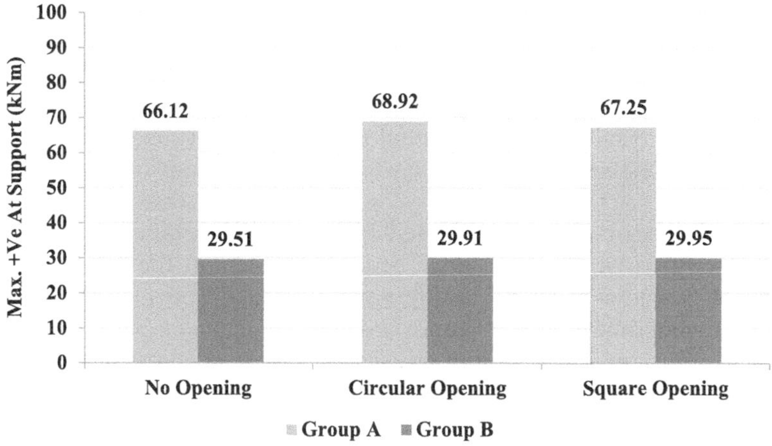

Figure 3.14 Peak positive bending moment at the support (kN·m) within the middle strip.

considerations must account for the size and location of openings to ensure structural integrity and prevent premature failure due to punching shear.

The inclusion of a drop panel in a flat slab, with a thickness of L/3, leads to a reduction in the punching shear value by a significant 73.53% compared to a flat slab without a drop panel, particularly under conditions where no openings are present. This reduction in punching shear strength highlights the structural benefit of incorporating drop panels in flat slab designs. The enhanced behavior is especially noticeable in situations where no openings are made in the slab, thereby ensuring a higher level of load distribution and resistance to shear failure. This finding emphasizes the effectiveness of drop panels in improving the punching shear capacity of flat slabs and offers important insights into their potential application in structural design, where enhanced shear performance is crucial.

The introduction of a drop panel with a size of L/3 in a flat slab results in a reduction of punching shear strength by 71.35% when compared to a flat slab without a drop panel, specifically under conditions involving a circular opening. This significant decrease in punching shear capacity emphasizes the effectiveness of the drop panel in mitigating shear failure, even in the presence of an opening. The drop panel contributes to a more efficient load distribution and enhances the slab's resistance to punching shear, thereby offering valuable insights for structural designs involving openings. The findings underline the potential advantages of incorporating drop panels in flat slab systems, particularly when openings are present.

In the case of a square opening, the incorporation of a drop panel with dimensions equal to L/3 in a flat slab leads to a substantial 70.52% reduction in punching shear value compared to a flat slab lacking a drop panel. This notable decrease illustrates the critical role that drop panels play in enhancing the shear resistance of flat slabs, even when geometric discontinuities such as square openings are present. The presence of the drop panel improves the slab's capacity to distribute loads more uniformly and resist localized shear stresses, thereby contributing to improved structural performance. These results provide essential evidence supporting the use of drop panels in flat slab systems, particularly in configurations involving square openings.

The presence of an opening located adjacent to the column face leads to a noticeable increase in the negative bending moment at the support region within the column strip. Specifically, when a circular opening is introduced, the negative moment rises by approximately 13.38% compared to the scenario without any opening. In the case of a square opening, the increase is even more pronounced, reaching 17.87% relative to the no-opening condition. These findings highlight the significant influence of opening geometry on the flexural behavior of flat slabs, particularly in the vicinity of critical support zones. The alterations in stress distribution caused by the openings necessitate careful consideration in structural design to ensure adequate performance and safety.

The introduction of an opening at the column face contributes to an increase in the negative bending moment at the support region within the middle strip of the slab. Specifically, a circular opening results in a 5.83% rise in the negative moment when compared to a slab without any openings, whereas a square opening leads to a 4.54% increase under the same conditions. These variations underscore the influence of opening geometry on the flexural response of flat slabs in critical regions. Furthermore, the incorporation of a drop panel with a dimension of L/3 in the flat slab significantly reduces the negative bending moment at the support in the middle strip across all scenarios. For slabs without openings, the use of such a drop panel decreases the negative moment by 57.97%. In the presence of a circular opening, the reduction is even more pronounced, reaching 60.23%, while for slabs with a square opening, the negative moment is reduced by 59.32%. These results clearly demonstrate the beneficial role of drop panels in mitigating negative moment intensities at support zones, thereby enhancing the structural efficiency and performance of flat slab systems regardless of opening conditions.

The presence of an opening located at the column face slightly elevates the positive bending moment at the support region within the column strip of a flat slab. Specifically, a circular opening induces a 0.69% increase in the positive moment compared to a slab without any opening, while a square opening results in a slightly higher increase of 1.57% under the same condition. Although these increments are relatively modest, they indicate that the introduction of openings, regardless of shape, influences the flexural behavior of the slab in support regions.

In contrast, the implementation of a drop panel with a size of L/3 in the flat slab substantially reduces the positive bending moment at the support in the column strip for all cases studied. Under no-opening conditions, the drop panel leads to a 41.05% decrease in the positive moment. When a circular opening is present, the reduction is measured at 40.26%, and for square openings, the positive moment decreases by 40.7%. These findings emphasize the structural effectiveness of L/3-sized drop panels in minimizing positive bending moments at support zones, thereby enhancing the load-bearing capacity and overall performance of flat slab systems with or without openings.

The presence of an opening adjacent to the column face leads to an increase in the positive bending moment at the support region within the middle strip of a flat slab. For a circular opening, the positive moment increases by 4.23% compared to the case without any opening, while for a square opening, the increase is 1.71%. These results indicate that openings, particularly circular ones, modestly affect the flexural response in the middle strip's support zone.

In contrast, incorporating a drop panel with a depth of L/3 in the flat slab configuration significantly reduces the positive bending moment at the middle strip support. In the scenario without any openings, this design intervention leads to a 55.37% reduction. When circular openings are introduced, the positive moment decreases further by 56.6%, and for square openings, the reduction is 55.46%. These findings clearly demonstrate the efficiency of drop panels in mitigating support moments across varying opening conditions.

Despite these reductions, the rate of change in slab moments near the openings is observed to be substantial. However, these moment disturbances do not critically influence the overall structural behavior of the flat slab system. It is evident that the size and shape of openings primarily impact the negative bending moments in the column strip, while their influence on other moment zones remains relatively minor.

Furthermore, an increase in opening size leads to a corresponding rise in the rate of change of slab moments, suggesting a nonlinear interaction between opening dimensions and moment distribution. Among the geometries studied, slabs with square openings exhibit the most pronounced effect on the internal moment values compared to those with circular openings or no openings at all. These insights underscore the necessity of carefully considering opening characteristics during the design phase to ensure both structural efficiency and safety.

3.7 CONCLUSION AND RECOMMENDATIONS

- An increase in the dimensions of slab openings directly contributes to a rise in punching shear demand, which, in turn, results in a notable reduction in the ultimate punching load capacity of the flat slab system. The observed data clearly indicate that slabs without openings demonstrate superior punching shear resistance compared to those containing voids.

- Analysis further reveals that the positioning of openings plays a critical role in structural performance. Specifically, openings situated away from the column face are less detrimental to punching shear capacity than those located directly at the column face. Among the tested configurations, the presence of a square opening adjacent to the column significantly compromises the punching shear perimeter of the resisting section. This leads to the highest recorded punching shear demand and a consequent reduction in load-bearing capacity, reinforcing the understanding that punching shear is the dominant failure mode in flat slab systems.
- Conversely, circular openings placed at the column face result in a comparatively lower increase in punching shear. This is attributed to a relatively minor reduction in the effective punching shear perimeter, highlighting the geometric advantage of circular shapes in preserving the integrity of the resisting section.
- For flat slabs designed without drop panels or column capitals, the introduction of a square opening at the column face increases the punching shear by approximately 48% compared to the baseline slab without openings. In the case of a circular opening under the same configuration, the increase is more moderate at 13%.
- However, when a drop panel and column capital are incorporated into the design, the structural behavior improves markedly. Under this configuration, a square opening at the column face leads to a 64% increase in punching shear, while a circular opening results in a 22% increase, demonstrating the mitigating effect of structural enhancements.
- The implementation of a drop panel with a depth equal to L/3 significantly improves punching shear resistance. Specifically, the inclusion of this structural element reduces the punching shear values by approximately 74% in slabs without openings and by 71% in slabs with either circular or square openings, relative to slabs without drop panels.
- Similar benefits are observed in terms of flexural behavior. The addition of an L/3-sized drop panel results in a reduction in negative bending moments at the middle strip support by 58%, 60%, and 59% for slabs with no openings, circular openings, and square openings, respectively. Likewise, positive bending moments at the support in the column strip are reduced by 41%, 40%, and 41% across the same conditions. In the middle strip support region, the positive bending moments are lowered by 55%, 57%, and 55%, respectively.
- Finally, it is evident that relocating slab openings away from the column face reduces the severity of the impact on the punching shear perimeter, thereby enhancing the structural efficiency of the slab. This improvement is attributed to the preservation of a more continuous and effective shear-resisting boundary around the column.

BIBLIOGRAPHY

Ahmed, E., Eltaly, B., El-Zhraa, F., &Tayel, M. (2018). Resisting punching shear stress in reinforced concrete slabs. *MATEC Web of Conferences*, *162*, 04025. https://doi.org/10.1051/matecconf/201816204025

Al-Jaberi, L., Abbas, Z. W., & Jaffar, E. K. (2021). Effect of volume of steel fibers on the punching shear behavior of hybrid reinforced concrete flat slab. *Journal of Engineering and Sustainable Development*, *25*(4), 1–10. https://doi.org/10.31272/jeasd.25.4.1

Beutel, R. (2001). 3D numerical punching analysis of shear reinforced flat slabs—variation of quantity and arrangement of stirrups. In *Punching of Structural Concrete Slabs* (pp. 79–85). CEB/fib International Federation for Structural Concrete.

Birkle, G., & Dilger, W. H. (2008). Influence of slab thickness on punching shear strength. *ACI Structural Journal*, *105*(2), 180–188.

Bompa, D. V., &Elghazouli, A. Y. (2020). Nonlinear numerical simulation of punching shear behavior of reinforced concrete flat slabs with shear-heads. *Frontiers of Structural and Civil Engineering*, *14*(2), 331–356. https://doi.org/10.1007/s11709-019-0596-5

Broms, C. E. (2000). Elimination of flat plate punching failure mode. *ACI Structural Journal*, *97*(1), 94–101.

Choi, J. W., & Song, J. K. (2007). The punching shear strength of lightly reinforced thick flat plates without shear reinforcement. *Journal of the Architectural Institute of Korea*, *23*(8), 67–74.

Corley, W. G., & Hawkins, N. M. (1968). Shearhead reinforcement for slabs. *ACI Journal Proceedings*, *65*(10), 811–824.

Deng, P., & Matsumoto, T. (2020). Fracture mechanics-based fatigue life prediction method for RC slabs in punching shear failure mode. *Journal of Structural Engineering*, *146*(1), 04019186. https://doi.org/10.1061/(ASCE)ST.1943-541X.0002504

Dilger, W. H., & Ghali, A. (1981). Shear reinforcement for concrete slabs. *Journal of Structural Division*, *107*(12), 2403–2420.

Eder, M. A., Vollum, R. L., Elghazouli, A. Y., & Abdel-Fattah, T. (2010). Modelling and experimental assessment of punching shear in flat slabs with shear-heads. *Engineering Structures*, *32*(12), 3911–3924. https://doi.org/10.1016/j.engstruct.2010.08.015

Einpaul, J., Ruiz, M. F., & Muttoni, A. (2015). Influence of moment redistribution and compressive membrane action on punching strength of flat slabs. *Engineering Structures*, *86*, 43–57. https://doi.org/10.1016/j.engstruct.2014.12.032

El-Salakawy, E. F., & Polak, M. A. (2005). Punching shear behavior of flat slabs reinforced with fiber-reinforced polymer bars. *ACI Structural Journal*, *102*(5), 708–715.

Elstner, R. C., & Hognestad, E. (1956). Shearing strength of reinforced concrete slabs. *Journal of the American Concrete Institute*, *53*(7), 29–58.

Eom, T. S., Kang, S. M., Choi, T. W., & Park, H. K. (2018). Punching shear tests of slabs with high-strength continuous hoop reinforcement. *ACI Structural Journal*, *115*(5), 1295–1305. https://doi.org/10.14359/51702231

Faridmehr, I., Nehdi, M. L., & Baghban, M. (2022). Novel informational bat-ANN model for predicting punching shear of RC flat slabs without shear reinforcement. *Engineering Structures*, *256*, 114030. https://doi.org/10.1016/j.engstruct.2022.114030

Ferreira, M. P., Melo, G. S., Regan, P. E., & Vollum, R. L. (2014). Punching of reinforced concrete flat slabs with double-headed shear reinforcement. *ACI Structural Journal*, *111*(2), 363–374. https://doi.org/10.14359/51686535

Ghoreishi, M., Bagchi, A., & Sultan, M. A. (2013). Review of the punching shear behaviour of concrete flat slabs in ambient and elevated temperature. *Journal of Structural Fire Engineering*, *4*(4), 259–279. https://doi.org/10.1260/2040-2317.4.4.259

Ghoreishi, M., Bagchi, A., & Sultan, M. A. (2015). Punching shear behavior of concrete flat slabs in elevated temperature and fire. *Advances in Structural Engineering*, *18*(5), 659–673. https://doi.org/10.1260/1369-4332.18.5.659

Guandalini, S., & Muttoni, A. (2004). *Punching tests on concrete slabs without shear reinforcement*. Swiss Federal Institute of Technology, Lausanne.

Hallgren, M., & Kinnunen, S. (1996). Increase of punching shear capacity by using high-strength concrete. In *4th International Symposium on Utilization of High-strength/High-performance Concrete* (pp. 1037–1046).

Hawkins, N. M., & Mitchell, D. (1979). Punching shear of reinforced concrete slabs. *Journal of the Structural Division*, *105*(11), 2471–2489.

Hegger, J., Kueres, D., & Schmidt, P. (2010). Punching shear of reinforced concrete slabs with shear reinforcement. *Structural Concrete*, *11*(1), 19–30.

Jasim, H. H. (2023). Parametric study for behavior of the punching shear capacity of flat slab without shear reinforcement by the selected design codes. *Eurasian Journal of Science and Engineering*, *9*(1), 57–66. https://doi.org/10.23918/eajse.v9i1p57

Kinnunen, S., & Nylander, H. (1960). Punching of concrete slabs without shear reinforcement. *Transactions of the Royal Institute of Technology, Stockholm*, 158.

Melo, G. S., Ferreira, M. P., & Regan, P. E. (2014). Punching of reinforced concrete flat slabs with double-headed shear reinforcement. *ACI Structural Journal*, *111*(2), 363–374.

Menétrey, P., &Khelidj, A. (2001). Punching strength of reinforced concrete slabs. *ACI Structural Journal*, *98*(5), 626–630.

Micallef, K., Sagaseta, J., Fernández Ruiz, M., &Muttoni, A. (2014). Assessing punching shear failure in reinforced concrete flat slabs subjected to localised impact loading. *International Journal of Impact Engineering*, *71*, 17–33. https://doi.org/10.1016/j.ijimpeng.2014.04.003

Muttoni, A., & Fernández Ruiz, M. (2008). Shear strength of members without transverse reinforcement as function of critical shear crack width. *ACI Structural Journal*, *105*(2), 163–172.

Regan, P. E., & Braestrup, M. W. (1985). Punching shear in reinforced concrete slabs. *ACI Structural Journal*, *82*(4), 543–551.

Schimanowski, E. R. Z., Campos Filho, A., Lazzari, B. M., & Lazzari, P. M. (2024). An investigation of punching shear in flat slabs with spherical voids using the finite element method. *Structural Concrete*. https://doi.org/10.1002/suco.202400489

Schmidt, P., Kueres, D., & Hegger, J. (2020). Punching shear behavior of reinforced concrete flat slabs with a varying amount of shear reinforcement. *Structural Concrete*, *21*(1), 235–246. https://doi.org/10.1002/suco.201900017

Setiawan, A., Vollum, R. L., Macorini, L., & Izzuddin, B. A. (2021). Numerical modelling of punching shear failure of reinforced concrete flat slabs with shear reinforcement. *Magazine of Concrete Research*, *73*(4), 172–188. https://doi.org/10.1680/jmacr.19.00562

Shatarat, N. K., & Salman, D. (2022). Investigation of punching shear behavior of flat slabs with different types and arrangements of shear reinforcement. *Case Studies in Construction Materials*, *16*, e01028. https://doi.org/10.1016/j.cscm.2022.e01028

Shwalia, A. S. I., Al-Salim, N. H. A., & Al-Baghdadi, H. M. (2019). Enhancement punching shear in flat slab using mortar infiltrated fiber concrete. *Civil Engineering Journal*, *5*(6), 1273–1284. https://doi.org/10.28991/cej-2019-03091327

Yang, Y., & Ashour, A. F. (2008). Punching shear strength of steel-fiber reinforced concrete slabs. *ACI Structural Journal*, *105*(5), 605–612.

Chapter 4

Finite element analysis of Sisal fiber and human hair composites

A review

Joeeta Mukherjee Ghosh

4.1 INTRODUCTION

The demand for sustainable and biodegradable materials has led to increased interest in natural fiber composites. Sisal fiber, derived from the Agave sisalana plant, and human hair, a readily available waste material, are both known for their excellent mechanical properties. This review focuses on the use of finite element analysis (FEA) to study the behavior of these composites under various loading conditions, aiming to optimize their use in engineering applications. Natural fibers' excellent mechanical qualities, accessibility, environmental friendliness, and crucial handling advantages are making their usage in composites increasingly important. In addition to being more economical and environmentally friendly, natural fibers are also infinitely renewable and lighter. However, the biological circumstances in which natural fibers are found, as well as the methods used for extraction and treatment, affect their mechanical qualities, giving them unique characteristics. Such fluctuation must be measured in order to make reasonable expectations about the fibers' qualities if natural fiber is to replace synthetic fiber. Studies on the tensile behavior of sisal and other natural fibers have previously been led by a few analysts.

In many non-industrialized countries around the world, natural fibers are sustainable resources since they are less costly, pose no health risks, and, finally, provide a solution to natural contamination by finding new uses for waste materials. Furthermore, another class of materials that seems to hold a lot of promise for the future as a replacement for wood-based and scarce wood materials in underlying applications are natural fiber-reinforced polymer composites.

4.2 FEATURES OF SISAL FIBERS BONDED WITH EPOXY RESINS

The use of sisal fiber as a building material in polymer-based composites has been reviewed by a group of scholars who looked into the construction and properties of sisal fiber, changes to the fiber surface, the physical

 DOI: 10.1201/9781003642152-4

and mechanical properties of sisal fiber-based polymer composites, and the status and future assumptions for regular filaments generally. Sisal strands offer a lot of potential as reinforcements in composites made of polymers, such as rubber, thermosets, and thermoplastics. Composites made from sisal strands may have generally good effects in the automotive and transportation industries due to their low thickness and high specific properties. These composites will be more attractive because of their higher quality, less scraped area of gear, and thus lower re-tooling costs. In addition to providing a sustainable resource, the use of sisal strands as a natural resource in the plastics sector may also generate non-food sources of economic growth for farming and rural areas. Given that Brazil is one of the world's leading producers of sisal fiber, sisal fiber-based polymer composites and their ensuing uses would be highly attractive from a financial standpoint. It became evident from the aforementioned illustrations that more modern composites made of readily available sisal strands are becoming more popular; this incorporates contemporary trends in composite materials. With an improvement in the predicted number of strands and a reduction in mechanical characteristics of more than 1.5%, the paper is thus closed. The strands will be added in small amounts to increase the strength. The mechanical characteristics of the sisal fiber primarily depend on the assembly, condition, and size of the sisal fiber, which will determine the trademark properties, as well as on the standard boundaries, such as the fiber's size, length, strain rate, and review climate. Regular sisal strands are familiar with an increase in the flexural strength and crack strength of the substance despite the creation without regular filaments.

4.3 FEATURES OF HUMAN HAIR WITH EPOXY RESIN

The following conclusions are drawn from this experimental analysis of the mechanical characteristics of human hair fiber-reinforced epoxy composites: It has been shown that the 5% treated fiber has a major impact on the mechanical properties of the composites, such as rigidity, flexural strength, and sway strength. According to the tractable test, the 5% treated fiber had the highest elasticity at 36.994 N/mm^2, while the 10% treated fiber had the lowest at 14.392 N/mm^2.

4.4 METHODOLOGIES

4.4.1 Finite element modeling

FEA involves creating a detailed numerical model of the composite material, incorporating the properties of sisal fiber, human hair, and the matrix material. The model is subjected to various loading conditions to simulate real-world scenarios.

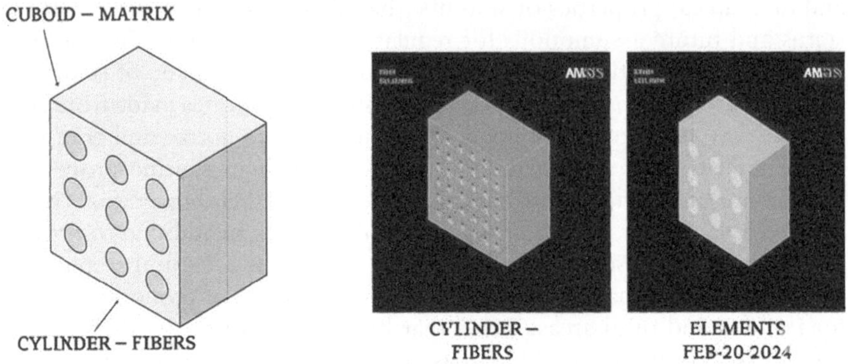

Figure 4.1 ANSYS model of natural fiber composites.

4.4.2 Material properties

The mechanical properties of sisal fiber and human hair, such as tensile strength, Young's modulus, and Poisson's ratio, are critical inputs for the FEA model. These properties are obtained from experimental studies and literature.

4.4.3 Simulation tools

Software such as ANSYS and ABAQUS is commonly used for FEA of natural fiber composites. These tools provide advanced capabilities for modeling complex geometries and analyzing the behavior of materials under different conditions (Figure 4.1).

4.5 RESULTS AND DISCUSSION

4.5.1 Mechanical behavior

FEA studies have shown that sisal fiber and human hair composites exhibit good tensile strength and flexibility. The combination of these fibers can enhance the overall mechanical performance of the composite (Figure 4.2).

4.5.2 Stress distribution

The analysis of stress distribution within the composite material reveals areas of high stress concentration, which are critical for understanding failure mechanisms. FEA helps in identifying these regions and optimizing the composite design (Figure 4.3).

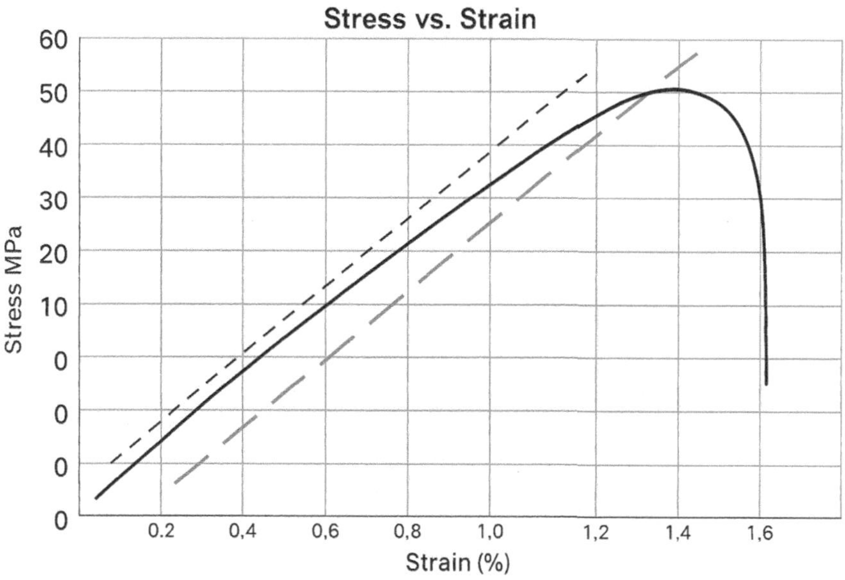

Figure 4.2 Mechanical behavior test graph of sisal fiber and human hair composites.

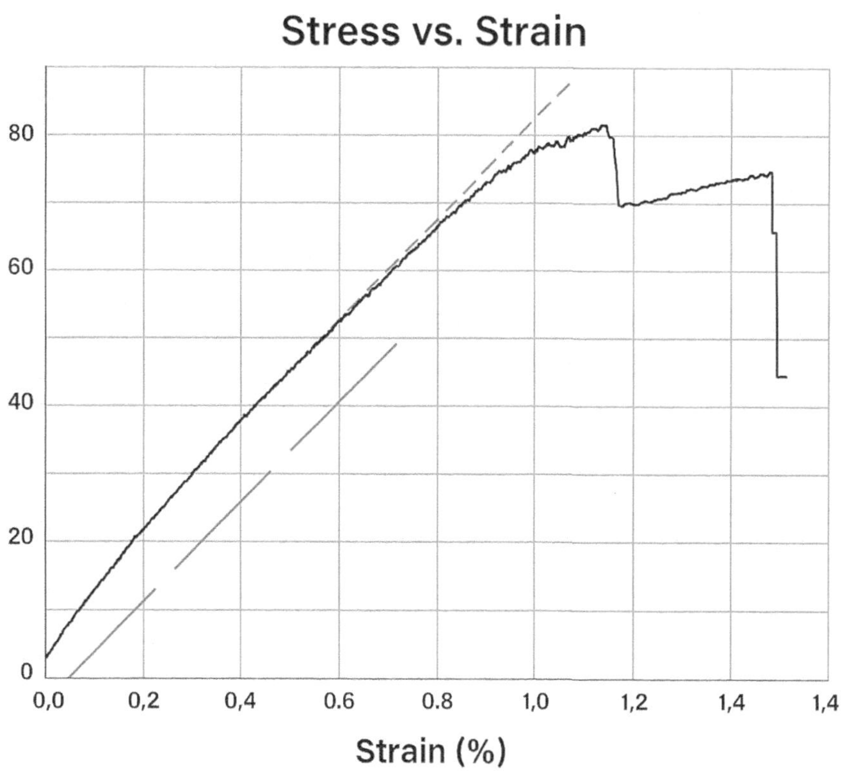

Figure 4.3 Stress distribution test graph of sisal fiber and human hair composites.

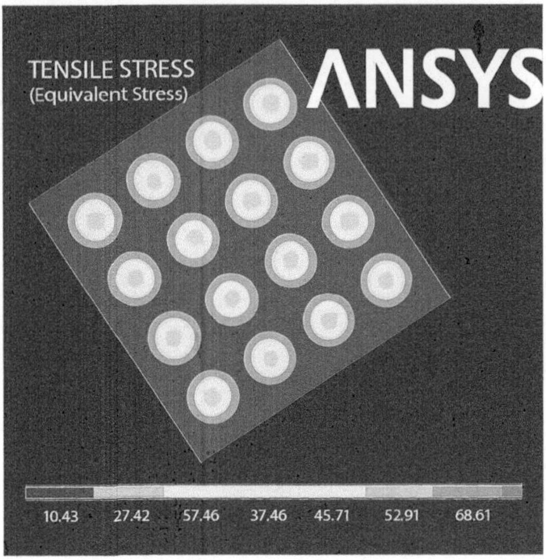

Figure 4.4 Deformation and failure of sisal fiber and human hair composites.

4.5.3 Deformation and failure

FEA provides insights into the deformation behavior of the composite under various loading conditions. The failure analysis helps in understanding the limits of the material and improving its performance through design modifications (Figure 4.4).

4.6 CHALLENGES AND FUTURE DIRECTIONS

4.6.1 Technological challenges

The accuracy of FEA depends on the quality of the input data and the complexity of the model. Ensuring accurate material properties and boundary conditions is essential for reliable results.

4.6.2 Environmental impact

While natural fiber composites are environmentally friendly, the production and disposal processes need to be optimized to minimize their ecological footprint.

4.6.3 Future research

Future studies should focus on the development of hybrid composites, combining sisal fiber and human hair with other natural or synthetic fibers.

Additionally, more experimental data is needed to validate FEA models and improve their accuracy.

4.7 CONCLUSION

FEA is a valuable tool for studying the mechanical behavior of sisal fiber and human hair composites. The insights gained from FEA can help in optimizing the design and performance of these sustainable materials, making them suitable for a wide range of applications. Continued research and technological advancements will further enhance the potential of natural fiber composites in engineering.

BIBLIOGRAPHY

Joeeta Mukherjee Ghosh, A review on human hair and sisal fiber epoxy based reinforced composites. HBRP publication 2023.

Mohanty AK, Misra M, Drzal LT. Surface modifications of natural fibers and performance of the resulting Bio composites: an overview. *Comp Interfaces* 2001; 8(5):313–343.

Smith W; Resin Systems; Delaware Composites Design Encyclopedia (DCDE), Volume 3; 1990.

South Wales Asia, Singapore Advanced Mechanics of Composite Materials.

Stokes VK, part II Characterization of the tensile strength. *Polymer Composites* 1990; 11(6); 354–367.

Stokes VK, Random glass mat reinforced thermoplastic composites; part I Phenomenology of tensile.

Valery V. Vasiliev Russian State University of Technology, Moscow, Evgeny V. Morozov The University of New.

Chapter 5

Green synthesis of gold nanoparticles and their applications

A comprehensive review

Pallavi Jain and Prachi Tadge

5.1 INTRODUCTION

5.1.1 Background

In the last few decades, research and development efforts in the broad, multidisciplinary field of nanostructures and nanotechnology have grown rapidly throughout the world [1]. Recent assessments of recent advancements around the globe indicate a great deal of interest in the topic of nanotechnology. Numerous distinctive physical and chemical characteristics of materials at the nanoscale level, such as a high surface-to-volume ratio and an increase in surface activity in relation to their bulk counterparts in the same composite, sparked this research. These approve their use in mechanical, optical, antibacterial, anticancer, and drug delivery applications [2]. The synthesis of nanoparticles has made use of a wide range of conventional methods, including spray pyrolysis, solvothermal, thermal decomposition of single-source precursors, thermal reduction, ultrasound, solid-state reaction, gas-phase condensation, liquid-phase chemical precipitation, surfactant-assisted liquid–liquid reaction, and so forth. The foundation of these methods is the reduction of metallic ions using costly conventional stabilizing and reducing agents such as sodium dodecyl sulfate, sodium citrate, and sodium borohydrate [3].

AuNPs, or gold nanoparticles, are well known for their many uses in fields including biomedicine, electronics, photonics, sensing, antibacterial, catalysis, and drug delivery for the treatment of cancer, encompassing both chemical contaminants and biomolecules [4]. These uses are appropriate given the characteristics of AuNPs, particularly their size and shape. The uses of AuNPs manufactured chemically are limited due to their recognized spherical nature. However, AuNPs with a variety of morphologies, including nanospheres, nanorods, nanowires, nanoporous, nanotriangles, and nanocaps, have been generated using a variety of techniques. To some extent, nevertheless, it is conceivable to regulate the shapes with constrained sizes [5]. In many different sectors of application, the ability to manage the morphology (size, shape, and crystalline structure) of AuNPs during

 DOI: 10.1201/9781003642152-5

manufacture plays a significant role. Recent studies have concentrated on the environmentally friendly synthesis of AuNPs for a range of applications, necessitating the development of novel techniques for the large-scale assembly and synthesis of NPs with precise sizes. The present and future functions of a green system for the synthesis of AuNPs from biodegradable materials are the main topics of this review, along with their usage.

5.1.2 Relationship of research area with Indian knowledge tradition

The idea of creating nanoparticles by green synthesis is in line with traditional Indian knowledge systems, especially Ayurveda, which has long supported the use of natural remedies in healing and therapy. The Indian knowledge tradition encourages the sustainable use of natural resources for a range of purposes, emphasizing harmony between humans and the environment [6]. This way of thinking is consistent with the goals of green chemistry, which are to minimize or completely do away with the use of dangerous materials in the creation, production, and use of chemical products.

5.1.2.1 Use of gold in Ayurveda

Historical literature, such as the Charaka Samhita and Sushruta Samhita, describes the application of "Swarna Bhasma" (gold ash) in the treatment of a variety of diseases. The use of gold in traditional Indian medicine is well-documented. Swarna Bhasma is made via a painstaking procedure that includes purifying and burning gold to produce small particles that are thought to have medicinal qualities [7]. Though with an emphasis on scientific rigor and use in modern medicine, the synthesis of AuNPs using green technologies in modern times is reminiscent of this age-old activity.

5.1.2.2 Plant-based synthesis

The green synthesis of AuNPs, which makes use of plant extracts, is a direct extension of the Indian practice of using plant-based medicines. The potential of plants that have been used for ages in Ayurvedic medicine, such as tulsi (Ocimum sanctum), aloe vera (Aloe barbadensis), and neem (Azadirachta indica), to stabilize and decrease AuNPs is currently being investigated. These plants' phytochemicals, which include phenolic acids, terpenoids, and flavonoids, function as organic reducing and capping agents to help in the environmentally friendly synthesis of nanoparticles [8].

5.1.2.3 Ethical and sustainable activities

Ethical and sustainable activities are highly valued in the Indian knowledge system. These ideals are supported by the green synthesis of nanoparticles, which steers clear of harmful chemicals and encourages the use of renewable natural resources. As a fundamental tenet of Ayurveda and other conventional Indian therapies, this method not only lessens the impact on the environment but also guarantees a safer synthesis procedure for human health [9].

5.1.3 Integrating traditional knowledge with modern science

An all-encompassing approach to scientific research is to integrate modern nanotechnology with traditional Indian knowledge. Combining cutting-edge technology with age-old knowledge allows researchers to synthesize nanoparticles in sustainable, effective, and environmentally responsible ways. This fusion of the ancient and the new also highlights the continued applicability of classical approaches to the resolution of contemporary scientific issues. Harmony and mutual support characterize the interaction between Indian knowledge heritage and the green synthesis of AuNPs. The current paradigm of green nanotechnology is strongly influenced by the values that underpin traditional Indian traditions, including sustainability, ethical issues, and the utilization of natural resources. Research in this area has the ability to honor and preserve the wisdom of old Indian knowledge systems while also bringing new insights and innovations to sectors such as environmental science and medicine [10].

5.1.4 Literature review

Globally, research and development activities in the vast, diverse subject of nanostructures and nanotechnology have expanded quickly in recent decades. Because of the unique physical and chemical properties of materials at the nanoscale, recent evaluations of developments show that nanotechnology is of tremendous interest. These characteristics, which include a high surface-to-volume ratio and increased surface activity in comparison to bulk materials, have sparked interest in their potential uses in the domains of medicine administration and mechanical, optical, antibacterial, and anticancer research.

Several traditional techniques, including spray pyrolysis, solvothermal processes, thermal breakdown, ultrasound-assisted synthesis, and liquid-phase chemical precipitation, have been used in the creation of nanoparticles. These techniques frequently rely on expensive traditional stabilizing and reducing agents, which are hazardous and have the potential to contaminate the environment and human health. Alternatively, there has been a growing interest in the investigation of natural bio-agents, including plant extracts, sugars, fungi, bacteria, and biodegradable polymers, as stabilizing and reducing agents during the manufacture of nanoparticles. The utilization of a "green" methodology in the synthesis of noble metal nanoparticles presents several benefits, including economic viability, environmental friendliness, and the absence of hazardous or poisonous components. The wide range of uses of AuNPs in biomedicine, electronics, photonics, sensing, catalysis, and drug administration makes them especially noteworthy. A key factor in AuNPs' efficacy in a variety of applications is their synthesis-related morphology, or their size, shape, and crystalline structure. The ecologically friendly synthesis of AuNPs has been the subject of recent research, which has brought attention to the need for innovative methods for large-scale manufacturing with exact control over size and shape [11].

5.1.4.1 Gold nanoparticle synthesis mediated by plants:

The environmentally benign and economical characteristics of the approach have made the green synthesis of AuNPs using plant extracts a prominent topic of research. Utilizing the wide range of bioactive substances found in plants, including phenolic acids, terpenoids, and flavonoids, which naturally function as reducing and stabilizing agents, is known as "plant-mediated synthesis." Research has shown that a variety of plant extracts, including neem (Azadirachta indica), tulsi (Ocimum sanctum), and aloe vera (Aloe barbadensis), can be successfully used to synthesize AuNPs. For example, neem leaf extract was used by Kesarkar et al. [22] to successfully produce AuNPs. The research demonstrated how neem, a key component of conventional Indian medicine, may reduce gold ions to create nanoparticles with precise size and structure. Similar to this, Rajan et al. (2015) [11] showed that phytochemicals can function as efficient reducing agents in the creation of nanoparticles by using Hibiscus rosa-sinensis leaf extract for the manufacture of gold and silver nanoparticles [12].

According to these findings, plant-mediated synthesis can harness the tremendous biodiversity of plants, particularly those employed in traditional medicine, while simultaneously providing an environmentally friendly substitute for traditional approaches. Since the synthesis process is typically conducted at room temperature and pressure, it has a low impact on the environment and is energy-efficient.

5.1.4.2 Gold nanoparticle synthesis by microorganisms

Biological agents such as bacteria, fungi, and algae are used in the microbial synthesis of AuNPs to produce the particles. This method turns metal ions into nanoparticles by using the metabolic mechanisms that microbes naturally have. For instance, Ali et al. (2019) [21] used the bacteria Pseudomonas aeruginosa to demonstrate the production of AuNPs. According to the study, by adjusting the culture conditions, microbial synthesis might create nanoparticles in a range of shapes, such as triangles and spheres. Green synthesis of AuNPs has also been achieved through the employment of fungi like Fusarium oxysporum. Fungi use an extracellular manufacturing mechanism, which is particularly beneficial because it makes nanoparticle downstream processing easier. A sustainable and biocompatible way to make AuNPs is by microbial synthesis. In addition to lowering the requirement for hazardous chemicals, the employment of microbes enables the creation of nanoparticles with a variety of morphologies, which are crucial for a range of applications [13].

5.1.5 Green synthesized gold nanoparticle applications

Because of their surface functionalization abilities, customizable size and shape, and biocompatibility, green-produced AuNPs have found many uses in a variety of sectors. These nanoparticles are being investigated for use in photothermal therapy, medication delivery, and imaging in the biomedical field.

5.1.5.1 Biomedical applications

The treatment of cancer is one of the most promising uses for AuNPs. Drugs can be delivered directly to cancer cells using green-produced AuNPs, reducing negative effects on healthy tissue, by functionalizing them with targeting molecules. The use of AuNPs in drug delivery systems was proven by Ye et al. (2018) [19], underscoring their potential to increase the effectiveness of cancer treatment. AuNPs are employed as contrast agents in imaging procedures such as magnetic resonance imaging (MRI) and computed tomography (CT). AuNPs are perfect for boosting picture contrast because of their optical characteristics, especially their capacity to absorb and scatter light.

5.1.5.2 Environmental applications

Remediation of the environment using green-produced AuNPs has also been investigated. These nanoparticles' catalytic qualities enable them to break down environmental contaminants in wastewater, including heavy metals and dyes. The potential of these nanoparticles to promote environmental sustainability was demonstrated by Sadiq et al. [20], who reported on the usage of AuNPs manufactured using Mentha longifolia extract for the photocatalytic destruction of organic contaminants [14].

5.1.5.3 Sensing and detection

AuNPs are a great option for biosensors due to their special optical characteristics. Green-synthesized nanoparticles have been used to detect a number of biomolecules, such as cholesterol and glucose. In order to provide a noninvasive approach for early cancer detection, Ye et al. (2018) investigated the use of AuNPs in the construction of biosensors for detecting cancer biomarkers [19].

5.1.5.4 Research prospects

The body of research on the ecologically benign and economically viable development of green synthesis techniques for producing AuNPs is noteworthy. Even though a lot of progress has been accomplished, there are still a number of untapped or underdeveloped research areas that present fascinating possibilities for further investigation. The literature review reveals the following areas for future research.

5.1.5.5 Green synthesis protocol optimization

Systematic optimization of synthesis techniques is necessary, notwithstanding the success of green synthesis employing microbial agents and plant extracts. The size, shape, and yield of AuNPs are greatly influenced by variables like pH, temperature, biological agent concentration, and reaction time. The development of standardized procedures that can reliably yield nanoparticles with the required properties should be the main goal of future research. Optimizing the synthesis also depends on knowing the precise functions that different phytochemicals and microbial metabolites play in the reduction and stabilization processes [16].

5.1.5.6 Investigating novel biological sources

For green synthesis, the richness of plants and microbes offers a huge, mainly unexplored resource. Future studies ought to look at a wider variety of biological sources, such as marine algae, uncommon medicinal plants, and extremophiles, which might have special qualities that make them useful for the production of nanoparticles. By examining these new sources, innovative reducing and stabilizing agents that improve the synthesis process' efficiency and specificity may be found [17].

5.1.5.7 Mechanism-based understanding of biosynthesis

The literature is replete with examples of successful green synthesis, but our understanding of the underlying mechanisms is still lacking. Subsequent investigations ought to focus on clarifying the molecular processes associated with the production of AuNPs. This entails pinpointing the precise molecular interactions and biochemical processes in charge of the stability of nanoparticles and the reduction of gold ions. To learn more about these processes, sophisticated methods like spectroscopy, proteomics, and molecular modeling could be used [15].

5.1.5.8 Expansion and market development

One essential component of green synthesis technologies' industrial use is their scalability. Subsequent investigations ought to tackle the obstacles linked with transferring these procedures from laboratory settings to commercial manufacturing. This entails creating large-scale synthesis techniques that are affordable, guaranteeing the consistency of nanoparticle quality, and preserving environmental sustainability. The conversion of green synthesis techniques into economically feasible technology will require cooperative efforts between academic researchers and industrial partners.

5.1.5.9 Combining conventional knowledge systems with integration

An exceptional chance for creativity exists in the interaction between traditional Indian knowledge systems and green synthesis. Subsequent investigations may examine the amalgamation of conventional Ayurvedic procedures with contemporary nanotechnology to formulate novel approaches for nanoparticle creation. This can entail conducting methodical research on conventional formulations—like Swarna Bhasma—and their possible uses in the manufacturing of nanoparticles. Such a multidisciplinary study may result in the creation of fresh, culturally appropriate methods for green synthesis.

5.1.5.10 Uses in environmental remediation and medicine

Green-produced AuNPs have demonstrated potential in a range of applications, especially in environmental remediation and biomedicine [18]. Subsequent investigations ought to concentrate on delving deeper into these uses, encompassing the following:Biomedical Applications: Examining photothermal treatment, targeted drug administration, and imaging with green-produced AuNPs. To evaluate these nanoparticles' safety, effectiveness, and biocompatibility in medicinal applications, preclinical and clinical research are required.Environmental Applications: Increasing the amount of time that environmentally harmful substances like heavy metals and organic dyes are broken down by using green synthetic AuNPs. The application of these nanoparticles in soil remediation, air filtration, and water purification should all be investigated further.

5.1.5.11 Extended-duration toxicity and environmental effect research

Even though green synthesis is thought to be environmentally benign, in-depth research is still required to determine the long-term toxicity and environmental effects of green-generated nanoparticles. Examining the effects of employing these nanoparticles in different applications on the environment and human health should be the main goal of future studies. This entails researching their possible environmental durability, biodegradability, and accumulation in living things. These kinds of investigations are necessary to guarantee the appropriate and safe application of AuNPs in practical applications.

5.2 CONCLUSION

As an innovative approach to nanotechnology, the green synthesis of AuNPs provides a sustainable, economical, and environmentally benign substitute for traditional chemical processes. Through the use of natural bio-agents like microbes, plant extracts, and biodegradable polymers, scientists have created AuNPs with improved functions and a variety of morphologies. These green-synthesized nanoparticles have a great deal of potential for use in biomedicine, environmental remediation, catalysis, and sensing, among other domains. Even though the environmentally friendly synthesis of AuNPs has already shown a great deal of promise, more study and development are needed to fully realize this technology's potential. Future research can help improve sustainable nanotechnology, advancing progress in a variety of scientific and industrial sectors while minimizing environmental effects, by tackling current obstacles and pursuing novel routes.

REFERENCES

[1] Lloyd-Hughes, H., *et al.* (2015). *Current and Future* Nanotechnology Applications in the Management of Melanoma: A Review. *Journal of Nanomedicine & Nanotechnology*. 6(6): 1–12.

[2] Rangari, V., S. Dey, and S. Jeelani. (2016). In-situ Synthesis of Cu Nanoparticles on MWCNTS using Microwave Irradiation. *NSTI-Nanotech*. 13(1): 140–143.

[3] Shukri, W. N. W. *et al.* (2015). Synthesize of Gold Nanoparticles with 532 nm and 1064 nm Pulsed Laser Ablation. *Jurnal Teknologi*. 78(3): 267–270.

[4] Bishop, P., *et al.* 2016. Flame Spray Pyrolysis for the Production of Nanoparticles. *NSTI-Nanotech*. 13: 15.

[5] Vaquero, F., R. Navarro, and J. Fierro. (2016). Evolution of the Nanostructure of CdS using Solvothermal Synthesis at Different Temperature and Its Influence on the Photoactivity for Hydrogen Production. *International Journal of Hydrogen Energy*. 41(27): 11558–11567.

[6] Umar, M. I. A. *et al.* (2013). Formation of Gold-coated Multilayer Graphene via Thermal Reduction. *Materials Letters*. 106: 200–203.

[7] Noruzi, M. 2015. Biosynthesis of Gold Nanoparticles Using Plant Extracts. *Bioprocess and Biosystems Engineering*. 38(1): 1–14.

[8] Ahmed, S., *et al.* (2016). A Review on Plants Extract Mediated Synthesis of Silver Nanoparticles for Antimicrobial Applications: A Green Expertise. *Journal of Advanced Research*. 7(1): 17–28.

[9] Patra, S., *et al.* 2015. Green Synthesis, Characterization of Gold and Silver Nanoparticles and Their Potential Application for Cancer Therapeutics. *Materials Science and Engineering: C.* 53: 298–309.
[10] Jannathul, F.M. and P. Lalitha. (2022). Biogenic Green Synthesis of Gold Nanoparticles and Their Applications – A Review of Promising Properties. *Inorganic Chemistry Communications.* 143, 109800.
[11] Rajan, A., V. Vilas, and D. Philip. (2015). Studies on Catalytic, Antioxidant, Antibacterial and Anticancer Activities of Biogenic Gold Nanoparticles. *Journal of Molecular Liquids.* 212: 331–339.
[12] González-Rubio, G., *et al.* (2015). Femtosecond Laser-controlled Tip-to-tip Assembly and Welding of Gold Nanorods. *Nano Letters.* 15(12): 8282–8288.
[13] Santhosh, P.B., J. Genova and H. Chamati. (2022) Green Synthesis of Gold Nanoparticles: An Eco-Friendly Approach. *Chemistry.* 4(2), 345–369.
[14] Usman, A. I., A. A. Aziz, and O. A. Noqta. (2018). Bio-synthesis of Triangular and Hexagonal Gold Nanoparticles Using Palm Oil Fronds' Extracts at Room Temperature. *Material Research Express.* 5(1), 015042.
[15] Dheyab, M. A., J. H. Tang, A. A. Aziz, S. H. Nowfal, M. S. Jameel, M. N. Alrosan Oladzadabbasabadi and M. Ghasemlou. (2024) Green synthesis of gold nanoparticles and their emerging applications in cancer imaging and therapy: a review. *Reviews in Inorganic Chemistry*, 48, 1–23.
[16] Dikshit, P.K., J. Kumar, K. A. Das, S. Sadhu, S. Sharma, S. Singh, P. K. Gupta, and B. S. Kim (2021) Green Synthesis of Metallic Nanoparticles: Applications and Limitations. *Catalysts.* 11, 902.
[17] He, W., F. Yang, K. Chen and Q. Zeng (2024) Targeted gold nanoparticles for ovarian cancer (Review). *Oncology Letters.* 28: 589.
[18] Khan, M. A. R., Al Mamun, M. S., Habib, M. A., Islam, A. N., Mahiuddin, M., Karim, K. M. R., ... & Ara, M. H. (2022) A review on gold nanoparticles: Biological synthesis, characterizations, and analytical applications. *Results in Chemistry*, 4, 100478.
[19] Ye, D., Zuo, X., & Fan, C. (2018). DNA nanotechnology-enabled interfacial engineering for biosensor development. *Annual Review of Analytical Chemistry*, *11*(1), 171–195.
[20] Sadiq, Z., Safiabadi Tali, S. H., Hajimiri, H., Al-Kassawneh, M., & Jahanshahi-Anbuhi, S. (2024). Gold nanoparticles-based colorimetric assays for environmental monitoring and food safety evaluation. *Critical Reviews in Analytical Chemistry*, *54*(7), 2209–2244.
[21] Ali, J., Ali, N., Wang, L., Waseem, H., & Pan, G. (2019). Revisiting the mechanistic pathways for bacterial mediated synthesis of noble metal nanoparticles. *Journal of Microbiological Methods*, *159*, 18–25.
[22] Kesarkar, R., Sangar, V., Oza, G., Sawant, T., Kothari, S., Sharon, M., & Chowdhary, A. (2014). Synthesis, characterization and hepatoprotective activity of Neem gold nanoparticles for improved efficacy and sustained drug release profile of Azidothymidine. *International Journal of Pharmaceutical Sciences Review and Research*, *26*, 117–122.

Chapter 6

Potential flow analysis of Rayleigh-Taylor instability in nanofluid layer

Mukesh Kumar Awasthi, Ashwani Kumar, Nitesh Dutt, and Dhananjay Yadav

6.1 INTRODUCTION

The viscosity correction within the viscous potential flow (VPF) framework, termed VCVPF, extends traditional VPF by incorporating viscous pressure to address irrotational flows. Unlike classical VPF, which neglects tangential viscous stresses, VCVPF integrates viscous effects through normal and tangential stress corrections. However, its limitations in capturing vorticity generation and dissipation restrict its utility in modeling turbulence and vortex dynamics. Wang et al. [1] applied VCVPF to capillary instability between a highly viscous fluid and a low-viscosity gas, showing that viscous pressure integration yields growth rates asymptotically consistent with exact solutions. In a follow-up study, Wang et al. [2] analyzed viscous contributions to pressure in capillary instability between two viscous fluids. Awasthi et al. [3, 4] further advanced VCVPF by evaluating viscous corrections for Rayleigh-Taylor (RT) instability at a viscoelastic-viscous fluid interface with heat and mass transfer.

RT instability, a hydrodynamic phenomenon driven by density contrasts under gravitational acceleration, has been widely studied across astrophysical systems (e.g., supernovae and accretion disks) and engineering applications (e.g., inertial confinement fusion and oceanography). Sharp [5] established a foundational theoretical framework for RT instability, while Jacobs and Catton [6] expanded this to three-dimensional non-viscous fluid-gas systems. Kull [7] developed theories for accelerated fluids, including nonlinear bubble dynamics, and Milkaelian [8] derived hybrid solutions for Richtmyer-Meshkov (RM) and RT instabilities.

For viscous fluids, irrotational flow theory has proven effective. Joseph et al. [9] examined RT instability between fluids of differing viscosities, while Asthana et al. [10] demonstrated viscosity's stabilizing role in cylindrical geometries using nonlinear perturbations. Awasthi et al. [11–13] extended this work to nonlinear analyses and rotational effects in cylindrical configurations.

Recent research has expanded RT instability studies to nanofluids—base fluids infused with nanoparticles that enhance thermal and transport

DOI: 10.1201/9781003642152-6

properties. These advancements have implications for energy systems, electronics cooling, and nanomedicine. Awasthi et al. [14–17] applied VPF theory, to nanofluid-viscous fluid interfaces. Moatimid and Hassan [18] investigated linear stability in viscoelastic nanofluids, while Moatimid and Gaber [19] analyzed magnetohydrodynamic effects in conducting nanofluid layers. Moatimid et al. [20] incorporated Marangoni effects into temporal instability models using VPF theory.

This work presents a critical synthesis of RT instability in nanofluids, integrating theoretical, numerical, and experimental insights into nanoparticle size, concentration, and fluid properties. We employ irrotational flow theory to model linear perturbations at a nanofluid-viscous fluid interface, incorporating viscous corrections to both shear and normal stresses—a departure from classical VPF. By deriving a dispersion relation dependent on viscosity, density, surface tension, and nanoparticle characteristics, we establish neutral stability criteria and analyze interfacial dynamics through parametric plots. The study highlights the interplay between buoyancy-driven instabilities and nanoparticle dispersion, offering design principles for nanofluid-based systems with optimized thermal and mechanical performance.

6.2 MODELING

We examine a stratified system consisting of a nanofluid layer overlying a viscous incompressible fluid, as illustrated in Figure 6.1. The two layers exhibit distinct thermophysical properties: the nanofluid is defined by its density ρ_{nano} and viscosity μ_{nano}, while the viscous fluid occupies the region $z \in [-h_v, 0]$, separated from the nanofluid layer (confined within $z \in [0, h_{nano}]$ by an interfacial boundary denoted as Γ).

The governing equations for the nanofluid phase are:

$$\left.\begin{aligned} &\nabla \cdot q_{nano} = 0 \\ &\rho_{nano}\left(\frac{\partial q_{nano}}{\partial t} + (q_{nano} \cdot \nabla) q_{nano}\right) = -\nabla p_{nano} + \rho_{nano} g + \mu_{nano} \nabla^2 q_{nano} \end{aligned}\right\} \quad (6.1)$$

where q_{nano} and p_{nano} represent the velocity and pressure fields in the nanofluid, respectively.

The nanofluid's viscosity μ_{nano} is influenced by multiple factors, including the nanoparticle volume fraction, base fluid viscosity, interfacial nano-layer effects, and nanoparticle geometry. This relationship is modeled as:

$$\mu_{nano} = \mu_{eff}\left(1 - \frac{\phi_{ag}}{\phi_m}\right)^{-[\alpha]\phi_m} \quad (6.2)$$

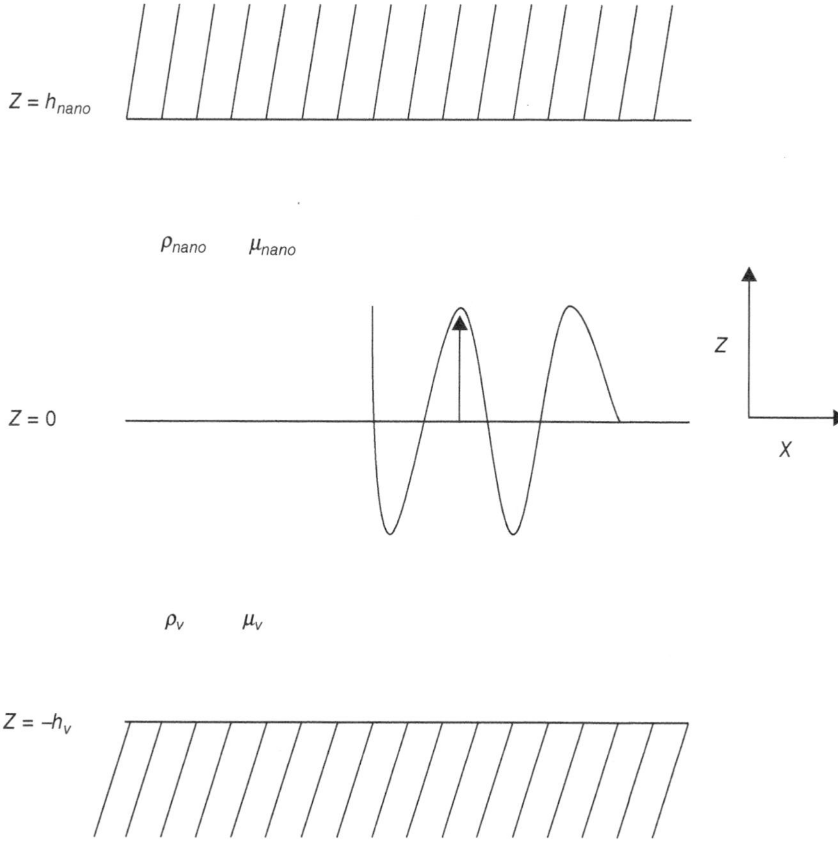

Figure 6.1 Schematic of a stratified two-layer fluid system.

where μ_{eff} is the base fluid viscosity, ϕ_m (ranging between 0.68 and 0.75 for prolate spheroids) denotes the maximum packing fraction, and α is a shape parameter accounting for nanoparticle rotation. The aggregate volume fraction ϕ_{ag} is given by:

$$\phi_{ag} = \phi_{\mathrm{mod}}\left(\frac{r_a}{r}\right)^{3-d} \tag{6.3}$$

where ϕ_{mod} denotes the equivalent volume fraction defined by

$$\phi_{\mathrm{mod}} = \phi\left(1+\frac{\delta}{r}\right)\left(1+\frac{\delta}{R}\right)^2 \tag{6.4}$$

Here, δ is the interfacial layer thickness, r and R are the semi-minor and semi-major axes of the nanoparticles, respectively, and d is the fractal index quantifying spatial variations in packing density.

The nanofluid's density ρ_{nano} is expressed as a weighted combination of the metal nanoparticle density ρ_p and the base fluid density ρ_f:

$$\rho_n = \phi\rho_p + (1-\phi)\rho_b \tag{6.5}$$

For the underlying viscous fluid phase, the dynamics are governed by the incompressible Navier-Stokes equations:

$$\left.\begin{aligned} &\nabla \cdot q_v = 0 \\ &\rho_v\left(\frac{\partial q_v}{\partial t} + (q_v \cdot \nabla)q_v\right) = -\nabla p_v + \rho_v g + \mu_v \nabla^2 q_v \end{aligned}\right\} \tag{6.6}$$

where q_v, p_v, and μ_v denote the velocity, pressure, and constant viscosity of the viscous fluid, respectively.

6.3 STABILITY ANALYSIS

6.3.1 Base state

In the unperturbed (base) state, both fluid layers remain motionless, yielding $q_{nano} = q_v = 0$. The interface is flat, resulting in constant pressure distributions $p_{nano} = p_{nano,0} = 0$ and $p_v = p_{v,0}$ within each layer.

6.3.2 Perturbed state

A small-amplitude perturbation is introduced to the base state, displacing the interface to a position $z = \Gamma(x,t)$. In the perturbed configuration, the velocities and pressures in the two layers are modified to $q_{nano} = 0 + q'_{nano}$ and $q_v = 0 + q'_v$ and $p_{nano} = p_{nano,0} + p'_{nano}$ and $p_v = p_{v,0} + p'_v$, respectively. The governing linearized equations for the perturbed quantities are:

$$\left.\begin{aligned} &\nabla \cdot q'_{nano} = 0 \\ &\rho_{nano}\frac{\partial q'_{nano}}{\partial t} = -\nabla p'_{nano} + \rho_{nano} g + \mu_{nano} \nabla^2 q'_{nano} \end{aligned}\right\} \tag{6.7}$$

$$\left.\begin{aligned} &\nabla \cdot q'_v = 0 \\ &\rho_v\frac{\partial q'_v}{\partial t} = -\nabla p'_v + \rho_v g + \mu_v \nabla^2 q'_v \end{aligned}\right\} \tag{6.8}$$

Assuming irrotational flow, the velocity fields can be expressed as gradients of potential functions, reducing the first part of (6.7) and (6.8) to Laplace equations:

$$\nabla^2 \varphi_{nano} = 0; \nabla^2 \varphi_v = 0 \tag{6.9}$$

6.3.3 Interfacial and boundary conditions

At the interface $z = \Gamma(x,t)$ between the two fluids, the kinematic condition is expressed as:

$$\frac{\partial \Gamma}{\partial t} = q_{nano} \cdot \hat{n} = q_v \cdot \hat{n} \quad \text{at} \quad z = 0 \tag{6.10}$$

The impermeable boundary ensures no vertical flow at the rigid walls, giving:

$$\left.\begin{aligned} q_{nano} \cdot \hat{n} = 0 \quad \text{at} \quad z = h_n \\ q_v \cdot \hat{n} = 0 \quad \text{at} \quad z = -h_f \end{aligned}\right\} \tag{6.11}$$

Finally, the stress balance at the interface, accounting for surface tension effects, is formulated as:

$$p_{nano} - p_v - 2\mu_{nano} \frac{\partial^2 \varphi_{nano}}{\partial z^2} + 2\mu_v \frac{\partial^2 \varphi_v}{\partial z^2} = \sigma \frac{\partial^2 \Gamma}{\partial x^2} \tag{6.12}$$

6.3.4 Inclusion of shearing stresses

Wang et al. [1] introduced a viscous adjustment mechanism to resolve the inconsistency in irrotational pressure predictions at unconfined interfaces of sustained flows, addressing the mismatch between non-zero and zero shear stress conditions. In this work, we derive the pressure correction terms associated with RT instability directly from the mechanical energy balance.

Let $\hat{n}_1$ and $\hat{n}_2$ denote the unit outward normal at the interface for the lower (fluid 1) and upper (fluid 2) layers, respectively, with $\hat{t}$ representing the unit tangent vector. Using subscripts:

a. i for irrotational components,
b. v for viscous components,
c. 2 and 1 for the upper and lower fluids,
d. the viscous stress tensor is decomposed into normal (τ^n) and shear (τ^s) components.

The mechanical energy equations for the upper and lower fluids are:

$$\frac{d}{dt}\int_{V_1}\frac{\rho_{nano}}{2}\left|q_{nano}\right|^2 dV = \int_{V_1}\rho_{nano}g q_{nano}dV + \int_{A}\left[q_{nano}\cdot \mathbf{T}\cdot \mathbf{n}_2\right]dA$$
$$-\int_{V_1}2\mu_{nano}\mathbf{D}_1 : \mathbf{D}_1\, dV$$
$$= \int_{V_1}\rho_{nano}g q_{nano}dV - \int_{A}\left[q_{nano}\cdot \mathbf{n}_1\left(-p_1^i + \tau_1^n\right) + q_{nano}\cdot \mathbf{t}\tau_1^s\right] dA$$
$$-\int_{V_1}2\mu_{nano}\mathbf{D}_1 : \mathbf{D}_1 dV \tag{6.13}$$

$$\frac{d}{dt}\int_{V_1}\frac{\rho_v}{2}\left|q_v\right|^2 dV = \int_{V_1}\rho_v g q_v dV + \int_{A}\left[q_v\cdot \mathbf{T}\cdot \mathbf{n}_1\right]dA - \int_{V_2}2\mu_v\mathbf{D}_2 : \mathbf{D}_2\, dV$$
$$= \int_{V_1}\rho_v g q_v dV + \int_{A}\left[q_v\cdot \mathbf{n}_1\left(-p_2^i + \tau_2^n\right) + q_v\cdot \mathbf{t}\tau_2^s\right]dA - \int_{V_1}2\mu_v\mathbf{D}_2 : \mathbf{D}_2 dV \tag{6.14}$$

By imposing the continuity of normal velocity at the interface $(q_{nano}\cdot \hat{n}_2 = q_v\cdot \hat{n}_1 = q_n)$ and summing Equations (6.13) and (6.14), we obtain:

$$\frac{d}{dt}\int_{V_1}\frac{\rho_{nano}}{2}\left|q_{nano}\right|^2 dV + \frac{d}{dt}\int_{V_1}\frac{\rho_v}{2}\left|q_v\right|^2 dV = \int_{V_1}\rho_{nano}g q_{nano}dV$$
$$+\int_{V_1}\rho_v g q_v dV - \int_{V_1}2\mu_{nano}\mathbf{D}_1 : \mathbf{D}_1 dV - \int_{V_1}2\mu_v\mathbf{D}_2 : \mathbf{D}_2 dV$$
$$+\int_{A}\left[q_n\left(-p_2^i + \tau_2^n + p_1^i - \tau_1^n\right) + q_v\cdot \mathbf{t}\tau_2^s - q_{nano}\cdot \mathbf{t}\tau_1^s\right]dA \tag{6.15}$$

To address the discontinuity in shear stress and tangential velocity at the interface, we introduce two distinct viscous pressure corrections: p_{nano}^v (upper fluid) and p_v^v (lower fluid). Incorporating these into Equation (6.15) yields:

$$\frac{d}{dt}\int_{V_1}\frac{\rho_{nano}}{2}\left|q_{nano}\right|^2 dV + \frac{d}{dt}\int_{V_1}\frac{\rho_v}{2}\left|q_v\right|^2 dV = \int_{V_1}\rho_{nano}g q_{nano}dV + \int_{V_1}\rho_v g q_v dV$$
$$-\int_{V_1}2\mu_{nano}\mathbf{D}_1 : \mathbf{D}_1 dV - \int_{V_1}2\mu_v\mathbf{D}_2 : \mathbf{D}_2 dV$$
$$+\int_{A}\left[q_n\left(-p_2^i - p_2^v + \tau_2^n + p_1^i + p_1^i - \tau_1^n\right)\right]dA \tag{6.16}$$

By equating the expressions derived in equations (6.15) and (6.16), we obtain the compatibility condition:

$$\int_A \left[q_v \cdot \mathrm{t}\tau_2^s \llcorner q_{nano} \cdot \mathrm{t}\tau_1^s \right] dA = \int_A \left[q_n \left(-p_2^v + p_1^i \right) \right] dA \tag{6.17}$$

The governing equation for the pressure correction term is formulated as:

$$\nabla^2 p_2^v = \nabla^2 p_1^i = 0 \tag{6.18}$$

Solutions to equation (18) for the pressure correction are:

$$p_v^2 = -\left(M_1 \cosh(kz) + N_1 \sinh(kz) \right) \mathrm{Exp}(-i\varpi t + ikx) \tag{6.19}$$

$$p_v^1 = -\left(M_2 \cosh(kz) + N_2 \sinh(kz) \right) \mathrm{Exp}(-i\varpi t + ikx) \tag{6.20}$$

The viscous pressure difference at the interface $z = 0$, accounting for shear stresses, is expressed as:

$$-p_v^2 + p_v^1 = (M_2 - M_1)\mathrm{Exp}(-i\varpi t + ikx) \tag{6.21}$$

Incorporating this correction, the original interfacial condition in equation (6.12) is modified to:

$$p_{nano}^i + p_2^v - p_v^1 - p_1^v - 2\mu_{nano} \frac{\partial^2 \varphi_{nano}}{\partial z^2} + 2\mu_v \frac{\partial^2 \varphi_v}{\partial z^2} = \sigma \frac{\partial^2 \Gamma}{\partial x^2} \tag{6.22}$$

6.4 DISPERSION RELATIONSHIP

To analyze the stability of the system, we employ the classical normal mode analysis. The perturbed interface elevation is expressed as $\Upsilon(x,t) = \Upsilon_0 \mathrm{Exp}(-i\varpi t + ikx)$. The remaining perturbed quantities (velocities and pressures) are similarly decomposed into normal modes as $\Gamma(x,z,t) = \Gamma(z)\mathrm{Exp}(-i\varpi t + ikx)$.

The velocity potentials φ_{nano} and φ_v for the two fluid layers, which satisfy the boundary conditions (6.10) and (6.11), take the form:

$$\varphi_{nano} = \frac{i\varpi}{k} \frac{\cosh(k(z - h_{nano}))}{\sinh(kh_{nano})} \Upsilon_0 \, \mathrm{Exp}(-i\varpi t + ikx) \tag{6.23}$$

$$\varphi_v = -\frac{i\varpi}{k} \frac{\cosh(k(z + h_v))}{\sinh(kh_v)} \Upsilon_0 \, \mathrm{Exp}(-i\varpi t + ikx) \tag{6.24}$$

Linearizing the stress balance equation (6.12) about the base state yields, and using Bernoulli's equation for irrotational pressure:

$$\left[\rho_{nano}\left(\frac{\partial\varphi_{nano}}{\partial t}+g\Upsilon\right)+2\mu_{eff}\left(1-\frac{\phi_{ag}}{\phi_m}\right)^{-[\alpha]\phi_m}\frac{\partial^2\varphi_{nano}}{\partial z^2}-p_2^v\right]$$
$$-\left[\rho_v\left(\frac{\partial\varphi_v}{\partial t}+g\Upsilon\right)+2\mu_v\frac{\partial^2\varphi_v}{\partial z^2}-p_1^v\right]=-\sigma\frac{\partial^2\Upsilon}{\partial x^2} \tag{6.25}$$

Substituting the expressions for φ_{nano} and φ_v (Equations 6.23--6.24) into (6.25) results in the dispersion relation:

$$b_2\varpi^2+ib_1\varpi+b_0=0 \tag{6.26}$$

Separating the real (ϖ_r) and imaginary (ϖ_i) components of Equation (6.26) gives:

$$b_2\left(\varpi_r^2-\varpi_i^2\right)-b_1\varpi_i+b_0=0 \tag{6.27}$$

$$b_2\left(2\varpi_r\varpi_i\right)+b_1\varpi_r=0\Rightarrow\varpi_r=0 \tag{6.28}$$

Then equation (6.26) takes the form as:

$$b_2\varpi_i^2+ib_1\varpi_i-b_0=0 \tag{6.29}$$

For neutral stability (ϖ_i), Equation (6.29) reduces to the critical condition:

$$b_0=0\Rightarrow\left(\rho_{nano}-\rho_v\right)gk-\sigma k^3=0 \tag{6.30}$$

6.5 NON-DIMENSIONAL FORM

Assuming U and $H=h_{nano}+h_v$ represent characteristic velocity and length scales, respectively, the governing equation (6.29) can be expressed in dimensionless form as:

$$\left[\coth\left(\hat{k}\hat{h}_{nano}\right)+\rho\coth\left(\hat{k}\hat{h}_v\right)\right]N^2$$
$$+\left[4\frac{\hat{k}^2}{\mathrm{Re}}\left(\left(1-\frac{\phi_{ag}}{\phi_m}\right)^{-[\alpha]\phi_m}\coth\left(\hat{k}\hat{h}_{nano}\right)+\mu\coth\left(\hat{k}\hat{h}_v\right)\right)\right]N^2+(1-\rho)\hat{k}Fr+\frac{\hat{k}^3}{We}=0 \tag{6.31}$$

where

$$\mathrm{Re} = \frac{\rho_{nano} U H}{\mu_{eff}}, \; We = \frac{\rho_{nano} U^2 H}{\sigma}, \rho = \frac{\rho_v}{\rho_{nano}}, \mu = \frac{\mu_v}{\mu_{eff}},$$

$$\hat{h}_{nano} = \frac{h_{nano}}{H}, \hat{h}_v = \frac{h_v}{H}, \; \mathrm{N} = \frac{\varpi_i H}{U}, \; Fr = \frac{gH}{U^2}$$

Here, *Fr* denotes the Froude number, Re denotes the Reynolds number, and *We* denotes the Weber number.

6.6 RESULTS AND DISCUSSION

This section presents a numerical solution of Equation (6.31) for a two-layer system consisting of a water–alumina nanofluid in the upper region and pure water in the lower region. The governing physical parameters for quantifying the perturbation growth rates, derived from Equation (6.31), are defined as follows:

$$\rho_v = 1.2, \mu_v = 0.000018, \rho_{H_2O} = 1000, \mu_{H_2O} = 0.001$$

$$\sigma_{water\&air} = 0.0473, r = R = 15^{*}10e-09, g = 9.8$$

Figure 6.2 compares the interfacial stability of VCVPF solutions between nanofluid and conventional viscous fluid interfaces [17]. The results demonstrate that VCVPF solutions for nanofluid interfaces exhibit significantly enhanced stability compared to those for viscous interfaces. This stability arises from the combined contributions of shear and normal viscous stresses in the nanofluid case. In contrast, the VPF model, which neglects shear stresses and accounts only for normal viscous effects, fails to capture this stabilizing mechanism for nanofluids.

Figure 6.3 investigates the influence of the Reynolds number on perturbation growth. Increasing the Reynolds number amplifies perturbation growth, indicating a destabilizing effect on the nanofluid-viscous interface. Since the Reynolds number is directly proportional to nanofluid density, higher nanofluid density exacerbates instability. Conversely, Reynolds number decreases with increasing effective nanofluid viscosity, highlighting the stabilizing role of viscous forces in suppressing interfacial disturbances.

Figure 6.4 analyzes the impact of upper fluid fraction on system stability. A higher upper fluid fraction elevates the perturbation growth rate,

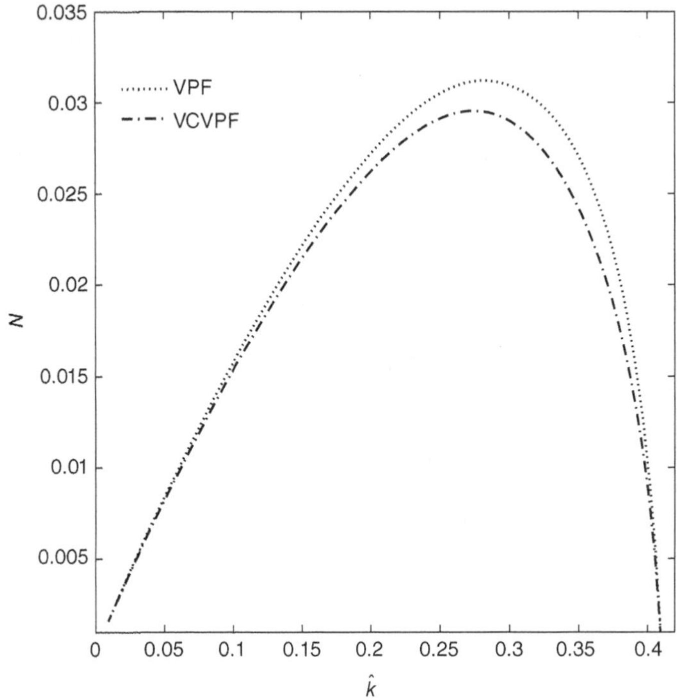

Figure 6.2 Comparison of interfacial stability between nanofluid and viscous fluid systems.

destabilizing the interface. This trend is attributed to the amplified gravitational influence on the denser upper layer, which accelerates interfacial deformations and intensifies instability.

Figure 6.5 explores the relationship between the Weber number and perturbation dynamics. Perturbation growth increases with the Weber number, underscoring its destabilizing role. As the Weber number is proportional to nanofluid density, higher density further destabilizes the interface. Conversely, surface tension (inversely related to the Weber number) exerts a stabilizing effect by resisting interfacial deformations.

Figure 6.6 examines the Froude number, which quantifies the influence of gravitational acceleration. Elevated Froude number values correlate with faster perturbation growth, confirming gravity's destabilizing role. This reinforces the conclusion that gravitational forces dominate over stabilizing mechanisms at higher Froude numbers.

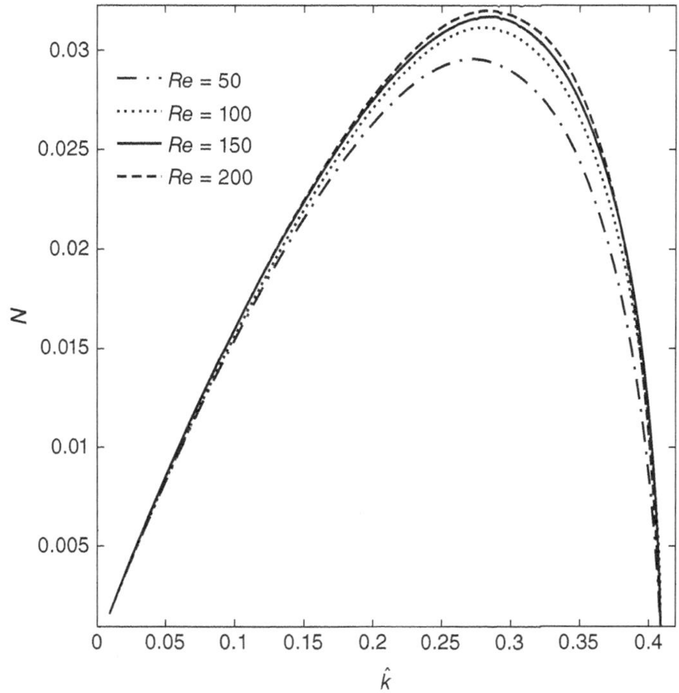

Figure 6.3 Effect of Reynolds number on perturbation growth.

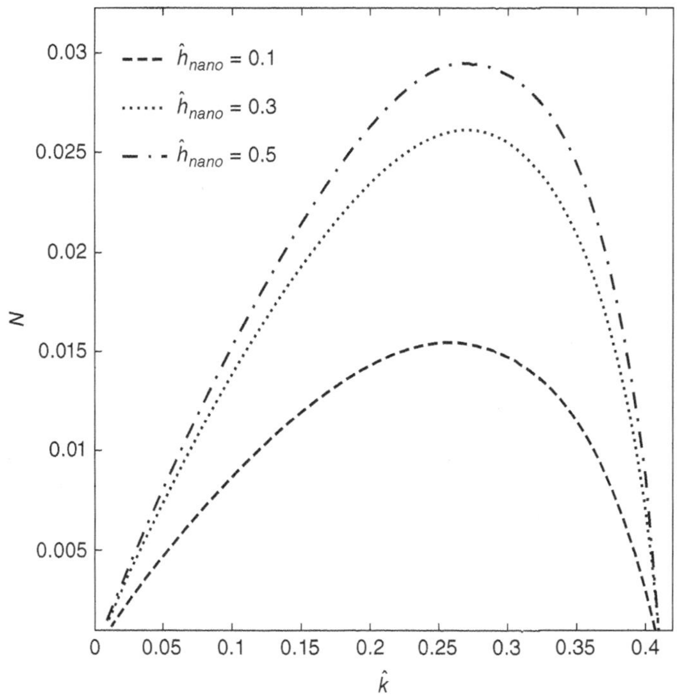

Figure 6.4 Influence of upper fluid fraction on interfacial instability.

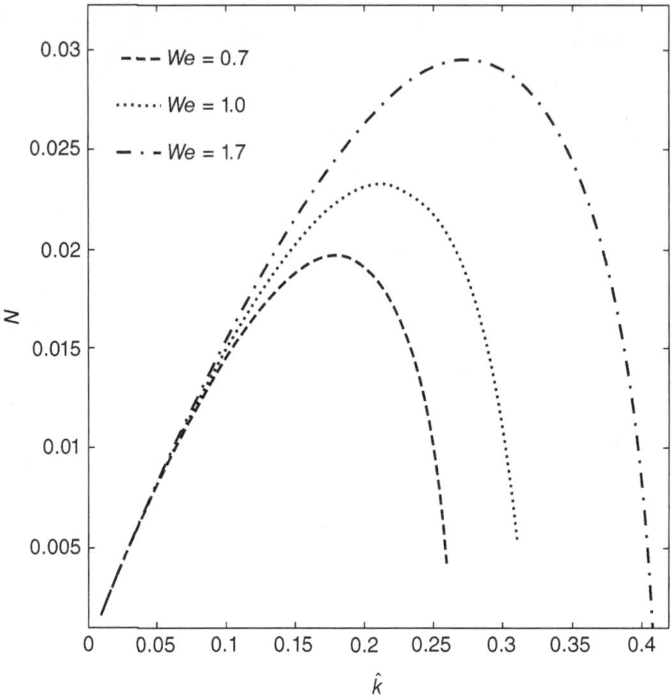

Figure 6.5 Weber number effects on interface stability.

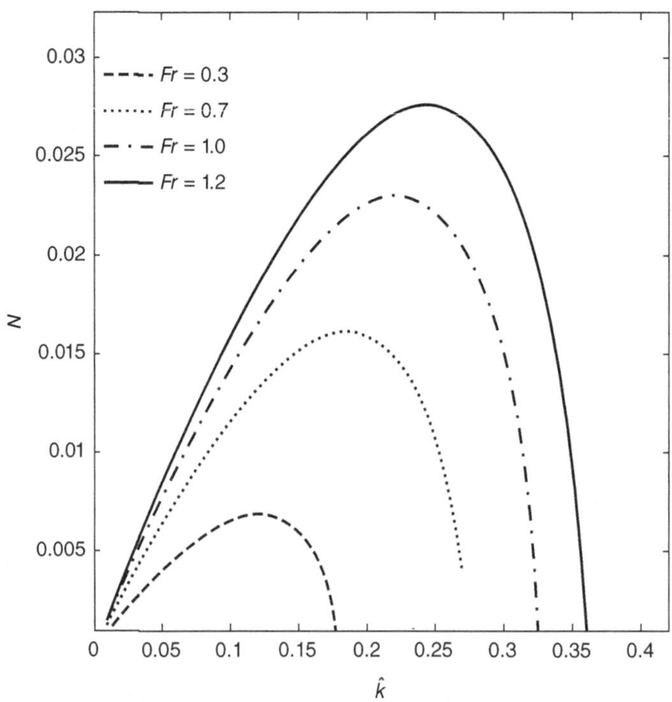

Figure 6.6 Froude number impact on interfacial stability.

6.7 CONCLUSIONS

This research explores RT instability at the interface between a viscous fluid and a nanofluid in a planar geometry. To analyze the system, a viscous correction is incorporated into the classical VPF theory. In the considered setup, the viscous fluid occupies the lower layer, while the nanofluid resides in the upper region. The dispersion relation for the perturbation growth rate exhibits a quadratic dependence. Results indicate that the solution obtained using the VCVPF model demonstrates greater stability compared to the standard VPF approach. Furthermore, interfacial stability improves with an increase in the viscosity of the viscous fluid, whereas a higher nanofluid density tends to destabilize the interface.

REFERENCES

1. J. Wang, D. D. Joseph, and T. Funada, "Pressure corrections for potential flow analysis of capillary instability of viscous fluids", *J. Fluid Mech.* 522 (2005), 383–394.
2. J. Wang, D. D. Joseph, and T. Funada, "Viscous contributions to the pressure for potential flow analysis of capillary instability of two viscous fluids", *Phys. Fluids* 17(5) (2005).
3. M. K. Awasthi, A. K. Shukla, A. Kumar, D. Yadav and N. Dutt, "Viscous correction to the potential flow analysis of Rayleigh–Taylor instability in a Rivlin–Ericksen viscoelastic fluid layer with heat and mass transfer", *Heat Transf.* 53(6), 3072–3088.
4. A. K. Shukla, M. K. Awasthi, D. Yadav, U. S. Mahabaleshwar, "Coupled effect of Pressure Corrections and Heat-Mass Transfer on Rayleigh-Taylor Instability in Casson Fluid Layer", *Chin. J. Phys.* (Accepted).
5. D. H. Sharp, "An overview of Rayleigh-Taylor instability", *Physica D*, 12, (1984), 3–18.
6. J. W. Jacobs and I. Catton, "Three-dimensional Rayleigh-Taylor instability Part 1. Weakly nonlinear theory", *J. Fluid Mech.* 187 (1988), 329–352
7. H. J. Kull, "Theory of the Rayleigh-Taylor instability", *Phys. Rep.* 206 (1991), 197–325.
8. K. O. Mikaelian, "Exact, approximate, and hybrid treatments of viscous Rayleigh-Taylor and Richtmyer-Meshkov instabilities", *Phys. Rev. E* 99 (2019), 023112.
9. D. D. Joseph, J. Belanger and G. S. Beavers, "Breakup of a liquid drop suddenly exposed to a high-speed airstream", *Int. J. Multiphase Flow* 25 (1999), 1263–1303.
10. R. Asthana, M. K. Awasthi, and G. S. Agrawal, "Viscous potential flow analysis of Rayleigh-Taylor instability of cylindrical interface", *App. Mech. Mat.* 110–116 (2012), 769–775.
11. M. K. Awasthi, R. Asthana and G. S. Agrawal, "Viscous potential flow analysis of nonlinear Rayleigh-Taylor instability with heat and mass transfer", *Microgravity, Sci. Technol.* 24 (2012), 351–363.

12. M. K. Awasthi, "Rayleigh–Taylor Instability of Swirling Annular Layer with Mass Transfer", *ASME-J. Fluid Engineering* 141 (2019), 071202 (5 pages).
13. M. K. Awasthi "Nonlinear analysis of Rayleigh-Taylor instability of cylindrical flow with heat and mass transfer", *ASME J. Fluid Engineering* 135 (2013), 061205 (7 pages).
14. M. K. Awasthi, Z. Uddin and R. Asthana, "Temporal instability of a Power-law viscoelastic nanofluid layer", *Eur. Phys. J. ST* 230 (2021), 1427–1434.
15. M. K. Awasthi, Dharamendra and D. Yadav, "Temporal instability of nanofluid layer in a circular cylindrical cavity", *Eur. Phys. J. ST* 231 (2022), 2773–2779.
16. S. Agarwal, M. K. Awasthi, A. K. Shukla, "Stability analysis of water-alumina nanofluid film at the spherical interface", *Proceedings of the Institution of Mechanical Engineers, Part E: Journal of Process Mechanical Engineering* 238(3) (2024), 1195–1205.
17. M. K. Awasthi, A. Kumar, N. Dutt, "Modeling Rayleigh-Taylor Instability in Nanofluid Layers", in *Computational Fluid Flow and Heat Transfer: Design, experiment and Application*, CRC Press (Taylor and Francis Group), USA.
18. G. M. Moatimid and M. A. Hassan, "Convection instability of non-Newtonian Walter's nanofluid along a vertical layer", *J. Egypt. Math. Soc.*, 25 (2017), 220–229.
19. G. M. Moatimid and M. Gaber, "Linear Instability of Water–Oil electrohydrodynamic Nanofluid Layers: Analytical and Numerical Study", *J. Comp. Theo. Nanoscience* 15 (2018), 1495–1510.
20. G. M. Moatimid, M. Gaber, and M. A. A. Mohamed, "Temporal instability of a confined nano-liquid film with the Marangoni convection effect: viscous potential theory", *Microsyst. Technol.* 26 (2020), 2123–2136.

Chapter 7

Materials Integration of Pressure Sensor Mechanism for Fluid Flow Control in MEMS Microfluidic Device Application

A Review

Ankur Saxena, Satyabrata Singha, Surendar Reddy, D. Lingamaia, and Md. Sadullah

7.1 INTRODUCTION

In the last few years, microfluidics has achieved advancements in research and development. The microfluidic system was developed whose width or height lies between 100 nm and 100 μm (Convery & Gadegaard, 2019). The microfluidic system and its applications are Bio-MEMS, microreactors, pressure sensors, and micro fuel cells. The microfluidics system has a crucial component, that is, the microchannel (Tamayol & Bahrami, 2010). The fluid flows in the microchannel; the properties of the fluid vary with the shape, size, and construction material of the microchannel. Mostly, a microchannel design approach with a high surface area to volume ratio (Koo & Kleinstreuer, 2003). The Naiver-Stokes equations are used to analyze the properties of fluid flow in microchannels. The fluidic properties vary along the length of an accountable microchannel (Papautsky et al., 2001). The monitoring of such types of properties for various types of applications is done by measuring instruments and sensors. The fluidic flow in microchannels exerts pressure; the pressure is varied according to the flow types of fluid corresponding to laminar flow, creeping flow, and turbulent flow. The analysis of behavior and monitoring of microfluidic pressure in microchannels for viscosity, density, and accuracy techniques is required.

Microfluidic pressure sensor senses pressure of fluid by various types of sensing mechanisms. The fluidic mechanics pressure is one key parameter of flow properties (Wu et al., 2012). A lot of methods are employed to detect the pressure inside a microfluidic chip. The piezoelectric, piezoresistive, capacitive, and optical pressure sensors are used to detect pressure through deflection of the membrane, and it is challenging to integrate them in microfluidics. The size of the MEMS pressure sensor in mm and microchannel is present in the microfluidic chip size of micrometers, so it is challenging for researchers to integrate a pressure sensor in a microchannel. These types of problems are removed when integrating a gas pressure sensor or PSP (pressure sensitive paints) layer in a microchannel. The monitoring of gaseous

DOI: 10.1201/9781003642152-7

fluid molecules through PSP is very attractive, and it provides a great global pressure profile (Hoera et al., 2018). The technology is extensively increasing in the chip-level market, with advancement and flexible device demand in the biomedical field increasing fast. The microfluidic pressure sensor is also involved in this flexible and wearable application. Flexible pressure sensors used flexible material as conductive material, carbon nanotubes, silicon nanowires, and some liquids, including galinstan indium, which is non-viscous (Yang et al., 2018).

This chapter discusses sensing layer integration in microchannel methodology, the material used in fluidic channels, and the pressure sensor application of microfluidics. The flexible pressure sensor is widely used in microfluidic biomedical applications.

7.2 MICROCHANNEL IN MICROFLUIDICS

The fluidic pressure in microchannels shows different types of behavior. The pressure not only depends on the shape and size of the microchannel, it also depends on the surface smoothness, the material of the channel, the velocity, and the flow profile of the fluid. The researchers investigate the properties of microchannels in the field of liquid and gas. The microchannel can be used in various types of applications, as shown in Table 7.1. The fluid flow velocity is directly related to the Reynolds' number in microchannel fluid flow. With a small velocity amount, the pressure exerted on the fluid due to the applied force on the walls is also a small amount. The high-viscosity fluid has high pressure and flow in a microchannel with low speed. The fluidic flow is divided into two categories according to the Reynolds number. If the Reynolds number is low, it is laminar flow; when the Reynolds number is high, then it is turbulent flow. The Reynolds number is the ratio of inertia to viscous force with respect to the velocity of the fluid. The Reynolds number is low, showing the viscous force is dominant because this fluid motion is smooth and constant, while a high Reynolds number shows the inertial force is dominant due to the production of eddies, vortices, and instability of fluid. The Reynolds number is defined by the following equation:

$$\mathrm{Re} = \frac{\rho U D}{\mu}$$

The Reynolds number describes the characteristics of fluid density, velocity, and viscosity with respect to length. The fluid in the microchannel has variable viscosity and compressible gas. The Newtonian fluid's viscous stress is shown to exhibit linear behavior with respect to fluid velocity. The Newtonian fluid follows Newton's law, and non-Newtonian fluids do not

Table 7.1 Various geometries of microchannels with their properties

Geometry	*Hydraulic diameter(m)*	*Material*	*Testing fluid*	*Reynolds number*	*Friction coefficient*	*Aspect ratio*	*References*
Circular	8μm	Silicon	DI water	1.0–2.0	64	0–1.0	(Tesař, 2008)
Trapezoidal	0.14	Silicon	DI water	0.1–2.0	64	0–1.0	(Tesař, 2008)
Triangular	24	Silicon	DI water	0.1–2.0	64	0–1.0	Tesař, 2008
Square	0.17	Silicon	DI water	0.1–2.0	64	0–1.0	Tesař, 2008
Rectangular	69.5-304.7	Silicon	DI water	112–9180	0.14	0.09-0.24	(Kundu et al., 2016)
Rectangular	NA	PDMS	DI water	1–35	14	0.13	(Zhang et al., 2005)
Circular	-	PDMS	DI water	1–35	12	0.17	Zhang et al., 2005
Trapezoidal	-	PDMS	DI water	1–35	4	0.4	Zhang et al., 2005
Triangular	-	PDMS	DI water	1–35	6	0.6	Zhang et al., 2005
Square	-	PDMS	DI water	1–35	2.5	0.76	Zhang et al., 2005
Trapezoidal	10mm	PDMS	DI water	32.5–265	0.09–2.07	-	(Song & Psaltis, 2010)

obey Newton's law. The microfluidic flow in a channel is divided into three categories: laminar flow, creeping flow, and turbulent flow.

7.2.1 Laminar fluid flow

The laminar fluid flow in a microchannel is a low-speed fluid flow in which particles move in a smooth layer without mixing with each other. The laminar fluid flow is an incompressible steady fluid flow that provides an exact solution for viscous fluids. This flow is known as fully developed, in which velocity is independent of the downstream coordinate (Kundu et al., 2016). The laminar flow produces a friction that depends on whether the surface is smooth or rough. The surface roughness is calculated through friction factor correlation (fRe). The microchannel has a size of 0.5 to 50 microns. The friction factor is different from the theoretical prediction. It is clear that the friction coefficient value depends on the size and shape of the microchannel (Tesař, 2008). For a steady laminar flow in a duct, the friction pressure loss is given by:

$$\Delta \mathrm{P} = \mathrm{f}\frac{L\rho v^2}{d2}$$

F: Darcy friction factor
L and d: Length of hydraulic diameter of duct
ρ: fluid density
v: mean velocity of ducts

The Darcy function of the Reynolds number of the flow and the relative surface roughness of the duct is defined as:

$$\mathrm{f} = \frac{8\tau_w}{\rho v^2} = \mathrm{F}\left(\mathrm{Re}, \frac{\varepsilon}{d}\right)$$

$\tau_{w:}$ Shear stress along the wall
ε: Wall roughness height

Laminar flow mostly depends on the Reynolds number, but some experiments investigated that the surface roughness of the duct friction factor also affects the **flow** (Jiang et al., n.d.). The researchers investigated fully developed vapor and liquid flow through a rectangular microchannel with a hydraulic diameter varying from 69.5 to 304.7 micrometers (Hrnjak & Tu, 2007). The rectangular cross section has been extensively studied and employed in many applications. The water flow through a different geometry of microchannel cross section includes triangular, trapezoidal, and rectangular. The diameter of microchannels ranges from 8 micrometers to 42 micrometers (Zhang et al., 2005). The five microchannels are taken with aspect ratios $0.13 < \varepsilon < 0.76$. The channel is designed for fully developed flow, so the author assumes the cross-section channel does not change under pressure-driven flow (Akbari et al., 2009). As experiments show, the friction factor decreases with the Reynolds number in a channel. The Poiseuille number does not depend on the Reynolds number; it only varies with respect to the aspect ratio.

7.2.2 Creeping flow

Stokes flow is also known as creeping flow; it is the motion of fluid in which advective inertial forces are small compared to viscous forces. Creeping fluid flow in a channel is where the Reynolds' number value is less than unity. The fluid flow developed resistance against the motion due to the applied drag force around a moving body. The steady incompressible creeping flow is given by Navier's-Stokes equation:

$$\eta\Delta 2\mathrm{v} = \Delta \mathrm{p}^*,$$

$$\Delta.\ v = 0$$

$$\mathrm{p}^* = \mathrm{p} + \rho_0 \phi \ \text{ - Effective pressure}$$

The research design investigates a microchannel with an integrated micropillar, which compares the mass transfer coefficient, thermal conductance, and surface area-to-volume ratio with and without a micropillar microchannel. Such a type of device is used in micro heaters, microreactors, micropumps, and microfilters. The array of micropillar cylinders is integrated into a rectangular microchannel. The staggered micropillar gives the same porosity as a square in a lower pressure drop (Tamayol et al., 2012). The small size of the channel reduced the Reynolds number flow in the system, which changed Navier's Stokes equation into Stokes equation for creeping flow. The author calculated different velocities and pressure drops through an experiment in a microchannel. The 3D orthogonal designed curvilinear model calculates the fluid velocity and pressure in the channel with the support of the Navier's-Stoke equation. The expansion of the contraction parameter in the channel increases the amount of fluid pressure (Lepchev & Weihs, 2010).

7.2.3 Turbulent flow

Turbulence in a microchannel. It is well known that the transition from a laminar to a turbulent flow takes place in a rigid tube at a Reynolds number of about 2100 and in a rigid channel at a Reynolds number of about 1200. The researcher studied the etching microchannels with fixed cross sections.

The microchannel is a linked transition from laminar to turbulent flow. The cross-section of microchannels is rectangular, hexagonal, trapezoidal, and double trapezoidal. The results are explained only by the conventional theory of microchannels (Wang et al., 2018). The researcher investigates the simulation study of turbulent flow in a microchannel having a micropost in an aligned and staggered arrangement. The author is investigating whether the slip velocity increases with the increase of the Reynolds number or cavity fraction. The Reynolds number is used to calculate flow mean velocity, slip velocity, and friction velocity (Kharati-Koopaee & Rezaee, 2017).

7.3 FLEXIBLE MICROCHANNEL IN MICROFLUIDICS

The flexible microfluidic sensor technology is rapidly developing, making life easier for human beings. The formation of flexible microchannels uses smart material for the development of a flexible microfluidic pressure sensor (An et al., 2019). The rigid microchannel is not suitable for wearable applications, and fluid passing through the channel shows weak structure interaction; therefore, flexible microchannels provide good moldability, high selectivity, and high sensitivity. In the earliest days, the first-choice microfluidic system was fabricated in glass or silicon, but such technology consumes a lot of time and needs sophisticated equipment. The silicon and glass materials are both rigid materials and are not suitable for flexible devices (Bollella et al., 2019). Flexible smart material has a property of bending, deforming, and stretching. Mostly, flexible microchannels are produced using polymer; polymer is selected due to its high elasticity so that it can return to its original shape (Dinh et al., 2018). The Young modulus shows the property of elasticity of the material in Figure 7.1. Various materials are shown with their elasticity sequence.

In the microfluidic chip, the size of the microchannel is under 100 μm where the pressure sensor is integrated. PDMS material is the most famous material for the fabrication of microchannels (Jia et al., 2008). PDMS material has soft walls, which undergo deformation when fluid flows through the microchannel.

In an observation of a flexible rectangular microchannel, an author found the nonlinear behavior of pressure with respect to the width-to-height ratio. The pressure characteristics with coated and non-coated microchannels vary when fluid passes through the channel (Adzima & Velankar, 2006). The PDMS microchannel also includes reservoir interdigitated electrodes for the healthcare application. Flexible microchannel investigates cardiovascular-related disease. The author investigates and resolves the problem of fluid structure interaction by reducing the thickness of the wall in a microchannel (Raj et al., 2018). The advancement in flexible microchannels continuously progresses, and there's a lot of future scope in the biomedical field.

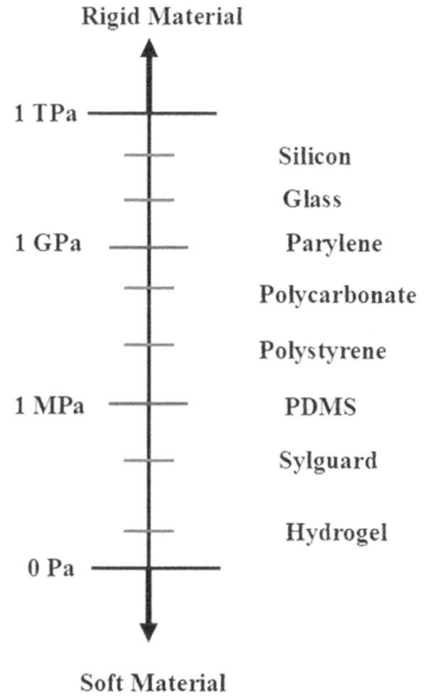

Figure 7.1 Pressure range for selection of material.

7.4 METHODOLOGY FOR PRESSURE SENSING IN MICROCHANNEL

7.4.1 PSP sensing layer pressure

The surface pressure measurement is an important method in microfluidics. The traditional surface pressure measurement required an instrument with several taps of pressure drilled into the model, making the model costly and the process time-consuming. It is very difficult to drill at thin edges and sharp corners where calculated pressure is required. To reduce the bulk and cost of the model, PSP coating is used in the model. PSP is a highly demanding approach in optical methodology that measures the fluid pressure without including the large setup of the instruments. The fluidic device surface is coated with fluorescent material, which is used to generate light as the fluid interacts with the material. The luminophore molecules are supported to generate light with the longer wavelength. In addition to normal deactivation via thermal collisions, some luminophores are also quenched through collisions with oxygen molecules. Since the relative concentration of oxygen in air remains constant, this oxygen quenching effect can be used to indicate air pressure above a surface. As pressure decreases, the amount of emitted light increases. The theory indicates that pressure is inversely proportional to the emitted light intensity.

The first experimental method was developed using a pressure sensor with PSP coating in 1980 (Gregory et al., 2008). The PSP technique measures continuous pressure measurement on the global surface. This method is non-intrusive, which means that there are no probes required, and provides qualitative and quantitative pressure profiles. Luminance is decreased continuously by increasing pressure or increased by decreasing pressure. The intensity of light depends on the pressure profile. The PSP method measures the pressure of interaction of molecules or atoms inside a microchannel. The molecules emit light; a photodetector collects light and transfers it to image processing equipment. The author measures pressure in the range of 0.001 psi–30 psi (Matsuda et al., 2007). The PSP method is mostly used in aerospace engineering, which calculates air pressure on the surface. This method is not able to detect high-pressure and low-speed fluid. The range of pressure measurement is in Pascal; if measuring the pressure of pores in microchannels, then PSP is not able to detect pressure because the pores exist in kPa–MPa (Zarikos et al., 2018).

7.4.2 Deformation of PDMS and sensing pressure

It is difficult to measure in a microfluidic chip due to the limitation of the pressure sensor. The PDMS structure is highly flexible and compliant, with fluid flow in the microchannel with a flow rate of 15–1550 μL/min in the laminar region. A microchannel becomes shallow and deformed at high pressure. It is challenging to reduce large deformation at high pressure and increase the stiffness of PDMS (Sollier et al., 2011). To increase the stiffness of the material, modify the ratio of mixing and thermal treatment of PDMS. The curing agent can enhance stiffness by up to 50%, allowing for improved channel deformation control under fluidic pressure (Kim et al., 2014).

A novel method for the detection of pressure inside a microchannel in a microfluidic device. An author sensed the pressure of fluid flow in a channel by observing deformation of PDMS (Roh et al., 2016). An author compares the result of pressure with its experimental and theoretical study. A positive pressure or negative pressure is applied on the microchannel, and the PDMS chamber is deformed into a convex shape or deformed back into a concave shape. A sensing layer senses a pressure, color fluid intensity according to pressure at the POI (point of contact) (Tsai et al., 2015). The PDMS microchannel includes a sensing chamber, which is filled with colored fluid. The colored fluid flows inside and outside of the sensing chamber; the sensing chamber senses the pressure of fluid at different intensities. The colored fluid shows various pressure levels of the deformation sensing chamber. The pressure applied was 50 kPa and 100 kPa to the sensing chamber, which shows 0.024 and 0.048 displacement (Tai et al., 2007).

Polydimethylsiloxane in microfabrication offers low cost, optical transparency, biocompatibility, and contour accuracy. The new approach to PDMS microchannels integrated an electronic obstacle in the flow of fluid

Table 7.2 Pressure sensing materials with pressure range

Pressure-sensing material	*Microchannel size (μm)*	*Pressure range*	*Flow rate*	*Ref.*
(Polystyrene + PtTFPP)	170 × 27	Up to 9 mbar	1 μL min^{-1} to 3 μL min^{-1}	(Ko et al., 2008)
SU-8	500–4000 × 23–55	Up to 140Kpa	0–1200 mg min^{-1}	(Tang et al., 2017)
Electrochemical	200 × 100	120 mmHg	0–10 sec	(Shen et al., 2020)
Parylene-C bond Si_3N_4	-	Up to 16 Mpa	0.1 M Pa s^{-1}	(Ciftlik & Gijs, 2012)
Self-powered Sensing	100–500 μm	Up to 1bar	-	(Goh et al., 2021)

and pressure drop under steady fluid flow. The PDMS microchannel swelling is detected when the fluid comes into contact with obstacles. The author studies quasi-steady flow and steady-state flow under low flow rates. The fluid flow in different microfluidic channels, that is, 15 μm and 100 μm, was compared to the theoretical study model (Roh et al., 2016). The deformation of the microchannel also depends on the aspect ratio of the channel. The rectangular channel required a low aspect ratio, which produced bulging in the channel under the pressure (Miralles et al., 2013). The comparative analysis of pressure-sensing materials is represented in Table 7.2.

7.5 MATERIALS PRESSURE SENSING IN MICROFLUIDICS

7.5.1 PSP pressure sensor in microfluidics

The pressure-sensitive paint method was introduced in 1980 to measure the pressure of the surface in aerodynamic applications. Later, the PSP technique is used for low-speed flow application (Quinn et al., 2013). Basically, PSP is an optical pressure method which measures the distributed pressure of the entire surface. Recently, researchers improved the PSP coating method to make it a cheap and efficient method to measure the pressure of complex 3D structures (Roh et al., 2016).

An author investigated the PSP formulation, in which luminophore and polymer binder are composed. Different types of photoluminescent probe molecules exhibit distinct emission characteristics for temperature and oxygen sensing. Polymer-based PSP without embedded particles features a uniform polymer binder layer. It interacts only with air molecules present at the surface of the binder layer. The response time of such PSP systems depends

Table 7.3 Types of PSP

PSP	*Luminophore*	*Polymer binder*	*Reference*
PSP-1	PtTFPP	RTV-118	(Roh et al., 2016)
PSP-2	PtTFPP	Poly(TMSP)	(Roh et al., 2016)
PSP-3	PtOEP	RTV-118	(Roh et al., 2016)
PSP- 4	PtOEP	Poly(TMSP)	(Roh et al., 2016)
PSP-5	PtTFPP	PC-PSP	(Roh et al., 2016)

on the oxygen diffusivity and the thickness of the binder layer. In comparison, PC-PSP exhibits a faster response time than the polymer binder layer, primarily due to the concentration of solvents for luminophores, even when the binder layer thickness remains the same (Cho et al., 2024). The thickness of the PSP layer is around 25–30 μm with a white paint layer. Table 7.3 shows five types of PSP formulation.

The polymer-based PSP, without particles has a uniform polymer binder layer. It contacts with air particles only, which are available on the surface of the binder layer. The response time of PSP depends on oxygen diffusivity and the thickness of the binder layer. As in PC-PSP, it shows a faster response time compared to the polymer binder layer, depending on the concentration of solvents for luminophores, while the thickness is the same as the binder layer (Huang et al., 2015).

7.5.2 Parylene pressure sensor in microfluidic

Parylene was founded in 1965 for commercial applications. Parylene is a semi-crystalline polymer, which is deposited as a thin film through the CVD process. The four types of Parylene commonly used are Parylene N, Parylene C, Parylene D, and Parylene HT. Parylene coating is important in electronics, as it provides waterproof insulation and is used in biomedical devices. Parylene has special features such as optical transparency, mechanical flexibility, biocompatibility, and low intrinsic stress. Parylene can be used on both rigid and flexible substrates (Dai et al., 2009; Ortigoza-Diaz et al., 2018).

The pressure sensor was coated with micro-coating, which provides a better environment with improved accuracy of the device. The micro-coating is done by a post-processing method, and the size of the coating is approximately 1 μm. The micro-coating is a very low-cost, high-accuracy method (Yao et al., 2019).

Parylene-based pressure MEMS sensor is designed to sense intraocular pressure. An author designed a pressure sensor, which senses without changes in electrical or power signal. It measures the pressure difference between the applied pressure in the microchannel and the ambient

Table 7.4 Types of Parylene with their function

Parylene	*Chemical structure*	*Function cells*
Parylene N	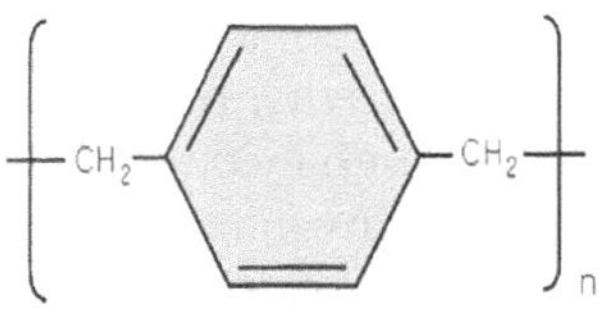	Used as a dielectric The deposition rate is very slow
Parylene C	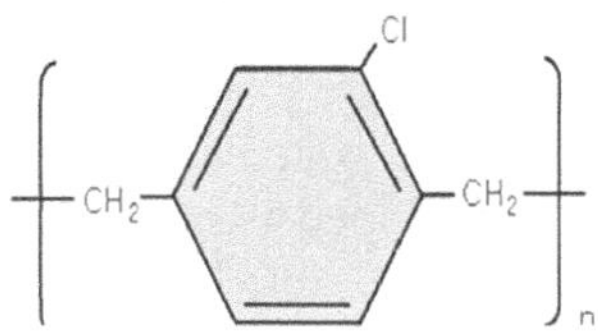	Chemical inertness, electrical resistivity, low moisture permeability, and biocompatibility
Parylene D	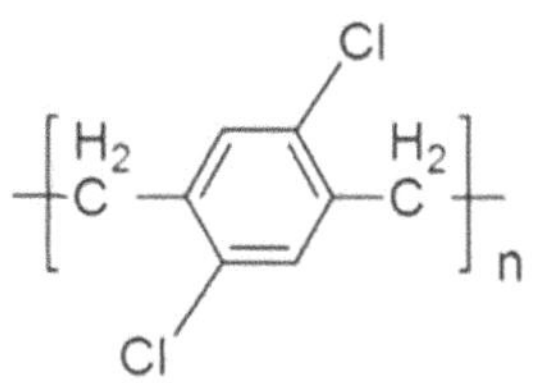	Higher temperature resistance Similar function as parylene C
Parylene HT	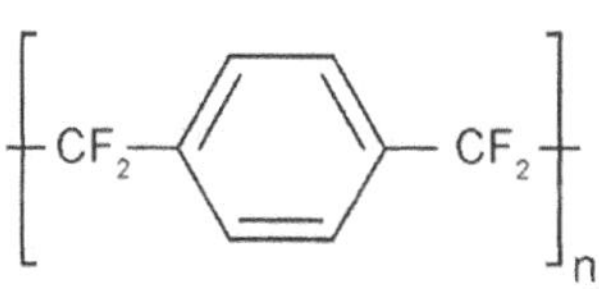	High temperature stability

temperature. The IOP sensor is designed with a protective parylene membrane. The channel is formed by parylene material with a Young modulus of 4 GPa and biocompatibility. The size of the channel is approximately 20 um, and there is a parylene C layer in a spiral pattern coating in the channel (Chen et al., 2007). Table 7.4 represents different types of parylene configurations that are used under pressure of fluid for different types of applications.

7.5.3 Galinstan pressure sensor in microfluidic

Galinstan has the property of a liquid and a metal; it is an alloy, which is a mixture of 65% gallium, 21.5% indium, and 10% tin. Gallium liquid has much better properties compared to mercury and DI water. Its melting point is -19°C, which can be used as a supercool environment. Galinstan is widely used in microfluidic devices and their applications. A Galinstan liquid metal is embedded in a microfluidic structure, which is made from PDMS, and the majority of it acts as pressure sensors (J. Wang et al., 2012).

The thin film pressure sensor is designed using Galinstan and investigates the pressure of DI water, which flows through a microchannel. Recently, eutectic gallium–indium is widely used in microfluidic applications for the detection and measurement of different aspects or parameters. The eutectic gallium–indium alloy has 75% gallium and 25% indium.

In addition to its high electrical conductivity, EGaIn can function as an optical reflector and can be incorporated into metamaterials (Khoshmanesh et al., 2017). Owing to its negligible solubility in most liquids, EGaIn can be integrated into devices alongside other fluids without risk of dissolution. Its low melting point of 15.5°C ensures it remains liquid at room temperature (Kwang-Seok Yun et al., 2002).

The design process of the microfluidic system involves several steps to integrate the sensor. The researchers developed a Galinstan-based integrated sensor microfluidic device consisting of multiple layers: a bottom layer for the fluidic channel, a middle layer with a PDMS thin membrane, and a top layer for sensor integration. Pressure measurement within the fluidic channel is performed using the Wheatstone bridge methodology. Pressure is measured in terms of terminal or node voltage difference or change in terminal voltage (Jung & Yang, 2015).

The wearable microfluidic diaphragm pressure sensor is also designed for health and tactile touch monitoring. The wearable pressure sensor's basic requirement is that it should be flexible. A diaphragm pressure sensor designed and embedded in a Galinstan microchannel. The flexible pressure sensor in microfluidics applications is a wristband, tactile sensing glove, or tissue detection device (Gao et al., 2017).

7.5.4 Polystyrene pressure sensor in microfluidic

Polystyrene is frequently found in standard biological labware, including multiwell plates and cell culture dishes. Because of its chemical resistance to the majority of common solvents, polystyrene (PS) was chosen as the polymer material for the sensor layer.

The trend is highlighted by various biological research industries that offer specific plastic microfluidic cell-based assays. The polystyrene is integrated with PDMS and makes a sensing layer. An author compared the PS microfabrication method for cell-based arrays in microfluidics with PDMS-based soft lithography in terms of cost, time expenditure, and reliability (G. Song et al., 2023). An author designed a polystyrene–PDMS bonding microfluidic device that studies the fluid flow under pressure and stress (Reddy et al., 2012). It developed a sensing device for diagnostics for health care applications. The device has functionalized electrodes in a microfluidic channel; these electrodes have the ability for sensing of chemical or label-free analytes. PDMS and PS, both substrates, are used as microfluidic channels and electrodes. A silane layer and plasma oxidation are used to link two substrates. An innovative method for creating a microfluidic system is represented by this bonding substrate formation. Electrochemical sensing capabilities are made possible by fabricating three electrodes inside a Y-shaped microfluidic channel. Strong bonds that withstand delamination at gaseous pressures of up to 150 kPa and fluid flow rates of up to 5.5 ml/min are produced via PS-PDMS bonding.

7.6 SPECIFIC CONSIDERATION OF THE PRESSURE SENSOR AND ITS APPLICATION

The integrated pressure sensor in a microfluidic chip is designed for various types of applications. This chapter discusses some specific designs, materials, and methodologies. The microfluidic pressure sensor is mostly applied in the field of biomedical and optical applications.

Human finger motion can be detected by the self-powered liquid triboelectric microfluidic pressure sensor. The sensing mechanism includes two mechanisms: capacitive and triboelectric. The pressure sensor measures magnitude and frequency. The pressure sensor measures the bending angle and frequency of the finger. Whereas the capacitive mechanism works in both static and dynamic water situations, the triboelectric mechanism produces signals when water flows over the electrodes. The output voltage's magnitude indicates the degree of bending, while the time difference between two bending angles establishes the bending rate (Sun et al., 2019). The 3D force is determined through a tactile electronic skin sensor. The CNT and PDMS materials are used for the detection of 3D force. The pressure sensor works on the principle of piezoresistivity. The pressure sensor is sensing finger bending, wrist pulse, and object grasping (Samoei & Jayatissa, 2020). The piezoresistive pressure sensor is highly attracted to material AZO (aluminum-doped zinc oxide). Its cost is lower than other materials. The AZO thin film is used as a diaphragm; as the AZO layer bends, the resistance increases. The AZO sensor investigates pressure as applied pressure on the diaphragm. The AZO pressure sensor measures pressure in both vacuum and gauge. The disposable electronic device was created without additional toxic and hazardous by-products. The PDMS and ZnO layer is used for a nontoxic and nonabsorbent surface material (Sinar & Knopf, 2020b). The cardiovascular disorder, immune diseases, and rheumatoid arthritis are diagnosed through blood plasma viscosity. The viscosity range for healthy persons lies between 1.1 cP and 1.3 cP. The microcantilever is designed to sense the viscosity of blood with high resolution. The pressure difference along the channel at two different locations is measured. The viscosity is measured with glycerol solution and variation in temperature (Cakmak et al., 2013). Table 7.5 represents the comparison of specific integrated pressure sensors in a microfluidic device.

Table 7.5 Comparison Study for specific integrated pressure sensor in microfluidic

Sensor	*Material*	*Mechanism*	*Application*	*Reference*
Flexible pressure sensor	PET, Cu, PDMS, SU-8	Tribloelectric mechanism and capacitive mechanism	Finger motion detection	(Goh et al., 2021)
Flexible Tactile Sensor	PDMS, CNT, Parylene –C	Piezoresistive	Human body monitoring and robot grasping	(Sun et al., 2019)
Pressure sensor	Al, ZnO, PE, PMMA, AZO	Piezoresistive	Flexible sensor	(Samoei & Jayatissa, 2020)
Vibration Sensor	Graphene nanoparticles, polymer, carboxymethyl cellulose, PDMS, ZnO	Piezoelectric	Wearable sensor	(Sinar & Knopf, 2020a)
Viscosity Sensor	$Ni(SO_3NH_2)_2$ $4H_2O$), Ni, Si, PMMA, Au	Magnetic actuation, optical sensing, and Temperature control	Serum and Blood Plasma	(Cakmak et al., 2013)

7.7 CONCLUSION

This chapter studies various fields of microfluidics integrated pressure sensors. The microfluidic technology is involved in each field very extensively. The study is divided into some categories: types of sensors, material used in healthcare sensing applications, sensors, and also some specific designs with their application. Mostly, sensors interact with biotechnology for the detection of disease. Biosensors and biochemical sensors have a wide range of applications. This chapter discusses the microfluidic material for various applications, highlighting their strength and application. The future scope of microfluidic devices in the field of healthcare applications is highly in demand.

REFERENCES

Adzima, B. J., & Velankar, S. S. (2006). Pressure drops for droplet flows in microfluidic channels. *Journal of Micromechanics and Microengineering*, *16*(8), 1504–1510. https://doi.org/10.1088/0960-1317/16/8/010

Akbari, M., Sinton, D., & Bahrami, M. (2009). Pressure drop in rectangular microchannels as compared with theory based on arbitrary cross section. *Journal of Fluids Engineering, Transactions of the ASME*, *131*(4), 0412021–0412028. https://doi.org/10.1115/1.3077143

An, H., Chen, L., Liu, X., Zhao, B., Zhang, H., & Wu, Z. (2019). Microfluidic contact lenses for unpowered, continuous and non-invasive intraocular pressure monitoring. *Sensors and Actuators A: Physical*, *295*, 177–187. https://doi.org/10.1016/j.sna.2019.04.050

Bollella, P., Sharma, S., Cass, A. E. G., & Antiochia, R. (2019). Microneedle-based biosensor for minimally-invasive lactate detection. *Biosensors and Bioelectronics*, *123*, 152–159. https://doi.org/10.1016/j.bios.2018.08.010

Cakmak, O., Elbuken, C., Ermek, E., Mostafazadeh, A., Baris, I., Erdem Alaca, B., Kavakli, I. H., & Urey, H. (2013). Microcantilever based disposable viscosity sensor for serum and blood plasma measurements. *Methods*, *63*(3), 225–232. https://doi.org/10.1016/j.ymeth.2013.07.009

Chen, P.-J., Rodger, D. C., Agrawal, R., Saati, S., Meng, E., Varma, R., Humayun, M. S., & Tai, Y.-C. (2007). Implantable micromechanical parylene-based pressure sensors for unpowered intraocular pressure sensing. *Journal of Micromechanics and Microengineering*, *17*(10), 1931–1938. https://doi.org/10.1088/0960-1317/17/10/002

Cho, K., Lee, J., & Kim, J. (2024). Integrated high-throughput drug screening microfluidic system for comprehensive ocular toxicity assessment. *Toxicology in Vitro*, *98*, 105843. https://doi.org/10.1016/j.tiv.2024.105843

Ciftlik, A. T., & Gijs, M. A. M. (2012). Parylene to silicon nitride bonding for post-integration of high pressure microfluidics to CMOS devices. *Lab Chip*, *12*(2), 396–400. https://doi.org/10.1039/C1LC20727J

Convery, N., & Gadegaard, N. (2019). 30 years of microfluidics. *Micro and Nano Engineering*, 2, 76–91. https://doi.org/10.1016/j.mne.2019.01.003

Dai, C. L., Lu, P. W., Chang, C., & Liu, C. Y. (2009). Capacitive micro pressure sensor integrated with a ring oscillator circuit on chip. *Sensors*, *9*(12), 10158–10170. https://doi.org/10.3390/s91210158

Dinh, T., Phan, H. P., Kashaninejad, N., Nguyen, T. K., Dao, D. V., & Nguyen, N. T. (2018). An On-Chip SiC MEMS Device with Integrated Heating, Sensing, and Microfluidic Cooling Systems. *Advanced Materials Interfaces*, *5*(20). https://doi.org/10.1002/admi.201800764

Gao, Y., Ota, H., Schaler, E. W., Chen, K., Zhao, A., Gao, W., Fahad, H. M., Leng, Y., Zheng, A., Xiong, F., Zhang, C., Tai, L., Zhao, P., Fearing, R. S., & Javey, A. (2017). Wearable microfluidic diaphragm pressure sensor for health and tactile touch monitoring. *Advanced Materials*, *29*(39). https://doi.org/10.1002/adma.201701985

Goh, Q. L., Chee, P., Lim, E. H., & Liew, G. G. (2021). Self-powered pressure sensor based on microfluidic triboelectric principle for human–machine interface applications. *Smart Materials and Structures*, *30*(7), 075012. https://doi.org/10.1088/1361-665X/ac01a5

Gregory, J. W., Asai, K., Kameda, M., Liu, T., & Sullivan, J. P. (2008). A review of pressure-sensitive paint for high-speed and unsteady aerodynamics. *Proceedings of the Institution of Mechanical Engineers, Part G: Journal of Aerospace Engineering*, *222*(2), 249–290. https://doi.org/10.1243/09544100JAERO243

Hoera, C., Kiontke, A., Pahl, M., & Belder, D. (2018). A chip-integrated optical microfluidic pressure sensor. *Sensors and Actuators B: Chemical*, *255*, 2407–2415. https://doi.org/10.1016/j.snb.2017.08.195

Hrnjak, P., & Tu, X. (2007). Single phase pressure drop in microchannels. *International Journal of Heat and Fluid Flow*, *28*(1), 2–14. https://doi.org/10.1016/j.ijheatfluidflow.2006.05.005

Huang, C. Y., Matsuda, Y., Gregory, J. W., Nagai, H., & Asai, K. (2015). The applications of pressure-sensitive paint in microfluidic systems. *Microfluidics and Nanofluidics*, *18*(5–6), 739–753. https://doi.org/10.1007/s10404-014-1510-z

Jia, Y., Jiang, J., Ma, X., Li, Y., Huang, H., Cai, K., Cai, S., & Wu, Y. (2008). PDMS microchannel fabrication technique based on microwire-molding. *Science Bulletin*, *53*(24), 3928–3936. https://doi.org/10.1007/s11434-008-0528-6

Jiang, X. N., Zhou, Z. Y., Huang, X. Y., & Liu, C. Y. (n.d.). Laminar flow through microchannels used for microscale cooling systems. *Proceedings of the 1997 1st Electronic Packaging Technology Conference (Cat. No.97TH8307)*, 119–122. https://doi.org/10.1109/EPTC.1997.723897

Jung, T., & Yang, S. (2015). Highly stable liquid metal-based pressure sensor integrated with a microfluidic channel. *Sensors (Switzerland)*, *15*(5), 11823–11835. https://doi.org/10.3390/s150511823

Kharati-Koopaee, M., & Rezaee, M. (2017). Investigation of turbulent flow through microchannels consisting of different micropost arrangements. *Engineering Computations*, *34*(5), 1367–1392. https://doi.org/10.1108/EC-02-2016-0069

Khoshmanesh, K., Tang, S.-Y., Zhu, J. Y., Schaefer, S., Mitchell, A., Kalantar-Zadeh, K., & Dickey, M. D. (2017). Liquid metal enabled microfluidics. *Lab on a Chip*, *17*(6), 974–993. https://doi.org/10.1039/C7LC00046D

Kim, M., Huang, Y., Choi, K., & Hidrovo, C. H. (2014). The improved resistance of PDMS to pressure-induced deformation and chemical solvent swelling for microfluidic devices. *Microelectronic Engineering*, *124*, 66–75. https://doi.org/10.1016/j.mee.2014.04.041

Ko, H. S., Liu, C. W., Gau, C., & Jeng, D. Z. (2008). Flow characteristics in a microchannel system integrated with arrays of micro-pressure sensors using a polymer material. *Journal of Micromechanics and Microengineering*, *18*(7), 075016. https://doi.org/10.1088/0960-1317/18/7/075016

Koo, J., & Kleinstreuer, C. (2003). Liquid flow in microchannels: experimental observations and computational analyses of microfluidics effects. *Journal of Micromechanics and Microengineering*, *13*(5), 568–579. https://doi.org/10.1088/0960-1317/13/5/307

Kundu, P. K., Cohen, I. M., & Dowling, D. R. (2016). Laminar Flow. In *Fluid Mechanics* (pp. 409–467). Elsevier. https://doi.org/10.1016/B978-0-12-405935-1.00009-5

Lepchev, D., & Weihs, D. (2010). Low Reynolds Number Flow in Spiral Microchannels. *Journal of Fluids Engineering*, *132*(7). https://doi.org/10.1115/1.4001860

Matsuda, Y., Mori, H., Niimi, T., Uenishi, H., & Hirako, M. (2007). Development of pressure sensitive molecular film applicable to pressure measurement for high Knudsen number flows. *Experiments in Fluids*, *42*(4), 543–550. https://doi.org/10.1007/s00348-007-0259-5

Miralles, V., Huerre, A., Malloggi, F., & Jullien, M. C. (2013). A review of heating and temperature control in microfluidic systems: Techniques and applications. *Diagnostics*, *3*(1), 33–67. MDPI. https://doi.org/10.3390/diagnostics3010033

Ortigoza-Diaz, J., Scholten, K., Larson, C., Cobo, A., Hudson, T., Yoo, J., Baldwin, A., Weltman Hirschberg, A., & Meng, E. (2018). Techniques and considerations in the microfabrication of Parylene C microelectromechanical systems. *Micromachines*, *9*(9), 422. https://doi.org/10.3390/mi9090422

Papautsky, I., Ameel, T., & Frazier, A. B. (2001). A Review of Laminar Single-Phase Flow in Microchannels. *Micro-Electro-Mechanical Systems (MEMS)*, 495–503. https://doi.org/10.1115/IMECE2001/MEMS-23872

Quinn, M. K., Gongora-Orozco, N., Kontis, K., & Ireland, P. (2013). Application of pressure-sensitive paint to low-speed flow around a U-bend of strong curvature. *Experimental Thermal and Fluid Science*, *48*, 58–66. https://doi.org/10.1016/j.expthermflusci.2013.02.008

Raj, A., Suthanthiraraj, P. P. A., & Sen, A. K. (2018). Pressure-driven flow through PDMS-based flexible microchannels and their applications in microfluidics. *Microfluidics and Nanofluidics*, *22*(11), 128. https://doi.org/10.1007/s10404-018-2150-5

Reddy, K., Guo, Y., Liu, J., Lee, W., Khaing Oo, M. K., & Fan, X. (2012). Rapid, sensitive, and multiplexed on-chip optical sensors for micro-gas chromatography. *Lab on a Chip*, *12*(5), 901. https://doi.org/10.1039/c2lc20922e

Roh, C., Lee, J., & Kang, C. (2016). The deformation of polydimethylsiloxane (PDMS) microfluidic channels filled with embedded circular obstacles under certain circumstances. *Molecules*, *21*(6). https://doi.org/10.3390/molecules21060798

Samoei, V. K., & Jayatissa, A. H. (2020). Aluminum doped zinc oxide (AZO)-based pressure sensor. *Sensors and Actuators, A: Physical*, *303*. https://doi.org/10.1016/j.sna.2019.111816

Shen, L., Zhang, G., & Etzold, B. J. M. (2020). Paper-Based Microfluidics for Electrochemical Applications. *ChemElectroChem*, *7*(1), 10–30. https://doi.org/10.1002/celc.201901495

Sinar, D., & Knopf, G. K. (2020a). Disposable piezoelectric vibration sensors with PDMS/ZnO transducers on printed graphene-cellulose electrodes. *Sensors and Actuators A: Physical*, *302*, 111800. https://doi.org/10.1016/j.sna.2019.111800

Sinar, D., & Knopf, G. K. (2020b). Disposable piezoelectric vibration sensors with PDMS/ZnO transducers on printed graphene-cellulose electrodes. *Sensors and Actuators, A: Physical*, *302*. https://doi.org/10.1016/j.sna.2019.111800

Sollier, E., Murray, C., Maoddi, P., & Di Carlo, D. (2011). Rapid prototyping polymers for microfluidic devices and high pressure injections. In *Lab on a Chip* (Vol. 11, Issue 22, pp. 3752–3765). Royal Society of Chemistry. https://doi.org/10.1039/c1lc20514e

Song, G., Weicheng, Y., & Yong, L. (2023). Rapid prototyping of microfluidics devices using novel thermoset polydicyclopentadiene. *Journal of Micromechanics and Microengineering*, *33*(7), 075002. https://doi.org/10.1088/1361-6439/acd25c

Song, W., & Psaltis, D. (2010). Optofluidic pressure sensor based on interferometric imaging. *Optics Letters*, *35*(21), 3604. https://doi.org/10.1364/OL.35.003604

Sun, X., Sun, J., Li, T., Zheng, S., Wang, C., Tan, W., Zhang, J., Liu, C., Ma, T., Qi, Z., Liu, C., & Xue, N. (2019). Flexible tactile electronic skin sensor with 3D force detection based on porous CNTs/PDMS nanocomposites. *Nano-Micro Letters*, *11*(1). https://doi.org/10.1007/s40820-019-0288-7

Tai, C.-H., Hsiung, S.-K., Chen, C.-Y., Tsai, M.-L., & Lee, G.-B. (2007). Automatic microfluidic platform for cell separation and nucleus collection. *Biomedical Microdevices*, *9*(4), 533–543. https://doi.org/10.1007/s10544-007-9061-7

Tamayol, A., & Bahrami, M. (2010). Laminar flow in microchannels with non-circular cross section. *Journal of Fluids Engineering*, *132*(11). https://doi.org/10.1115/1.4001973

Tamayol, A., Gunda, N. S. K., Akbari, M., Mitra, S. K., & Bahrami, M. (2012). Creeping flow through microchannels with integrated micro-pillars. *ASME 2012 10th International Conference on Nanochannels, Microchannels, and Minichannels*, 411–417. https://doi.org/10.1115/ICNMM2012-73199

Tang, L., Liu, J., Hong, W., Wang, T., Wang, X., Chen, X., Yang, B., & Yang, C. (2017). A smart microfluidic system integrated with pressure sensors and flow sensor based on electrochemical impedance methods. *2017 IEEE 30th International Conference on Micro Electro Mechanical Systems (MEMS)*, 1300–1303. https://doi.org/10.1109/MEMSYS.2017.7863657

Tesař, V. (2008). *Pressure-Driven Microfluidics*.

Tsai, C.-H. D., Nakamura, T., & Kaneko, M. (2015). An on-chip, electricity-free and single-layer pressure sensor for microfluidic applications. *2015 IEEE/RSJ International Conference on Intelligent Robots and Systems (IROS)*, 165–170. https://doi.org/10.1109/IROS.2015.7353369

Wang, J., Liu, S., Vardeny, Z. V., & Nahata, A. (2012). Liquid metal-based plasmonics. *Optics Express*, *20*(3), 2346. https://doi.org/10.1364/OE.20.002346

Wang, M. T. M., Tien, L., Han, A., Lee, J. M., Kim, D., Markoulli, M., & Craig, J. P. (2018). Impact of blinking on ocular surface and tear film parameters. *The Ocular Surface*, *16*(4), 424–429. https://doi.org/10.1016/j.jtos.2018.06.001

Wu, C.-Y., Lu, J.-C., Liu, M.-C., & Tung, Y.-C. (2012). Integrated electrofluidic circuits: pressure sensing with analog and digital operation functionalities for microfluidics. *Lab on a Chip*, *12*(20), 3943. https://doi.org/10.1039/c2lc40436b

Yang, X., Wang, Y., & Qing, X. (2018). A flexible capacitive pressure sensor based on ionic liquid. *Sensors (Switzerland)*, *18*(7). https://doi.org/10.3390/s18072395

Yao, J.-L., Chen, Y.-X., Qiang, W.-J., Wang, X.-Z., Wei, H., Gao, B.-H., & Yang, X. (2019). A simple, low-cost micro-coating method for accuracy improvement and its application in pressure sensors. *Sensors*, *19*(20), 4601. https://doi.org/10.3390/s19204601

Yun, Kwang-Seok, Cho, Il-Joo, Bu, Jong-Uk, Kim, Chang-Jin, & Yoon, Euisik. (2002). A surface-tension driven micropump for low-voltage and low-power operations. *Journal of Microelectromechanical Systems*, *11*(5), 454–461. https://doi.org/10.1109/JMEMS.2002.803286

Zarikos, I. M., Hassanizadeh, S. M., van Oosterhout, L. M., & van Oordt, W. (2018). Manufacturing a micro-model with integrated fibre optic pressure sensors. *Transport in Porous Media*, *122*(1), 221–234. https://doi.org/10.1007/s11242-018-1000-y

Zhang, Y., Gu, X.-J., Barber, R. W., & Emerson, D. R. (2005). Influence of the electric double layer on induced pressure fields and development lengths in electroosmotic flows. *Modern Physics Letters B*, *19*(28n29), 1655–1658. https://doi.org/10.1142/S0217984905010141

Chapter 8

Mathematical and computational analysis of a two-fluid nonlinear model for pulsatile blood flow in a stenosed artery

Mona P. Dange, Bhawna Agrawal, and Sanjit Kumar

8.1 INTRODUCTION

Atherosclerosis is a dangerous disorder in which plaques or cardiovascular blockages in arteries cause them to thicken and narrow. This can harm organs and keep them from working properly. The endothelium's damage brought on by smoking, high cholesterol, or high blood pressure (BP) is the first step in stenosis. Flow of blood via a stenosed artery can be depicted using a variety of fluid models, such as single-fluid and two-fluid, Newtonian, and non-Newtonian flow of blood. In blood flow modeling, body acceleration is important because it affects normal blood flow and results in issues like headaches, stomachaches, and raised BP. Understanding cardiovascular conditions like atherosclerosis requires awareness of the dynamics of blood flow in stenosed arteries. The rheological properties of blood are affected by plasma and cellular components, which cause non-Newtonian behaviors. Computational and mathematical models are essential for examining flow properties under different physical conditions. The mathematical and computational evaluation of a two-fluid nonlinear model for steady flow of blood in a stenosed artery is the main purpose of this work. The model represents blood as the accumulation of red blood cells (RBCs) in plasma, with a layer of Newtonian blood plasma close to the arterial wall and a non-Newtonian fluid that represents RBCs. Essential flow parameters are analyzed through computational simulations employing numerical techniques such as the finite difference method (FDM), which provide information on how arterial stenosis affects blood circulation and indicate the complications of vascular disease. The findings may lead to improved methods of identifying and dealing with cardiovascular diseases.

The heart and artery systems are affected by the blood flow in the area, even though the precise causes of stenosis are still unknown. For high shear rate flow, Newtonian blood activity is a reasonable assumption. This is especially relevant for pulsative blood flow in arteries that are stenosed (Tu and Deville, 1996). The Newtonian blood flow model has been used with an applied magnetic field to study the effects related to temperature and velocity fields (Kumar and Diwakar, 2012). The H-B model (non-Newtonian

DOI: 10.1201/9781003642152-8

fluid model) and power legislation were used in the analysis of blood flow via elastic arteries (Kumar, 2015). A fluid flow mathematical model through an elastic artery with non-Newtonian pulses and elasticity effects was developed (Chitra and Karthikeyan, 2017). The blood flow under various other external factors that influence it, including exposure to magnetic fields, abrupt shocks, and many more, was studied in similar ways. The research focused on the effects of blood thickening, stenosis percentage, and curvature and length ratios on the blood flow properties (Lakzian and Akbarzadeh, 2020). The findings highlight the fact that the stenosed neck is where the highest shear stress on the wall occurs and that, as predicted, when the stenosis length decreases, the maximum wall shear stress (WSS) rises. A vertically symmetric function, like a cosine or exponential function, can explain symmetrical vertical stenosis, while a function that is vertically asymmetric, like the union of two different stenosis shapes, can representvertically asymmetric stenosis (Pinyo and Somachai, 2021). Both Newtonian and non-Newtonian properties depend on the shear rate. Aside from that, the blood vessel is believed to have mild stenosis and saturated porous walls (Fahim et al., 2024).

8.2 MATHEMATICAL FORMULATION

We examine entirely developed, axially symmetric, laminar, and pulsating flow of blood via a round tube that depicts an artery with mild axisymmetric stenosis, as seen in Figure 8.1. Assuming that the blood is an incompressible and two-phase fluid, a Newtonian fluid is used to describe the outermost layer, which represents plasma, while the core region is composed of a

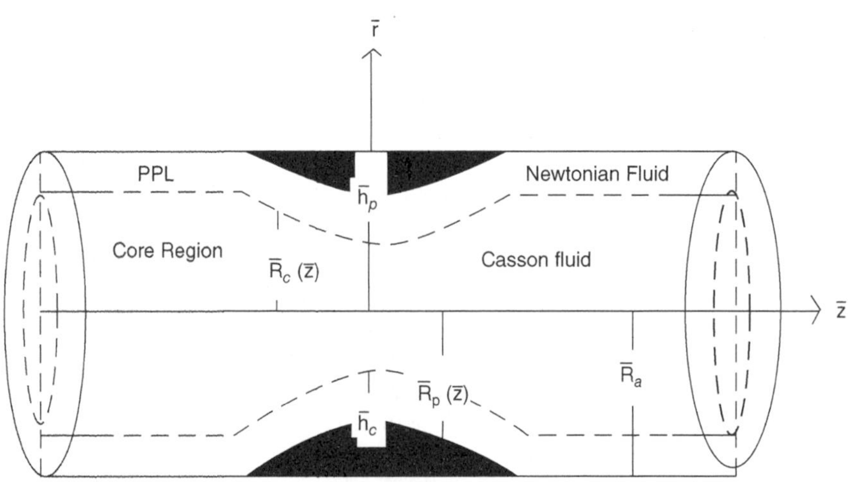

Figure 8.1 The geometry of an axisymmetric arterial stenosed.

Casson fluid, which simulates the suspension of erythrocytes. Because the artery length is assumed to be much greater than the artery radius, entrance and exit effects are disregarded, and the arterial wall is assumed to be rigid.

$$\bar{R}_p(\bar{z}) = \begin{cases} \bar{R}_n - \left(\bar{h}_p/2\right)\left(1+\cos\left(\dfrac{\pi\bar{z}}{\bar{z}_1}\right)\right), \text{ if } |\bar{z}| \le \bar{z}_1 \\ \bar{R}_n, \text{ if } |\bar{z}| > \bar{z}_n \end{cases} \tag{8.1}$$

The core region's stenosis geometry is provided by

$$\bar{R}_c(\bar{z}) = \begin{cases} \alpha\bar{R}_n - \left(\bar{h}_c/2\right)\left(1+\cos\left(\dfrac{\pi\bar{z}}{\bar{z}_1}\right)\right), \text{ if } |z| \le z_l \\ \alpha\bar{R}_n, \text{ if } |z| > z_n \end{cases} \tag{8.2}$$

where $\bar{R}_p$ is the artery stenosis radius with a layer of peripheral, $\bar{R}_c$ is the core region's stenosed artery radius, such that $\bar{R}_c(\bar{z}) = \propto \bar{R}_p(\bar{z})$, $\bar{R}_n$, and $\alpha\bar{R}_n$ are the radii and core regions of the normal artery, respectively. $\bar{h}_p$ and $\bar{h}_c$ are the highest stenosis heights in the peripheral and the core regions, respectively, such that $\bar{h}_c = \propto \bar{h}_p$, $\bar{z}_l$ is the stenosis half length. According to reports, with mild stenosis in a tube, when the flow has a low Reynolds number, the radial velocity is very, very little.

The fluid flow's equation of motion is provided by

$$\bar{\rho}^{(c)}\frac{\partial \bar{v}_c}{\partial \bar{t}} = -\frac{\partial \bar{P}}{\partial z} - \frac{1}{\bar{r}}\frac{\partial}{\partial \bar{r}}\left(\bar{r}\bar{\tau}^{(c)}\right) + \bar{f}(\bar{t}), if\, 0 \le \bar{r} \le \bar{R}_c(\bar{z}) \tag{8.3}$$

$$\bar{\rho}^{(N)}\frac{\partial \bar{v}_N}{\partial \bar{t}} = -\frac{\partial \bar{P}}{\partial \bar{z}} - \frac{1}{\bar{r}}\frac{\partial}{\partial \bar{r}}\left(\bar{r}\bar{\tau}^{(N)}\right) + \bar{f}(\bar{t}), if\, \bar{R}_c(z) \le \bar{r} \le \bar{R}_p(\bar{z}) \tag{8.4}$$

The shear stresses for the Newtonian and the Casson fluids are denoted by $\bar{\tau}^{(N)}$ and $\bar{\tau}^{(c)}$, respectively, while the velocities of the peripheral and core regions fluid are represented by $\bar{v}_N$ and $\bar{v}_c$, respectively, along with $\bar{\rho}^{(N)}$ and $\bar{\rho}^{(c)}$ are the Newtonian and Casson fluid densities, respectively, where $\bar{f}(\bar{t})$ is the body acceleration and $\bar{P}$ is the pressure.

The following formulas are provided for Newtonian fluid and Casson fluid, respectively:

$$\bar{\tau}^{(N)} = -\bar{u}_N\frac{\partial \bar{v}_N}{\partial \bar{r}}, if \bar{R}_c(z) \le \bar{r} \le \bar{R}_p(\bar{z}) \tag{8.5}$$

$$\sqrt{\bar{\tau}^{(c)}} = \sqrt{-\bar{u}_c \frac{\partial \bar{v}_c}{\partial \bar{r}}} + \sqrt{\bar{\tau}^{(y)}}, if\ \bar{\tau}^{(c)} \geq \bar{\tau}^{(y)}\ and\ \bar{r}_p \leq \bar{r} \leq \bar{R}_c(\bar{z})$$

$$\frac{\partial \bar{v}_c}{\partial \bar{r}} = 0, if\ \bar{\tau}^{(c)} \geq \bar{\tau}^{(y)}\ and\ 0 \leq \bar{r} \leq \bar{r}_p \tag{8.6}$$

where the plug flow region's radius is denoted by $\bar{r}_p$.

The axial periodic body acceleration is determined by

$$\bar{f}(\bar{t}) = A\cos(\bar{\omega}_B \bar{t} + \theta) \tag{8.7}$$

Here, the amplitude is denoted by A, $\bar{\omega}_B = 2\pi \bar{f}_B$, frequency in hertz is denoted by $\bar{f}_B$, and in relation to the heart action, θ is the lead angle of $\bar{f}(\bar{t})$. In order to ignore the wave effect, it is assumed that the body acceleration frequency, $\bar{f}_B$, is low.

The following is a representation of the gradient of pressure at $\bar{t}$ and $\bar{z}$.

$$\frac{\partial \bar{P}}{\partial \bar{z}}(\bar{z}, \bar{t}) = A_s + A_f \cos \bar{\omega}_P \bar{t} \tag{8.8}$$

where A_s represents the pressure gradient's steady component, A_f represents the fluctuating component's amplitude, and $\bar{\omega}_P = 2\pi \bar{f}_p$, and the pulse frequency is represented by $\bar{f}_p$. We introduce the non-dimensional variables listed below, where A_s and A_f are both functions of $\bar{z}$.

$$\begin{gathered} z = \frac{\bar{z}}{R_n}, R_p(z) = \frac{\bar{R}_p(\bar{z})}{R_n}, R_c(z) = \frac{\bar{R}_c(\bar{z})}{R_n}, r = \frac{\bar{r}}{R_n}, t = \bar{\omega}_P \bar{t}, \omega = \frac{\bar{\omega}_B}{\bar{\omega}_P}, \\ h_p = \frac{\bar{h}_p}{R_n}, h_c = \frac{\bar{h}_c}{R_n}, B = \frac{A}{A_s}, v_c = \frac{\bar{v}_c}{A_s R_n^2 / 4\bar{\eta}_c}, v_N = \frac{\bar{v}_N}{A_s R_n^2 / 4\bar{\eta}_N}, v_s = \\ \frac{\bar{v}_s}{A_s R_n^2 / 4\bar{\eta}_N}, \tau^{(c)} = \frac{\bar{\tau}^{(c)}}{A_s R_n^2 / 2}, \tau^{(N)} = \frac{\bar{\tau}^{(N)}}{A_s R_n^2 / 2}, \varepsilon = \frac{A_f}{A_s} \end{gathered} \tag{8.9}$$

where β_c and β_N represent the pulsatile Reynolds numbers for Newtonian and Casson fluids, respectively.

Equations (8.1) and (8.2) become dimensionless variables.

$$R_p(z) = \begin{cases} 1 - \left(h_p/2\right)\left(1 + \cos\left(\frac{\pi z}{z_l}\right)\right), if\ |z| \leq z_l \\ 1,\ if\ |z| > z_n \end{cases} \tag{8.10}$$

$$R_c(z) = \begin{cases} \propto - \left(\frac{h_c}{2}\right)\left(1 + \cos\left(\frac{\pi z}{z_l}\right)\right), \text{ if } |z| \leq z_l \\ \propto, \text{ if } |z| > z_n \end{cases} \tag{8.11}$$

Equations (8.3) and (8.4) provide the governing equation of motion, which is represented in the dimensionless form as:

$$\beta_c^2 \frac{\partial v_c}{\partial t} = 4g(t) - \frac{2}{r}\frac{\partial}{\partial r}\left(r\tau^{(c)}\right) \tag{8.12}$$

$$\beta_N^2 \frac{\partial v_N}{\partial t} = 4g(t) - \frac{2}{r}\frac{\partial}{\partial r}\left(r\tau^{(N)}\right) \tag{8.13}$$

where

$$g(t) = (1 + \varepsilon cost\,) + B\cos(\omega t + \theta). \tag{8.14}$$

Reducing equations (8.5) and (8.6) with dimensionless variables to

$$\sqrt{\tau^{(c)}} = \sqrt{-\frac{1}{2}\frac{\partial v_c}{\partial r}} + \sqrt{\varnothing} \tag{8.15}$$

$$\frac{\partial v_c}{\partial r} = 0 \tag{8.16}$$

$$\tau^{(N)} = -\frac{1}{2}\frac{\partial v_N}{\partial r} \tag{8.17}$$

The requirements for the boundary are

$$\begin{gathered} \overline{\frac{\partial v_c}{\partial \overline{r}}} = 0 \ \textit{when} \ \overline{r} = 0, \overline{\tau^{(c)}} \text{ is finite} \\ \overline{v_N} = 0 \ \textit{when} \ \overline{r} = \overline{R_p} \\ \overline{\tau^{(N)}} = \overline{\tau^{(c)}} \ \textit{and} \ \overline{v_N} = \overline{v_c} \ \textit{when} \ \overline{r} = \overline{R_c} \end{gathered} \tag{8.18}$$

According to the dimensionless form, the boundary conditions are,

$$\begin{gathered} \frac{\partial v_c}{\partial r} = 0 \ \textit{when} \ r = 0, \ \tau^{(c)} \text{ is finite} \\ v_N = 0 \ \textit{when} \ r = R_p \\ \tau^{(N)} = \tau^{(c)} \ \textit{and} \ v_N = v_c \ \textit{when} \ r = R_c \end{gathered} \tag{8.19}$$

The rate of dimensionless volumetric flow is determined by

$$Q_v = 4\int_0^{R_p(z)} v(z,r,t)\,dr \tag{8.20}$$

where $Q_v(t) = \dfrac{\overline{Q_v(t)}}{\pi R_n^4 A_s / 8u_c}$; $\overline{Q_v(t)}$ is the rate of volumetric flow.

8.2.1 Method of Solution Perturbation

Using the perturbation method, the simulated solution to the unknowns, v_c, v_N, $\tau^{(N)}$, $\tau^{(c)}$, is obtained because the system of nonlinear equations (8.12)–(8.17) cannot be solved exactly. The expansion of equations from (8.12) to (8.17) in terms of β_c^2 and β_N^2 is more appropriate, as these terms naturally arise when equations of momentum (8.3) and (8.4) are nondimensionalized.

Let's extend the plug core velocity v_p and the core region velocity v_c in the β_c^2 perturbation sequence as shown below, where $1 \gg \beta_c^2$.

$$v_p(z,t) = v_{0,p}(z,t) + \beta_c^2 v_{1,p}(z,t) + \ldots \tag{8.21}$$

$$v_c(z,r,t) = v_{0,c}(z,r,t) + \beta_c^2 v_{1,c}(z,r,t) + \ldots \tag{8.22}$$

$$r_p(z,t) = r_{0,p}(z,t) + \beta_c^2 r_{1,p}(z,t) + \ldots \tag{8.23}$$

$$v_N(z,r,t) = v_{0,N}(z,r,t) + \beta_c^2 v_{1,N}(z,r,t) + \ldots \tag{8.24}$$

The perturbation sequence is expanded by substituting in equations (8.12), (8.15), and (8.16), and the term constants and power of β_c^2 are compared.

$$\frac{\partial v_{0,c}}{\partial t} = -\frac{2}{r}\frac{\partial}{\partial r}\left(r\tau_1^{(c)}\right) \tag{8.25}$$

$$g(t) = \frac{1}{2r}\frac{\partial}{\partial r}\left(r\tau_0^{(c)}\right) \tag{8.26}$$

$$\frac{\partial v_{0,c}}{\partial r} = 4\sqrt{\varphi\tau_0^{(c)}} - 2\tau_0^{(c)} - 2\varphi \tag{8.27}$$

$$\frac{\partial v_{1,c}}{\partial r} = 2\sqrt{\varphi / \tau_0^{(c)}} - 2\tau_1^{(c)} \tag{8.28}$$

In the same way, equations (8.13) and (8.17) apply the perturbation sequence expansion, equating the term constants with the β_N^2 terms.

$$\begin{aligned}
&\frac{\partial v_{0,N}}{\partial t} = -\frac{2}{r}\frac{\partial}{\partial r}\left(r\tau_1^{(N)}\right)\\
&g(t) = \frac{1}{2r}\frac{\partial}{\partial r}\left(r\tau_0^{(N)}\right)\\
&\frac{\partial v_{0,N}}{\partial r} = -2\tau_0^{(N)}\\
&\frac{\partial v_{1,N}}{\partial r} = -2\tau_1^{(N)}
\end{aligned} \tag{8.29}$$

After replacing the perturbation sequence expanded in equation (8.19), we can evaluate the constant term along with β_c^2, β_N^2, and obtain,

$$\begin{aligned}
&\frac{\partial v_{0,p}}{\partial t} = \frac{\partial v_{0,p}}{\partial t} = 0,\ when\ r = 0, \text{ for } \tau_0^{(p)} \text{ and } \tau_1^{(p)} \text{ are finite}\\
&\tau_0^{(c)} = \tau_0^{(N)},\ \tau_1^{(c)} = \tau_1^{(N)},\ v_{0,c} = v_{0,N}\ and\ v_{1,c} = v_{1,N}\ when\ r = R_c\\
&v_{0,N} = 0\ and\ v_{1,N} = 0\ when\ r = R_p\\
&\tau_c^{(0)}\ and\ \tau_c^{(1)}\ are\ finite\ when\ r = 0
\end{aligned} \tag{8.30}$$

Regarding the solution of equations (8.25) to (8.29) for unknowns $\tau_0^{(c)}$, $\tau_0^{(N)}$, $\tau_0^{(p)}$, $v_{0,N}$, $v_{0,c}$, $v_{0,p}$, $\tau_1^{(c)}$, $\tau_1^{(N)}$, $v_{1,N}$, $v_{1,c}$, $v_{1,p}$ Equation (8.30) can be used to get,

$$\tau_0^{(c)} = g(t)r \tag{8.31}$$

$$\tau_0^{(N)} = g(t)r \tag{8.32}$$

$$\tau_0^{(p)} = g(t)r_{0,p} \tag{8.33}$$

$$v_{0,N} = g(t)R_p^2\left[1-\left(\frac{r}{R_p}\right)^2\right] \tag{8.34}$$

$$v_{0,c} = g(t)\left(R_p^2 - r^2\right) + 2\varphi(R_c - r) - \frac{8}{3}\sqrt{\varphi g(t)}\left(R_c^{\frac{3}{2}} - r^{\frac{3}{2}}\right) \tag{8.35}$$

$$v_{0,p} = g(t)\left(R_p^2 - r_{0,p}^2\right) + 2\varphi(R_c - r_{0,p}) - \frac{8}{3}\sqrt{\varphi g(t)}\left(R_c^{\frac{3}{2}} - r_{0,p}^{\frac{3}{2}}\right) \tag{8.36}$$

$$\tau_1^{(c)} = g'(t)\left[\frac{1}{8}\left(r^3 - 2R_p^2 r\right) + \frac{1}{21}\kappa\left(7rR_c^{\frac{3}{2}} - 4r^{\frac{3}{2}}\right)\right] \tag{8.37}$$

$$\tau_1^{(N)} = \frac{-1}{8} g'(t)\left(2R_p^2 r - r^3\right) + \frac{\kappa}{7r} g'(t) R_c^{\frac{7}{2}} \tag{8.38}$$

$$\upsilon_{1,N} = \frac{g'(t)}{4}\left[R_p^2 r^2 - r^4 - \frac{3}{4}R_p^4 - \frac{8}{7}\kappa R_c^{\frac{7}{2}} \ln\frac{r}{R_p}\right] \tag{8.39}$$

$$\upsilon_{1,c} = g'(t)\left[\begin{array}{c} \kappa\left(\frac{371}{2058} r^{\frac{7}{2}} - \frac{1}{3}\left(R_p^2 r^{\frac{3}{2}} + R_c^{\frac{3}{2}} r^2 - R_p^2 R_c^{\frac{3}{2}}\right) + \frac{315}{2058} R_c^{\frac{7}{2}}\right) \\ + \frac{4}{63}\kappa^2\left(7r^{\frac{3}{2}} R_c^{\frac{3}{2}} - 2r^3 - 5R_c^3\right) + \frac{1}{16}\left(R_c^4 - 4R_p^2 R_c^2 - r^4 + 4R_p^2 r^2\right) \\ + \frac{1}{4}\left(R_p^2 R_c^2 - R_c^4 - \frac{3}{4}R_p^4\right) - \frac{2}{7}\kappa R_c^{\frac{7}{2}} \ln\frac{R_c}{R_p} \end{array}\right] \tag{8.40}$$

$$\upsilon_{1,p} = g'(t)\left[\begin{array}{c} \kappa\left(\frac{371}{2058} r_{0,p}^{\frac{7}{2}} - \frac{1}{3}\left(R_p^2 r_{0,p}^{\frac{3}{2}} + R_c^{\frac{3}{2}} r_{0,p}^2 - r_{0,p}^2 R_c^{\frac{3}{2}}\right) + \frac{315}{2058} R_c^{\frac{7}{2}}\right) \\ + \frac{4}{63}\kappa^2\left(7r_{0,p}^{\frac{3}{2}} R_c^{\frac{3}{2}} - 2r_{0,p}^{\frac{3}{2}} - 5R_c^3\right) + \frac{1}{16}\left(R_c^4 - 4R_p^2 R_c^2 - r_{0,p}^4 + 4R_p^2 r_{0,p}^2\right) \\ + \frac{1}{4}\left(R_p^2 R_c^2 - R_c^4 - \frac{3}{4}R_p^4\right) - \frac{2}{7}\kappa R_c^{\frac{7}{2}} \ln\frac{R_c}{R_p} \end{array}\right] \tag{8.41}$$

where $\kappa^2 = \frac{\varphi}{g(t)}$,

Equation (8.23) ignores the higher power of β_c when the terms of $o\left(\beta_c^2\right)$ and the initial approximation plug core radius can be found as,

$$r_{0,p} = \kappa^2 = \frac{\varphi}{g(t)} \tag{8.42}$$

By applying equations (8.34), (8.35), (8.39), and (8.40), the velocities in the periphery and core are expressed as

$$\upsilon_N = g(t)\left(R_p^2 - r^2\right) + \beta_N^2 \frac{g'(t)}{4}\left[R_p^2 r^2 - r^4 - \frac{3}{4}R_p^4 - \frac{8}{7}\kappa R_c^{\frac{7}{2}} \ln\frac{r}{R_p}\right] \tag{8.43}$$

Equation (8.44) yields the value of v_c when the values of $v_{0,c}$ and $v_{1,c}$ are substituted.

$$v_c = v_{0,c} + \beta_c^2 v_{1,c} \tag{8.44}$$

One can obtain the expression for WSS $\tau^{(w)}$ as,

$$\tau^{(w)} = \left(\tau_0^{(N)} + \beta_N^2 \tau_1^{(N)}\right)_{r=R_p}$$

$$\tau^{(w)} = g(t) R_p - g'(t) \beta_N^2 \left(\frac{R_p^3}{8} - \frac{\kappa}{7R_p} R_c^{\frac{7}{2}} \right) \tag{8.45}$$

The rate of volumetric flow can be obtained from equations (8.20), (8.43), and (8.44) by,

$$\begin{aligned} Q_v = {} & 4\int_0^{r_{0,p}} r\left(v_{0,p} + \beta_c^2 v_{1,p}\right) dr + 4\int_{r_{0,p}}^{R_c} r\left(v_{0,c} + \beta_c^2 v_{1,c}\right) dr \\ & + 4\int_{R_c}^{R_p} r\left(v_{0,N} + \beta_N^2 v_{1,N}\right) dr \end{aligned} \tag{8.46}$$

8.3 RESULTS AND DISCUSSIONS

Using the pulsatile pressure gradient and body acceleration, a proposal has been provided to analyze some of the main components of blood flow via a stenotic artery in the current model. To highlight the biological significance and examine the model's validity, computer codes are created which evaluate the mathematical solution for rate of flow, velocity profile, WSS, and effective viscosity for various equation parameter values with the help of equations (8.31) to (8.46).

The study uses Womersley frequency parameters (0.5), body acceleration parameters (0–2), pressure gradient parameters (0.5), lead angle magnitude (0.2), peripheral stenosis height (0–0.5), and yield stress values (0, 0.1). The axial velocity is observed to vary with radial distance, with maximum and minimum values at the stenosed wall at r = 0 for fixed values of 0 and different body acceleration parameters. As body acceleration increases, axial velocity decreases. The maximum velocity occurs when blood is flowing to the start, and as time increases, velocity decreases; at t = 1 and 1.5, the velocity is lower than at t = 0. This indicates that time plays a vital role in blood flow modeling. The study also shows that the axial velocity increases with body acceleration and decreases with time, as shown in Figure 8.2.

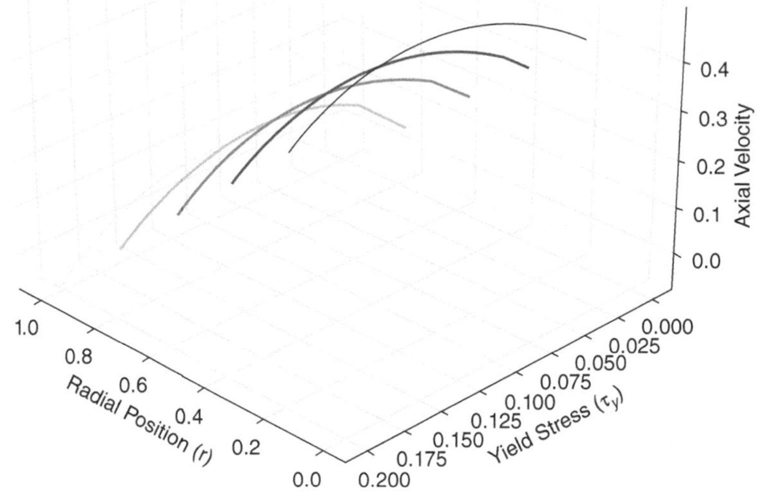

Figure 8.2 3D axial velocity versus radial position and yield stress.

The distribution of flow rates for the two fluid Casson models' r_0,= 0.1, ∝= 0.8, h_p = 0.2 and $t = 45^\circ$ at the stenosis's throat when $z = 0$ are shown in Figure 8.3. Similar to the axial velocity, the rate of flow rises as the body accelerates. The curves are linear when yield stress is absent, but they become nonlinear as yield stress φ increases. Additionally, it is observed that the rate of flow drops as the yield stress φ maximizes from $\varphi = 0$ tφ $\varphi = 0.1$ as the width of the plug flow region increases. When yield stress is absent, the relationship between axial velocity and the rate of flow is found to be linear, as shown in Figure 8.3. Axial velocity and flow rate are directly correlated, and this relationship is dependent on the artery's cross-sectional area. The axial velocity and flow rate both rise linearly as the pressure gradient increases. The pressure differential roughly doubles the rate of flow.

Figure 8.4 illustrates how WSS changes with time t; the entire time scale of t*f* or $0^\circ \leq t \leq 360^\circ$ has been taken into consideration. It has been discovered that body acceleration significantly affects shear stress in a stenosed artery wall. Shear stress on the wall decreases as body acceleration increases. A depicted value is one that reaches its maximum when t=0° and $t = 180^\circ$, and gets a minimum at $t = 90^\circ$ and $t = 270^\circ$. The impact of peripheral stenosis height cannot be overlooked because it is a crucial component of two-layered blood flow modeling. An increase in WSS is the result of peripheral stenosis height.

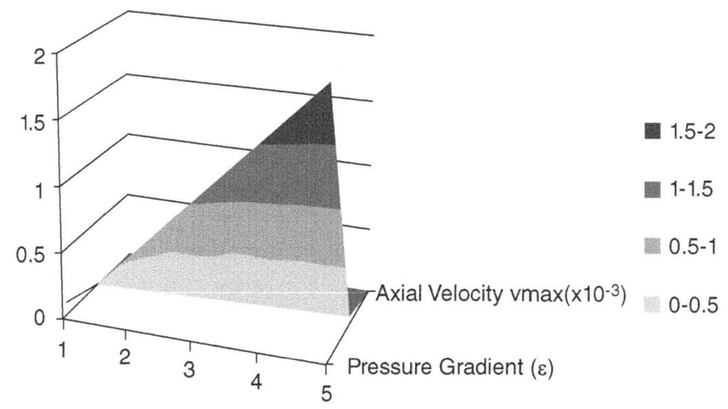

Figure 8.3 3D-pressure gradient versus flow rate and axial velocity.

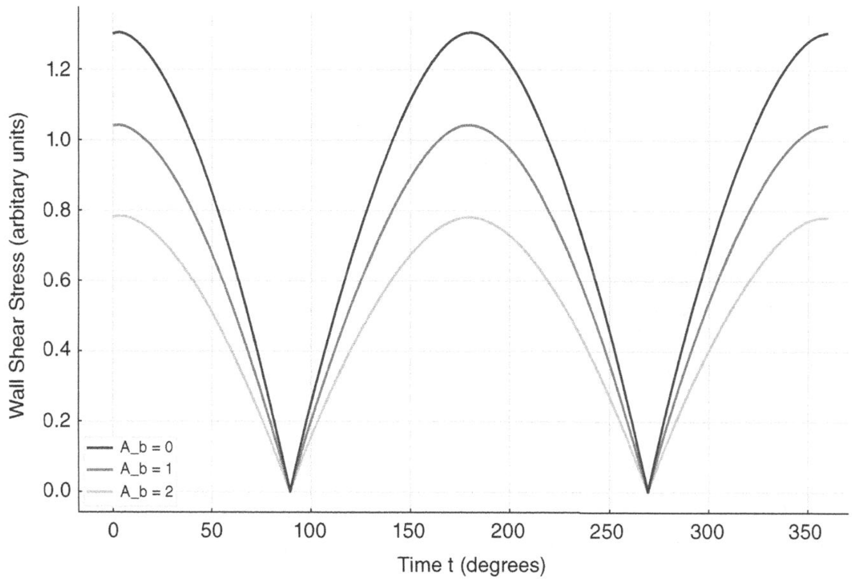

Figure 8.4 Time t° versus wall shear stress for various body accelerations.

8.4 CONCLUSIONS AND FUTURE SCOPE

We have taken into consideration that blood flows through a stenosed artery in two layers. In the core region, all of the erythrocytes are Casson fluids, while the outer layer of plasma is Newtonian fluid. It is shown by the current findings that the acceleration of the body plays a significant role in the flow of blood modeling. We have examined how body acceleration affects different flow parameters in this paper. According to our imagination, it is shown that while the rate of flow and axial velocity increase, WSS exhibits both rising and falling patterns as the body acceleration parameter increases. This model may be helpful in biomedical applications and can predict certain blood flow characteristics. This model has the potential to be helpful in biomedical applications and can predict certain blood flow properties. The current investigation can soon be expanded to take into account the permeability of walls and the magnetic effect of blood. The main objective of the study is to use high-performance computing analysis to produce correct clinical mathematical models with feasible arterial geometries, wall elasticity, and complicated blood flow behaviors.

BIBLIOGRAPHY

Ali, D., and Sen, S. (2018). Permeability and fluid flow induced wall shear stress of bone tissue scaffolds: computational fluid dynamic analysis using Newtonian and non-Newtonian blood flow models. *Computers in Biology and Medicine*, 99, 201–208.

Agrawal, B., Kumar, S., and Das, G., (2022). Mathematical model of blood flow through stenosed arteries with the impact of hematocrit on wall shear stress. *International Journal of Applied Research*, 8(3), 439–443.

Chakravarty, S., Mandal, P. K., and Sarifuddin (2005). Effect of surface irregularities on unsteady pulsatile flow in a compliant artery. *International Journal of Non-Linear Mechanics*, 40(10), 1268–1281.

Chakravarty, S., Sarifuddin, and Mandal, P. K. (2004). Unsteady flow of a two-layer blood stream past a tapered flexible artery under stenotic conditions. *Computational Methods in Applied Mathematics*, 4(4), 391–409.

Chitra, M., and Karthikeyan, D. (2017). Mathematical modelling of pulsatile flow of non-Newtonian fluid through an elasticartery: Effects of elasticity. *International Journal of Pure and Applied Mathematical Sciences*, 10(1), 29–40.

Das, G., Agrawal, B., and Kumar, S. (2023a). A Mathematical model of blood flow of a stenosed artery in variable shape. *Journal of Advanced Zoology*, 44(S7), 1193–1208.

Das, G., Agrawal, B.and Kumar, S. (2023b). Mathematical modeling of non-Newtonianblood flow through arterial stenosis. *Samdarshi*, 16(4), 3133–3139.

Das, G., Agrawal, B. and Kumar, S. (2024). Comparative analysis of mathematical models for non-Newtonian stress relaxation flow of blood in an artery with multiple stenosis. Futuristic Trends in Physical Sciences, IIP Series. 3(3), 139–148.

Fahim, M., Sajid, M., & Ali, N (2024). Pulsatile pressure-driven non-Newtonian blood flow through porous stenotic artery: A computational analysis. International Journal of computational and Methodology.

Gijsen, F.J., van de Vosse, F.N., and Janssen, J.D. (1999). The influence of the non-Newtonian properties of blood on the flow in large arteries: Steady flow in a carotid bifurcation model. *Journal of Biomechanics*, 32(6), 601–608.

Kumar, S., and Diwakar, C. S. (2012). A mathematical model for Newtonian blood flow in the presence of applied magnetic field. *Aryabhatta Journal of Math & Infomatics*, 4, 265–278.

Kumar, S. (2015). Study of blood flow using Power law and Harschel-Bulkley non-Newtonian fluid model through elastic artery. Proceedings of ICFM, 229–235.

Lakzian, E., and Akbarzadeh, P., (2020). Numerical investigation of unsteady pulsatile Newtonian/non-Newtonian blood flow through curved stenosed arteries. *Bio-Medical Materials and Engineering*, 30(5–6), 525–540.

Moore, S., David, T., Chase, J.G., Arnold, J., and Fink, J. (2006). 3D models of blood flow in the cerebral vasculature. *Journal of Biomechanics*, 39(8), 454–1463.

Nanda, S., Basu Mallik, B., Singh, C., Das, S., Ghosh, S., Bhattacharya, S., and Chatterjee, S.S., (2017). Study on the effect of non-Newtonian nature of blood flowing through an elastic artery in slip condition. *Journal of Chemical, Biological and Physical Sciences*, 7(4), 934–942.

Pinyo, O., and Somachai, S., (2021). Mathematical modeling of non-Newtonian fluid in arterial blood flow through various stenoses. *Advances in Differential Equations*, (340), 2–20.

Rakshit, S., Agrawal, B. and Kumar, S. (2024), Numerical and analytical study of unsteady arterial blood flow in time dependent stenosis using the non-Newtonian power-law blood fluid flow model. *Journal of Advanced Zoology*,45 (S-1), 297–310.

Roth, G.A., Abate, D., Abate, K.H. *et al.* (2018). Global, regional, and national age-sex-specific mortality for 282 causes of death in 195 countries and territories, 1980–2017.

Sinnott, M., Cleary, P.W. and Prakash, M., (2006). An investigation of pulsatile blood flow in a bifurcation artery using a grid-free method. Fifth International Conference on CFD in the Process Industries, CSIRO, Melbourne, Australia.

Saikrishna, C., Talwar, S., Gulati, G., Kumar, A.S., (2006). Normal coronary artery dimensions in Indians. *Indian Journal of Thoracic and Cardiovascular Surgery*, 22(3), 159–164.

Shukla, J.B., Parihar, R.S., and Rao, B.R.P. (1980). Effects of stenosis on non-Newtonian flow of the blood in an artery. *Bulletin of Mathematical Biology*, 42(3), 283–294.

Tu, C., Deville, M., Dheur, L., Vanderschuren, L. (1992). Finite element simulation of pulsatile flow through arterial stenosis. Journal of Biomechanics. 25(10), 1141–1152.

Tu, C., and Deville, M. (1996).Pulsatile flow of non-Newtonian fluids through arterial stenoses. *Journal of Biomechanics*, 29(7), 899–908.

Young, D.F. (1979). Fluid mechanics of arterial stenoses. *Journal Biomech Engineering*, 101(3), 157–175.

Chapter 9

Numerical simulation of unsteady blood flow in a multi-stenosed artery

Prashant Dange, Bhawna Agrawal, and Sanjit Kumar

9.1 INTRODUCTION

Important factors in the beginning and progression of arterial stenosis are cardiovascular stenotic flows and vascular wall deformability, a widespread disease affecting the cardiovascular system. Injury to the intima, the inner layer of the arteries, is thought to be the cause of the first stenosis, followed by regrowth resulting from balloon angiogenesis. The flow disorder that results from mild stenosis alters the regional rheology of blood, the deformability of the arteries, and the course of the disease. Research on hemodynamic characteristics through occluded vessels has shown that over the full range of Re (Reynolds numbers), the greatest pressure drop is seen across stenosis with a cosine shape as opposed to multiple irregular-shaped stenosis. Local geometry plays a significant role in local hemodynamics, especially at anastomotic sites, particularly at composite coronary artery grafts. Disease of the coronary arteries can develop at the locations of the lowest regions of shear stress, which are the bifurcations' lateral walls. A 3-dimensional (3-D) model of a porcine heart's coronary arteries was used by researchers to examine variations in blood flow and pressure. The study may help medical research in the diagnosis and treatment of heart diseases because it used both digital simulations and real arteries.

Numerous researchers claim that because blood flows through blood vessels with smaller diameters at low rates of shear and has a plasma layer free of erythrocytes next to the wall of the vessel, it cannot realistically be modeled as a single fluid in a non-Newtonian model. Therefore, blood up to 4–100 μm should be modeled as a two-fluid passing through arteries with small diameters and low shear rates. The use of numerical illustrations in their paper to effectively measure flux, graphical depictions of resistive impedance and WSS, along with comparisons to the outcomes of the current model (Chakravarty and Mandal, 2005). A comprehensive quantitative analysis of surface irregularities in non-staggered and non-uniform grids

DOI: 10.1201/9781003642152-9

with numerical calculations and mathematical models (Chakravarty et al., 2004). Through a number of graphical representations and a mathematical model that was quantitatively enhanced, the double-constricted flow's reaction to body acceleration was estimated using the finite difference approximation (Mustapha et al., 2008). About how these flow quantities changed in relation to peripheral layer thickness, amplitude, pulsatile Reynolds number ratio, catheter radius ratio, and yield stress. According to the calculations, the generalized Newtonian fluid has a smaller separation region, a reduced shear stress on the peak wall, and a higher pressure drop than the Newtonian model. An unsteady axisymmetric hemodynamics in a segment of a diseased porous artery, displaying areas of recirculation and separation for severe stenosis, and comparing experimental and numerical results. A numerical study on the influence of gravitational force on a two-dimensional nonlinear blood flow model via a stenosed artery (Mustapha et al., 2008). A steady pulsating flow when a magnetic field is present, flow of blood via a porous-saturated stenotic artery, results show that the permeability parameter increases flow velocity and resistance, while a magnetic field decreases flow and increases resistance. A non-Newtonian hemodynamics model via multiple arteries with smooth axisymmetric contraction result shows that increasing stenosis height increases the friction factor and Nusselt number, affecting flow impedance, axial velocity, and shear rate. Blood flow through vertically asymmetric stenosis in cardiovascular patients, comparing flow quantities in both symmetric and asymmetric stenoses, results show significant differences in flow quantities in bell and cosine shapes, with slight differences in single shape behavior. The interaction between the vascular fluid and tube in an asymmetrical nonlinear aneurysm, utilizing and simplifying the Lagrangian system, Navier-Stokes equations, and large deformation models. The Carreau fluid model and Darcy's Law are used to analyze nanoflow of blood in a porous artery with overlapping stenosis, enhancing understanding of hemodynamic flow and potential cardiovascular disease treatments. Analyzed the physiological variations in flow in stenosed arteries and external body acceleration, determining whether hyperviscosity and physiological conditions are related, and examining potential complications.

9.2 MODEL OF STENOSIS

An approximate (irregular) stenosis profile is the geometry of the stenosis under consideration, and it is created from the data. As the area of the cross-section variation of the human cadaver's left side circumflex coronary artery casting is preserved, simulating actual surface irregularities. The second smooth stenosis geometrical model that is evaluated is the cosine curve, which is commonly used.

$$\bar{R}(\bar{z},\tau)=\begin{cases}\alpha_1(\tau)\left\{1-\dfrac{\delta_1}{2\bar{R}_0}\left(1+\cos\left(\dfrac{\pi(\bar{z}-\sigma_1)}{\bar{Z}_1}\right)\right)\right\}, \text{ if } \sigma_1-\bar{Z}_1<\bar{z}<\sigma_1+\bar{Z}_1\\ \alpha_1(\tau)\left\{1-\dfrac{\delta_2}{2\bar{R}_0}\left(1+\cos\left(\dfrac{\pi(\bar{z}-\sigma_2)}{\bar{Z}_2}\right)\right)\right\}, \text{ if } \sigma_2-\bar{Z}_2<\bar{z}<\sigma_2+\bar{Z}_2\\ \alpha_1(\tau), \text{otherwise}\end{cases} \tag{9.1}$$

In the stenotic area, the artery's radius is $\bar{R}(\bar{z},\tau)$, and outside of the stenotic area, it is $\bar{R}_0$. This is the description of the time-variant parameter:

$$\alpha_1(\tau)=1+\kappa_{\bar{R}}\cos(\omega\tau-\theta) \tag{9.2}$$

where the angular frequency is $\omega=2\pi f_n$, the frequency of the pulse is f_n, and the constant is $\kappa_{\bar{R}}$. Second, in Figures 9.1 and 9.2, multiple instances of irregular stenoses are contrasted with a single, long, irregular stenosis with a comparable overall length L.

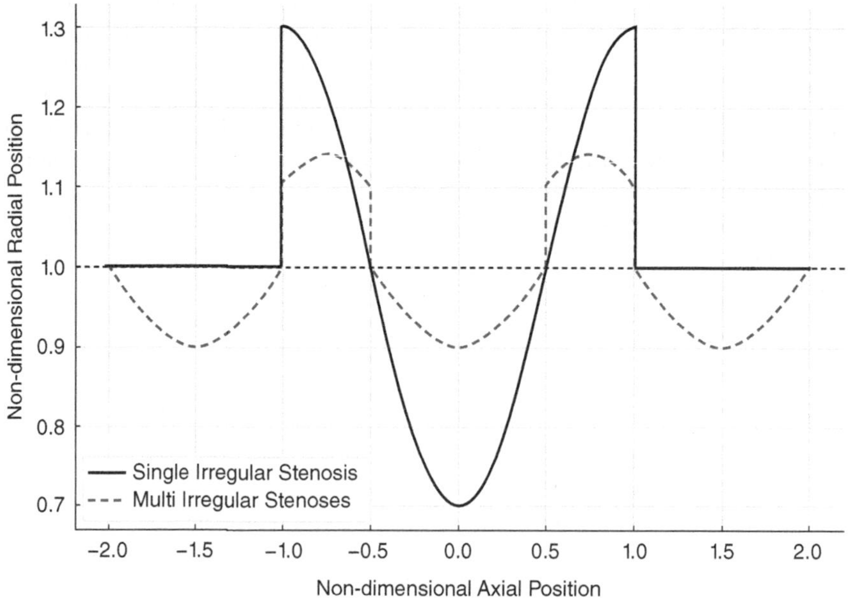

Figure 9.1 Profiles of single and multiple irregular stenosis.

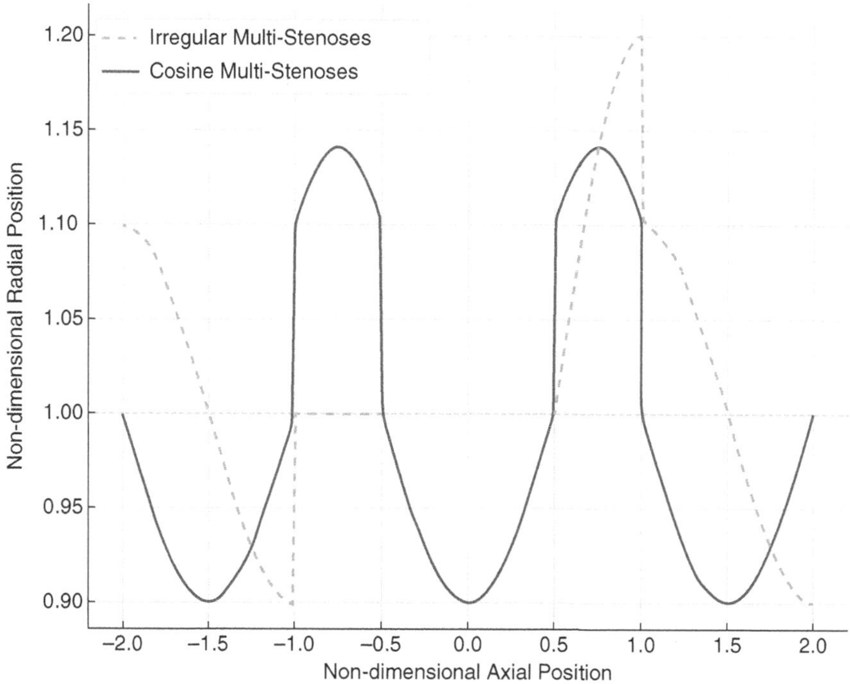

Figure 9.2 Profile of a multi-stenosis (irregular and cosine curve stenosis).

9.3 THE GOVERNING EQUATIONS

A homogeneous Newtonian fluid is used to describe the lumen of the artery, where blood is streaming. By adding a transformation of radial coordinates $\mathrm{r} = \bar{\mathrm{R}}(\bar{\mathrm{z}},\tau)\cdot x(\bar{\mathrm{z}},\tau)$, non-dimensional form can be used to express the equations for Navier-Stokes and continuity that govern the completely nonlinear and unsteady swirl-free blood flow in the following forms.

$$\frac{\partial v}{\partial \tau} = \frac{x}{\bar{R}}\left(\frac{\partial \bar{R}}{\partial \tau}\frac{\partial v}{\partial x} + \frac{\partial \bar{R}}{\partial \bar{z}}\frac{\partial v^2}{\partial x}\right) - \frac{vu}{x\bar{R}} - \frac{1}{\bar{R}}\frac{\partial (vu)}{\partial x} - \frac{\partial v^2}{\partial \bar{z}} - \frac{\partial p}{\partial \bar{z}} + \frac{x}{\bar{R}}\frac{\partial \bar{R}}{\partial \bar{z}}\frac{\partial p}{\partial x}\frac{1}{\mathrm{Re}\,\bar{R}^2}$$

$$\left\{\left[1 + \left(x\frac{\partial \bar{R}}{\partial \bar{z}}\right)^2\right]\frac{\partial^2 v}{\partial x^2} + \left[\frac{1}{x} + 2x\left(\frac{\partial \bar{R}}{\partial \bar{z}}\right)^2 - x\bar{R}\frac{\partial^2 \bar{R}}{\partial \bar{z}^2}\right]\frac{\partial v}{\partial x} + \bar{R}^2\frac{\partial^2 v}{\partial \bar{z}^2}\right\} \tag{9.3}$$

$$\frac{\partial u}{\partial \tau} = \frac{x}{\bar{R}}\left(\frac{\partial \bar{R}}{\partial \tau}\frac{\partial u}{\partial x} + \frac{\partial \bar{R}}{\partial \bar{z}}\frac{\partial (vu)}{\partial x}\right) - \frac{u^2}{x\bar{R}} + -\frac{\partial (vu)}{\partial \bar{z}} - \frac{1}{\bar{R}}\frac{\partial u^2}{\partial x} - \frac{1}{\bar{R}}\frac{\partial p}{\partial x}\frac{1}{\mathrm{Re}\,\bar{R}^2}$$

$$\left\{\left[1 + \left(x\frac{\partial \bar{R}}{\partial \bar{z}}\right)^2\right]\frac{\partial^2 u}{\partial x^2} + \left[\frac{1}{x} + 2x\left(\frac{\partial \bar{R}}{\partial \bar{z}}\right)^2 - x\bar{R}\frac{\partial^2 \bar{R}}{\partial \bar{z}^2}\right]\frac{\partial u}{\partial x} - \frac{u}{x^2} + \bar{R}^2\frac{\partial^2 u}{\partial \bar{z}^2}\right\} \tag{9.4}$$

$$x\bar{R}\frac{\partial v}{\partial \bar{z}} - x^2\frac{\partial v}{\partial x}\frac{\partial \bar{R}}{\partial \bar{z}} + \frac{\partial (ux)}{\partial x} = 0, \tag{9.5}$$

The non-dimensional coordinates, $\bar{z}$ and r, are scaled in reference to the z-axis, and r_0 is situated along the artery's axisymmetry. The entire velocity is determined by the non-dimensional axial and u and v, the radial components, scaled in relation to the cross-sectional average velocity U_0, since there is neither secondary nor rotational flow. Re = $U_0 r_0 \eta / \mu$, where η is the blood's density and μ is its viscosity, is one way to express the Re (Reynolds number). The initial and boundary conditions that have been modified are,

$$u(\bar{z},x,\tau) = \frac{\partial \bar{R}}{\partial t} \ and \ v(\bar{z},x,\tau) = 0, \ when \ x = 1, \tag{9.6}$$

$$u(\bar{z},x,\tau) = 0 \ and \ \frac{\partial v(\bar{z},x,\tau)}{\partial x} = 0 \ when \ x = 0, \tag{9.7}$$

$$u(\bar{z},x,\tau) = 0 \ and \ v(\bar{z},x,\tau) = 2\left(1 - x^2\right), \ when \ \bar{z} = 0, \tag{9.8}$$

$$\frac{\partial u(\bar{z},x,\tau)}{\partial \bar{z}} = 0 \ and \ \frac{\partial v(\bar{z},x,\tau)}{\partial \bar{z}} = 0 \ when \ \bar{z} = l \tag{9.9}$$

and

$$u(\bar{z},x,0) = v(\bar{z},x,0), p(\bar{z},x,0) = 0 \ when \ \bar{z} > 0 \tag{9.10}$$

9.4 SOLUTION METHODOLOGY

The MAC method, which uses a staggered grid and the method of finite differences, utilizes numerical methods to solve the set of initial conditions and boundary conditions (9.6)–(9.10) and the governing equations (9.3)–(9.5). In this kind of grid alignment, as shown in Figure 9.3, pressure and field of flow velocity are computed at various places of the control volumes, and (9.3) and (9.4), the axial and radial momentum, and (9.5), the continuity equation, are represented by the three distinct cells in which the difference equations have been derived.

To discretize the time derivative terms, a linear order exact two-step forward-in-time differencing formula is used, whereas the momentum equations' convective terms are subjected to a formula for hybrids that combines second-order upwinding and central differencing. On the other hand, a third-order exact three-point central difference formula is used to discretize the diffusive terms. Using this formulation of finite differences as shown

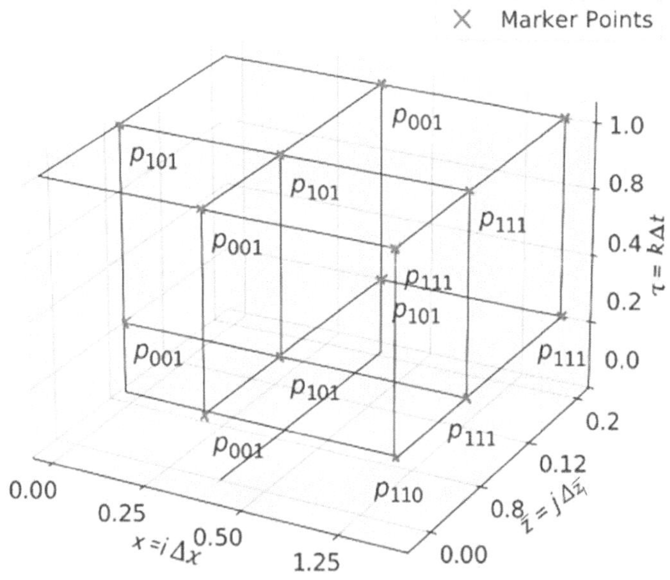

Figure 9.3 Typical marker and cell in 3D.

in Figure 9.3, we therefore define by $(x,\bar{z},\tau)=(i\Delta x, j\Delta\bar{z}_j, k\Delta\tau)$ *and* $p_{ij}^k = p(x,\bar{z},\tau) = p(i\Delta x, j\Delta\bar{z}_j, k\Delta\tau)$.

where Δx and $\Delta\bar{z}_j$ are the length and breadth $(i,j)^{th}$ control volumes, $\Delta\tau$ is the increment of time, and k is the time direction.

$$x_{il}\bar{R}_{lj}^k\left(\frac{v_{i,j}^k - v_{i,j-i}^k}{\Delta\bar{z}_j}\right) - (x_{il})^2\left(\frac{\partial\bar{R}}{\partial\bar{z}}\right)_{lj}^k\left(\frac{v_{a\tau}-v_{ab}}{\Delta x}\right)+\left(\frac{x_i u_{i,j}^k - x_{i-1}u_{i-1,j}^k}{\Delta x}\right)=0, \tag{9.11}$$

where

$$v_{a\tau} = \left(\frac{v_{i,j}^k}{4} + \frac{v_{i+1,j}^k}{4} + \frac{v_{i,j-1}^k}{4} \frac{+v_{i+1,j-1}^k}{4}\right), \tag{9.12}$$

$$v_{ab} = \left(\frac{v_{i,j}^k}{4} + \frac{v_{i-1,j}^k}{4} + \frac{v_{i,j-1}^k}{4} + \frac{v_{i-1,j-1}^k}{4}\right), \tag{9.13}$$

$$x_{il} = \frac{2x_i - \Delta x}{2},\ \bar{R}_{lj}^k = \bar{R}(\bar{z}_{lj}),\ \bar{z}_{lj} = \frac{2\bar{z}_j - \Delta\bar{z}_j}{2} \tag{9.14}$$

Here, the coordinates for the upper right corner and the control volume's cell center are denoted by $(\bar{z}_j, x_i)$ and $(\bar{z}_{lj}, x_{il})$, respectively. One way to express the axial momentum equation is as,

$$\left(\frac{v_{i,j}^{k+1} - v_{i,j}^{k}}{\Delta t^*}\right) = 2\left(\frac{p_{i,j}^{k} - p_{i,j+1}^{k}}{\Delta\bar{z}_j + \Delta\bar{z}_{j+1}}\right) + \frac{x_{il}}{\bar{R}_j^k}\left(\frac{\partial\bar{R}}{\partial\bar{z}}\right)_j^k\left(\frac{p_t - p_b}{\Delta x}\right) + (vme)_{i,j}^k, \tag{9.15}$$

$$\text{With } p_\tau = \frac{1}{2}\left[\frac{\left(p_{i+1,j}^k + p_{i,j}^k\right)\Delta\bar{z}_{j+1} + \left(p_{i,j+1}^k + p_{i+1,j+1}^k\right)\Delta\bar{z}_j}{\left(\Delta\bar{z}_j + \Delta\bar{z}_{j+1}\right)}\right], \tag{9.16}$$

$$p_b = \frac{1}{2}\left[\frac{\left(p_{i,j}^k + p_{i-1,j}^k\right)\Delta\bar{z}_{j+1} + \left(p_{i,j+1}^k + p_{i-1,j+1}^k\right)\Delta\bar{z}_j}{\left(\Delta\bar{z}_j + \Delta\bar{z}_{j+1}\right)}\right] \tag{9.17}$$

and

$$vme_{i,j}^k = conw_{i,j}^k + \frac{1}{Re}\left(Diffv_{i,j}^k\right), \tag{9.18}$$

For the purpose of conciseness, the expressions for the diffusive terms $\left(Diffv_{i,j}^k\right)$ and for the nth time step of the (i,j) cell, the convective terms $\left(conv_{i,j}^k\right)$ of the axial momentum is not provided. Similarly, the radial direction of the momentum equation is approximated by the finite difference equation, which is given by

$$u_{i,j}^{k+1} = \Delta\tau\left[\frac{1}{\bar{R}_j^k}\left(\frac{p_{i,j}^k - p_{i+1,j}^k}{\Delta x}\right) + ume_{i,j}^k\right]u_{i,j}^k, \tag{9.19}$$

where

$$ume_{i,j}^k = conu_{i,j}^k + \frac{1}{Re}\left(Diffu_{i,j}^k\right) \tag{9.20}$$

After being obtained from equations (9.11), (9.15), and (9.19), the Poisson pressure equation ultimately takes the following form:

$$\begin{aligned}\frac{Diw_{i,j}^{k+1} - Diw_{i,j}^k}{\Delta\tau} &= x_{il}\bar{R}_{lj}^k\left(\frac{vme_{i,j}^k - vme_{i,j-i}^k}{\Delta\bar{z}_j}\right) + \left(\frac{x_i ume_{i,j}^k - x_{i-1}ume_{i-1,j}^k}{\Delta x}\right)\\ &+A_{i,j}p_{i,j}^k + B_{i,j}p_{i+1,j}^k + C_{i,j}p_{i-1,j}^k + D_{i,j}p_{i,j+1}^k + E_{i,j}p_{i,j-1}^k + F_{i,j}p_{i+1,j+1}^k\\ &+G_{i,j}p_{i-1,j+1}^k + H_{i,j}p_{i-1,j-1}^k + T_{i,j}p_{i+1,j-1}^k\end{aligned} \tag{9.21}$$

Here, $Diw_{i,j}^{k}$ stands for the transformed form of the divergences in velocity fields for the cell (i, j), and the appendix contains the corresponding A, B, C, D, E, F, G, H, and T expressions. At the time level of kth, the SOR method obtains the intermediate pressure field by iteratively solving the pressure Poisson equation (9.21) with a predefined number of iterations. Here, 1.2 is used as the over-relaxation parameter value to maximize the SOR's rate of convergence.

9.5 RESULT AND DISCUSSION

In comparison to cosine-shaped stenosis, multi-irregular stenosis exhibits a greater dimensionless pressure drop across all Reynolds numbers, as shown in Figure 9.4. Pressure drop gradually decreases as the Reynolds number rises. Because there is less flow resistance and less swirl formation, the pressure drop profile of cosine-shaped stenosis is smoother and lower. Additionally, the cosine shape promotes more efficient flow and reduces flow separation. Because of the increased swirling, multi-irregular stenosis has a high pressure drop at low Reynolds numbers. Although the flow becomes more inertia-dominated as the Reynolds number rises, multi-irregular stenosis remains to show a greater pressure drop.

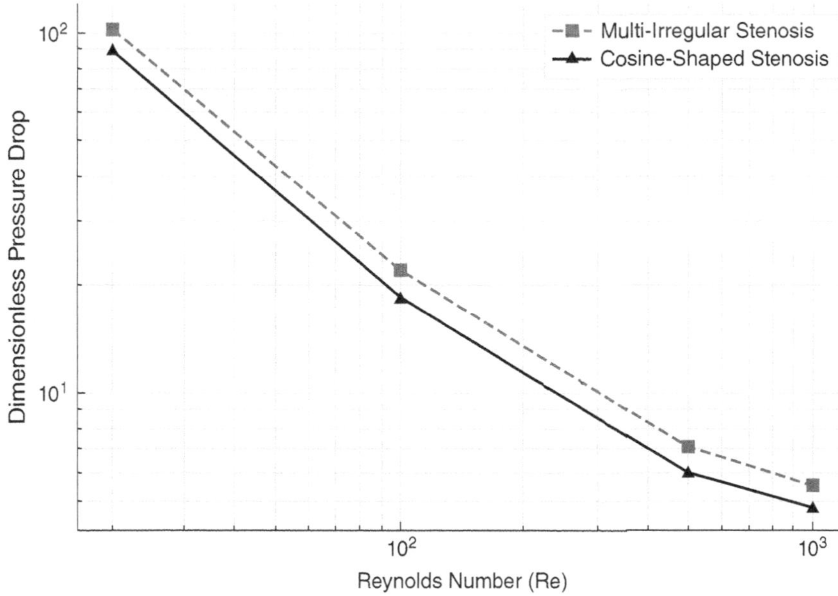

Figure 9.4 Comparison of multiple irregular stenoses versus cosine stenoses.

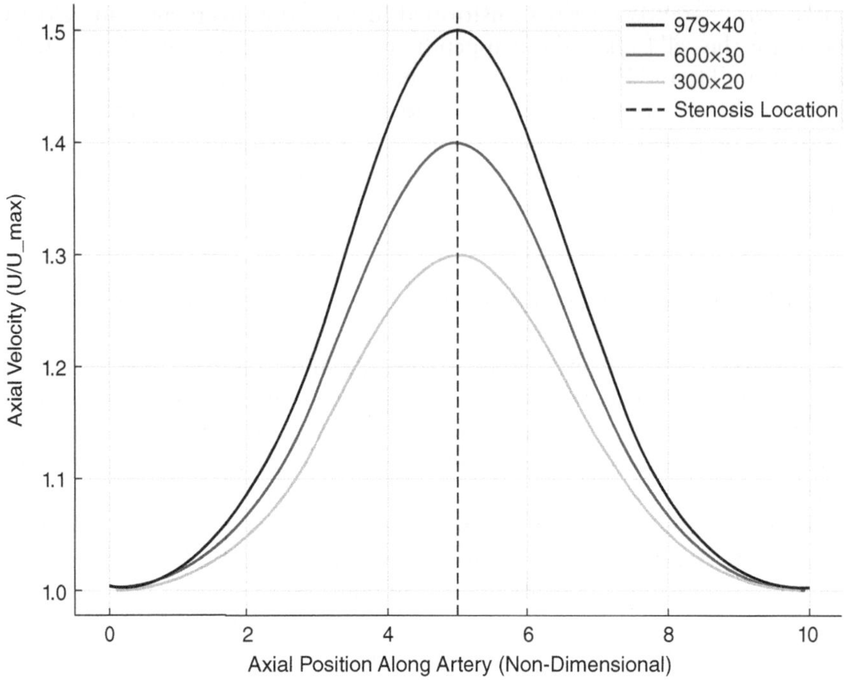

Figure 9.5 Variation of axial velocity along a stenosed artery for different grid sizes.

The study simulates the flow field in a stenosis using a non-dimensional arterial length of 60.6, staggered grids measuring 979 × 40, 512 MB of RAM, and a Pentium IV processor running at 2.8 GHz. Six hours are needed for the steady-state velocity field simulation, as shown in Figure 9.5. The fine grid displays a smooth transition prior to, during, and following narrowing, accurately capturing abrupt changes in velocity within the stenosis. Although it offers a fair approximation, the medium grid may marginally smooth out peak velocities inside the stenosis. Steep velocity gradients are difficult for the coarse grid to capture, which produces unreliable results in areas with fast acceleration and deceleration, especially inside and after the stenosis.

In Figure 9.6, when blood flow velocity rises and pressure drops suddenly, stenosis happens. Although there is some pressure recovery following stenosis, complete restoration is not possible due to energy losses. Increased flow resistance is indicated by a slight reduction in mild stenosis (30%), a significant drop in moderate stenosis (50%), and a drastic loss in severe stenosis (70%). In addition to increasing heart workload, more severe stenosis can result in arterial damage, ischemic attack, and decreased oxygen delivery. Diagnosing and treating cardiovascular diseases requires an understanding of pressure behavior.

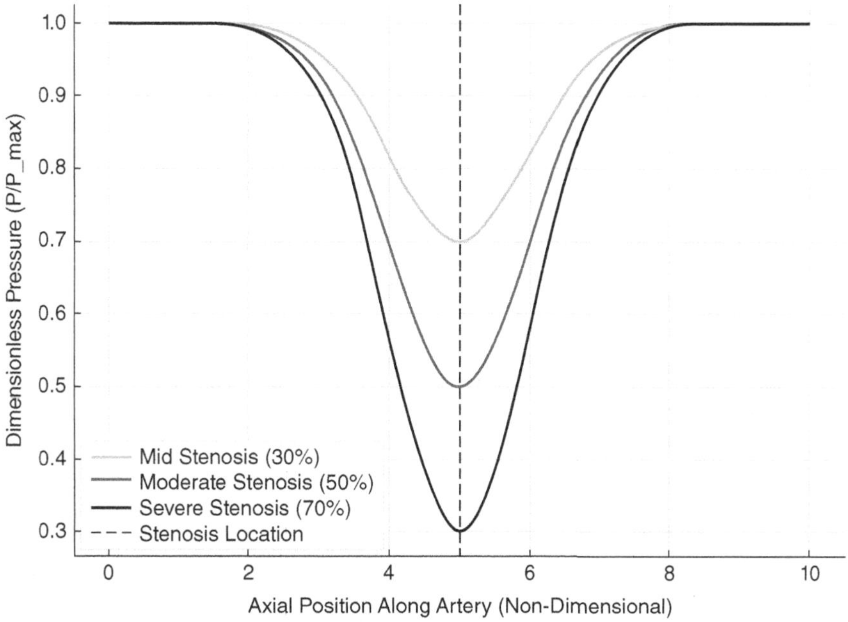

Figure 9.6 Comparison of non-dimensional pressure drop versus a single irregular stenosis.

9.6 CONCLUSIONS AND FUTURE SCOPE

Blood flow via arterial stenoses with multiple irregularities was examined in the study using the MAC method. According to the findings, a single stenosis causes localized constriction, whereas several irregular stenoses cause intricate flow disruptions. Higher velocity gradients and possible turbulence are caused by the irregularity of stenoses, which also alter flow characteristics. In order to accurately capture flow dynamics, the study also discovered that grid refinement is essential. In contrast to their cosine-based counterparts, irregular stenoses show larger pressure gradients, and the non-dimensional pressure drop rises with the severity of the stenosis. The outcomes confirm the accuracy of the MAC approach in simulating blood flow through stenosed arteries. Drug delivery, biomedical engineering, computational science, patient-specific modeling, machine learning, and AI integration can all benefit from the simulation of unsteady blood flow numerically in stenosed arteries.

BIBLIOGRAPHY

Agrawal, B., Kumar, S. and Das, G. (2022). Mathematical model of blood flow through stenosed arteries with the impact of hematocrit on wall shear stress. *International Journal of Applied Research*, 8 (3), 439–443.

Chakravarty, S., and Mandal, P. K. (2005). Effect of surface irregularities on unsteady pulsatile flow in a compliant artery. *International Journal of Non-Linear Mechanics*, 40 (10), 1268–1281.

Chakravarty, S., Sarifuddin, and Mandal, P. K. (2004). Unsteady flow of a two-layer blood stream past a tapered flexible artery under stenotic conditions. *Computational Methods in Applied Mathematics*, 4 (4), 391–409.

Das, G., Agrawal, B. and Kumar, S (2023a). A Mathematical model of blood flow of a stenosed artery in variable shape. *Journal of Advanced Zoology*, 44 (S7), 1193–1208.

Das, G., Agrawal, B. and Kumar, S. (2023b). Mathematical modeling of non-Newtonian blood flow through arterial stenosis. *Samdarshi*, 16 (4), 3133–3139,

Das, G., Agrawal, B. and Kumar, S. (2024). Comparative analysis of mathematical models for non-Newtonian stress relaxation flow of blood in an artery with multiple stenosis. *Futuristic Trends in Physical Sciences*, IIP Series, 3 (3), 139–148.

El-Shahed, M. (2003). Pulsatile flow of blood through a stenosed porous medium under periodic body acceleration. *Applied Mathematics and Computation*, 138 (2–3), 479–488.

El-Shehawey, E., Elbarbary, E., Afifi, N. and Elshahed, M. (2000). MHD flow of an elastic-viscous fluid under periodic body acceleration. *International Journal of Mathematics and Mathematical Sciences*, 23 (11), 795–799.

Hou, L. *et al.* (2020). Evaluation of the behavioral characteristics in a gas and heavy oil stratified flow according to the Herschel-Bulkley fluid model. *ACS Omega*, 5, 17787–17800.

Ikbal, M. A., Chakravarty, S., Wong, K. K. L., Mazumdar, J. and Mandal, P. K. (2009). Unsteady response of non-Newtonian blood flow through a stenosed artery in magnetic field. *Journal of Computational and Applied Mathematics*, 230 (1), 243–259.

Ismail, Z., Abdullah, I., Mustapha, N. and Amin, N. (2008). A power-law model of blood flow through a tapered overlapping stenosed artery. *Applied Mathematics and Computation*, 195 (2), 669–680.

Kuppalapalle, V., Sreedharamalle, S., Palluru, D. and Kerehalli, V. P. (2011). Mathematical model for a Herschel-Bulkley fluid flow in an elastic tube. *Central European Journal of Physics*, 9 (5), 1357–1365.

Magnon, E. and Cayeux, E. (2021). Precise method to estimate the Herschel-Bulkley parameters from pipe rheometer measurements. *Fluids*, 6, 157–167.

Mandal, P. K., Chakravarty, S., Mandal, A. and Amin, N. (2007). Effect of body acceleration on unsteady pulsatile flow of non-newtonian fluid through a stenosed artery. *Applied Mathematics and Computation*, 189 (1), 766–779.

Mekheimer, S. K. and El Kot, M. A. (2012). Mathematical modeling of unsteady flow of sisko fluid through an anisotropically tapered elastic arteries with time-variant overlapping stenosis. *Applied Mathematical Modelling*, 36 (11), 5393–5407.

Mustapha, N., Chakravarty, S., Mandal, P. K. and Amin, N. (2008). Unsteady response of blood flow through a couple of irregular arterial constrictions to body acceleration. *Journal of Mechanics in Medicine and Biology*, 8 (3), 395–420.

Nallapu, S. and Radhakrishnamacharya, G. (2016). A two-fluid model for Herschel-Bulkley fluid flow through narrow tubes. *Journal of Applied Science and Engineering*, 19 (3), 241–248.

Nallapu, S., Radhakrishnamacharya, G., Chamkha, A. J. (2015). Effect of slip on Herschel-Bulkley fluid flow through narrow tubes. *Alexandria Engineering Journal*, (54), 889–896.

Priyadharshini, S. and Ponalagusamy, R. (2015). Biorheological model on flow of Herschel-Bulkley fluid through a tapered arterial stenosis with dilatation. *Applied Bionics and Biomechanics* 2015, Article ID 406195, 1–12.

Rakshit, S., Agrawal, B. and Kumar, S. (2022). A mathematical study of constrained fluid movement in the arterial system due to the formation of multiple stenosis. *International Journal of Applied Research*, 8 (3), 455–464.

Rakshit, S., Agrawal, B. and Kumar, S. (2023). A mathematical model of flow of blood in a segment of an artery by a non-homogeneous approach. *Journal of Data Acquisition and Processing*, 38 (1), 5495–5504.

Chapter 10

Exploring the compatibility of just-in-time with circular economy principles

Sandip Das, Dinesh Kumar Soni, and Ankit Goyal

10.1 INTRODUCTION

Just-in-time (JIT) compilation is a runtime technique designed to enhance the performance of software applications by dynamically translating frequently executed code into native machine code during execution. Unlike traditional interpretation, which processes code line-by-line, or ahead-of-time (AOT) compilation, which compiles the entire program before running it, JIT strikes a balance by initially interpreting the code and then identifying and compiling "hot spots"—sections of code that are executed repeatedly—into optimized machine code. This approach combines the flexibility and portability of interpretation with the speed and efficiency of native execution. While JIT offers significant performance benefits, it does come with some trade-offs, such as initial startup delays due to the time required for compilation and increased memory usage to store both interpreted and compiled code. Despite these drawbacks, JIT is widely used in modern runtime environments, including the Java Virtual Machine (JVM), the .NET Common Language Runtime (CLR), and JavaScript engines like V8, to deliver faster and more efficient execution of programs across various platforms.

10.1.1 History and motivation

JIT was invented by Taiichi Ohno of Toyota shortly after World War II. Ohno's system was designed to handle large or small volumes of a variety of parts. Many people are intimidated by JIT because of its association with Japan. If these people take a broader look at JIT, they will see that it is nothing more than good, common-sense manufacturing.

Ohno learned the kanban (pull) system from our supermarket system in which customers pulled items from the shelves to fill their shopping carts, thereby creating an empty space on the shelf. The empty space is a signal for the stocker to replace that item. If an item was not bought that day, there was no need to replace it. When item quantities become low, that is the signal for the stockers to order more goods from their suppliers. Customers are

DOI: 10.1201/9781003642152-10

content to take just what they need, because they know that the goods will be there the next time they need them.

The JIT system continued to evolve, with the central thrust being the elimination of waste. Ohno's system has become a totally flexible system in which production rates are determined by the end user rather than the producer.

10.1.2 Theory and methodology

Maybe it is believed that a company is efficient enough and that the benefits of JIT are not worth the frustration and stress associated with change. At this point, we have a decision to make—we can adopt a new company motto such as "We're no worse than anybody else," or you can take positive steps toward improving the process. To strengthen the incentive for change, companies should identify the inefficiencies (wastes) in their present manufacturing processes.

To identify waste in your company, a value-added analysis should be performed. We must always be aware that any activity that does not add value to a product is waste. There are specific methods for performing a value-added analysis, but we will use a simplified approach for our purposes. Take a pad and a pencil and go out on the shop floor. Pick a product and follow it through the entire manufacturing process from raw materials to shipping. Note every activity performed on the product. Do not get a routing slip to see how the process is supposed to go, but accurately record the process, including delays, transportation, inspection, and storage. Everything from purchase orders to typing memos should be streamlined. How long does a purchase order sit on someone's desk awaiting a signature, only to be changed, retyped, and submitted again? It may then stall at a higher level in the organization and so on up the ladder. Not only will the techniques of JIT manufacturing help eliminate some of the nonvalue-added steps in your process, but in doing so, they will reduce lead time, improve throughput time, and increase quality.

10.2 THE CONCEPT OF WASTE

JIT is much more than a plan for decreasing inventory; it is a manufacturing philosophy for eliminating waste. For our purposes, waste can be defined as something other than the essential resources of people, machines, and material needed to add value to the product. Anything else, such as inventory, scheduling, meetings, warehousing goods, management, and moving stock, can be considered wasteful because these actions do not directly add value to the product. All waste cannot be purged from the system; however, we must strive toward that ideal goal. Above all, it must be ever present in the attitudes of our manufacturing system that cost without value is waste.

Traditional manufacturing processes build in safety stock at every station throughout the entire system, from extra raw materials to warehouses full of completed products. This superfluous work in progress provides manufacturers with a means to endure the problems, rather than solving the problems at the root cause.

The water shown in Figure 10.1 is inventory, and the rocks depict manufacturing problems. As the water level drops, more and more rocks begin to surface.

Do not make the mistake of raising WIP to allow the line to flow smoothly. We need the problems to surface so that we can solve them. Remember, the WIP is not the solution to the problems; it is only a way to wade through them. Inventory must first be reduced, and then you can solve the problems. Inventory must be decreased using a systematic approach. A methodical approach is to cut inventory by one half, then solve the problems, then cut half of that inventory and solve the problems.

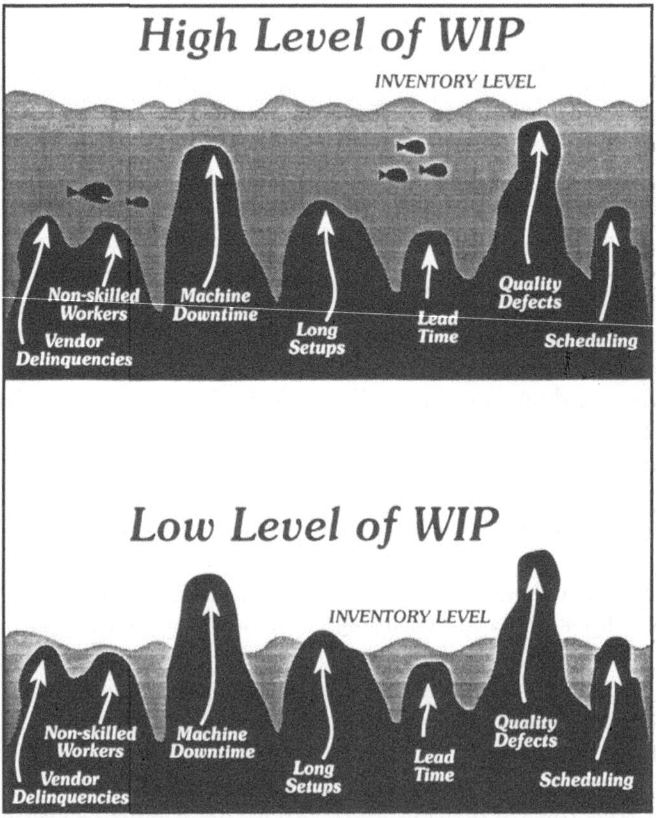

Figure 10.1 Inventory and the rocks depict manufacturing.

10.3 THE CONCEPT OF QUALITY

It is impossible for JIT to be successful until the company has drastically improved its attitude toward quality. In the language of the Malcolm Baldrige National Quality Award, quality is a "race with no finish line." The ultimate aspiration is to satisfy all customers (internal and external) all the time.

Quality is a two-way street; JIT is impossible without quality, but quality is directly enhanced by JIT. Although quality is possible without JIT, it requires the use of wasteful procedures such as inspection and rework. JIT proposes the idea of "do it right the first time" rather than inspecting for quality. To produce quality, you must install quality. Quality must evolve from both sides at the same time. A mathematics riddle known as Zeno's paradox asks if a person walks toward a wall, each step being one-half as large as the previous one, when will that person reach the wall? The answer is "never," but that person is continuously getting closer and closer to his or her goal.

Figure 10.2 gives an illustration of a restaurant which is facing a crisis due to bad service, which needs to be looked into deeper than the superficial causes of bad quality.

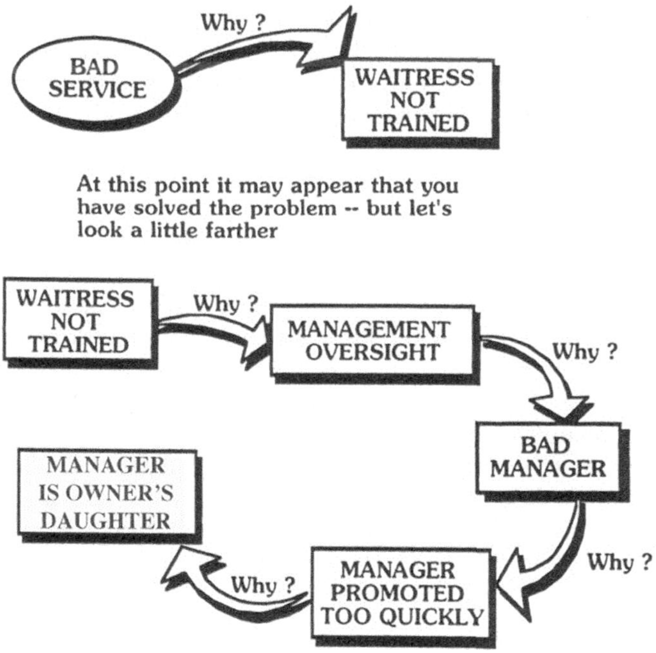

Figure 10.2 Illustration of a restaurant.

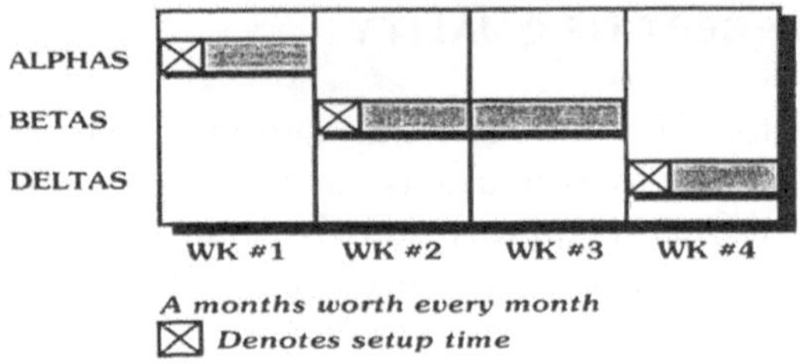

Figure 10.3 Equal distribution of load.

10.4 CONCEPT OF EQUAL DISTRIBUTION OF LOAD

This concept incorporates two radically different facets of production. They are the rate of production (cycle time) and frequency of production (level loading). It must be remembered that neither of these concepts will achieve maximum results until the process is under control and quality has been improved to world-class or near world-class standards.

Under JIT, cycle time is the total time required for a worker to complete one cycle of operations, including walking, loading/unloading, and inspecting. Cycle time should equal the customer requirement rate, or better stated, the sales rate.

Level loading suggests that if you sell a product every month, then make the product every month. Ideally, if you sell a product every day, then make the product every day. You must make your products as frequently as your customers require them. Figure 10.3 shows the same effect as discussed above.

10.5 CONCEPT OF REDUCTION IN SETUP TIME

Setup time is the interval between the production of one good part and the production of another good but dissimilar part. Setup reduction is a prerequisite to implementing many aspects of JIT by directly or indirectly influencing cycle time, level loading, work cells, pull systems, cost, WIP, purchasing, floor space, quality, operator numbers, and batch sizes. Our objective is to apply this hour to more frequent setups, thus giving us more flexibility to better implement JIT.

The new Setup Reduction Team should consist of the real experts—the operators and setup people. Contrary to traditional management's beliefs, these people have superior knowledge of their machines and the process.

One proven technique for optimizing setup time is the Single-Minute Exchange of Die (SMED) system. The SMED system was founded by Shigeo Shingo while consulting with Toyota in 1969. No man has revolutionized the setup reduction philosophy as much as Shingo. He has won numerous productivity improvement awards in Japan, the United States, and the world over. The basis of SMED is the performance of setup operations in under ten minutes, that is, in a number of minutes expressed in a single digit.

10.6 CONCEPT OF CELLULAR MANUFACTURING

The traditional method of arranging a manufacturing facility is by departmental specialty—that is, each department houses specialized equipment or technology. All the lathing, milling, drilling, grinding, and assembly would be done in completely separate departments. Production in large batch sizes is inevitable when factories are laid out in this manner. The JIT philosophy maintains that a manufacturing floor should be laid out by product rather than by function. All equipment should be dedicated to a product or family of products and organized logically in the order in which the various processes are performed on that family of products. Two characteristics must be fulfilled before a group of machines can be deemed the optimal JIT work cell. The first is whether the product is flowing one at a time from machine to machine, and, second, whether the cell has the flexibility to produce at different rates with varying crew sizes (cycle time) (Figure 10.4).

The most flexible work cell is the U-shaped layout. The U-shaped work cell should be large enough to allow operators to work side by side, back to back, but not so close as to inconvenience each other (Figure 10.5).

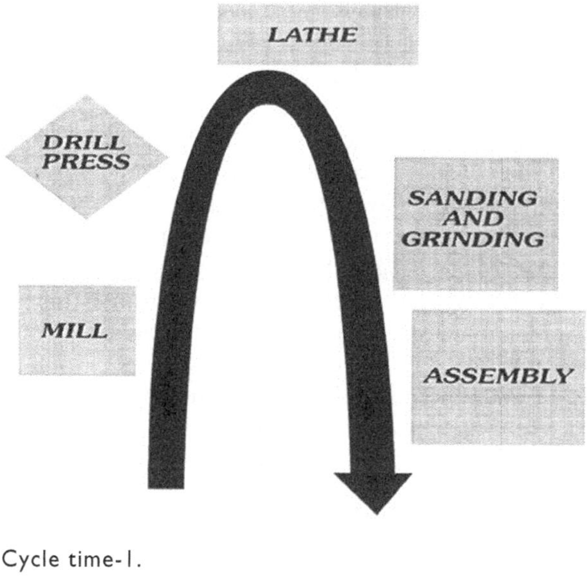

Figure 10.4 Cycle time-1.

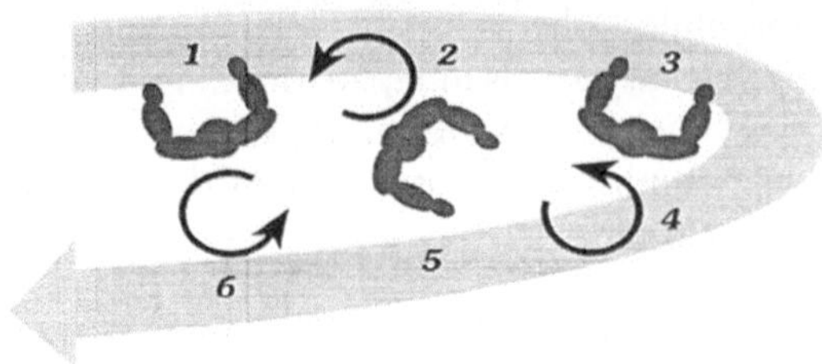

Figure 10.5 Cycle time-2.

10.7 CONCEPT OF THE PULL SYSTEM

Pull systems, sometimes known as kanban systems or supermarket systems. The United States is in the process of phasing out the use of the word kanban due to its association with Japan. Kanbán is not even a universally accepted Japanese term since some plants in Japan that compete with Toyota view it as a Toyota-coined word.

Before knowing further about the pull system, we must have a clear insight into why it came into existence. It was due to the push system. A push system originates with a forecast schedule. This forecast is forced upon the previous step (assembly) in the manufacturing process, all the while adjusting for lead times to predict which subassemblies are needed and by when. The subassembly forecasts are then forced upon the various component levels, still maintaining lead time, thus completing the cycle through the total manufacturing system down to raw materials. Paperwork accompanies this system, informing operators what is needed and by when (Figure 10.6).

The Japanese visited some of our supermarkets in their spare time. What they learned and took back to Japan has revolutionized manufacturing. A supermarket is managed very differently from a factory. Shoppers come to a supermarket knowing that there will always be a small stock of needed inventory. Customers feel no pressure to buy large quantities because they know that the goods will be available when they need them. Every night, a "stocker" replenishes the inventory that has been removed. Empty spaces on the shelves are the stocker's signal to produce more goods (restock). Exactly what has been taken is what is replaced. The customers have directly told the store what to replace with what has been purchased. No operation can produce goods until it has received a signal from its customers. When the operator gets a signal from the customer, he then has authorization to produce a certain number of parts in a specific time period. The most effective pull (kanban) signals are visual indicators such as empty containers or empty floor space. If you have an empty container, fill it up; if you have no container, then do not produce that part. The most commonly used signal is a Kanban card. The card tells what type of part to build, what to put the

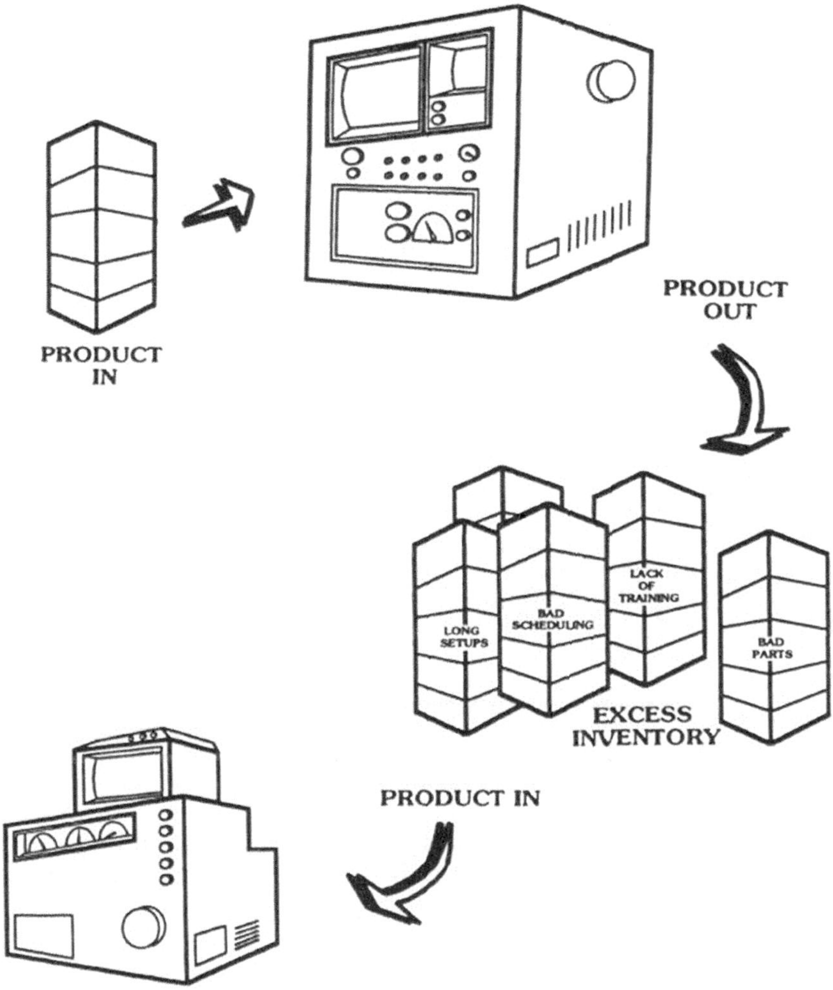

Figure 10.6 Product process.

parts in, how many parts to build, where to send the parts, and how many cards of these parts are required to maintain a smooth flow. Paperwork is limited in a pull system (Figure 10.7).

10.8 CONCEPT OF PURCHASING

JIT purchasing is as different from traditional purchasing as JIT manufacturing is from traditional manufacturing. The goal of a purchasing department is the same as that for the manufacturing floor—the elimination of

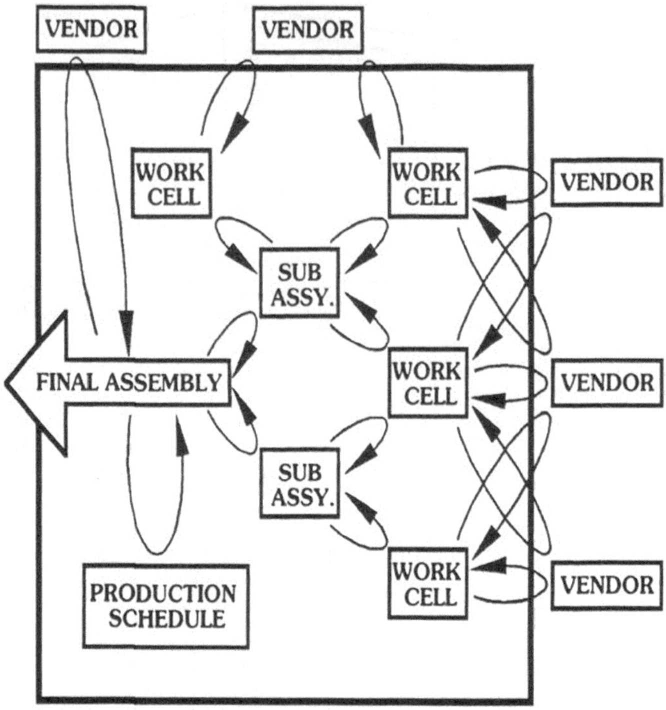

Figure 10.7 Production process flow.

waste. Purchasing cost is a critical factor for a JIT manufacturing system, but it lags behind quality and delivery lead time in importance. Vendors must deliver quality products on time (just-in-time) before a JIT system can work, regardless of cost. JIT purchasing offers a framework for a true partnership between vendors and companies that helps to solve the problems of cost, quality, and lead time.

The new JIT partnership that we are striving for is a long-term, mutually beneficial relationship with fewer but better vendors. Mutual trust must be developed between companies and vendors. This cannot be accomplished if vendors change every time new bids are sent out. For this reason, a company should have few suppliers (preferably one) for each purchased material or component.

Three areas of waste need to be recognized before a company can successfully implement JIT. First is the waste in a company's own manufacturing process: the moving, counting, rework, storing, scheduling, setup times, and inspection. Second, waste within the purchasing process itself should be eliminated. A third area that is often overlooked is the manufacturing philosophy of a company's vendors.

10.9 CONCEPT OF CONJURING JIT WITH MRP

Many MRP software companies fail to comprehend the overall picture, resulting in the misconception that JIT will cut into their market share. The MRP gurus tend to view only parts of the whole, so that they fail to address the manufacturing process and master the usefulness of both systems. JIT and MRP can and do work well together, but we must be conscious of the fact that MRP wants to work with the scheduling system, while JIT wants to radically alter the manufacturing process.

10.10 MODELS FOR JIT

10.10.1 Classification

There is a large number of factors that have to be considered explicitly in describing a manufacturing environment. D is the set of external parameters.

Krajewski et al. [1987] and, subsequently, Chu and Shih [1992] identify 41 system characteristics that appear as factors in analyzing manufacturing environments. They summarize these factors under eight major categories: inventory policy, process characteristics, product structure, customer influence, vendor influence, buffer mechanisms, facility design, and inescapably, others.

In somewhat abstract terms, we can classify the literature using the following elements:

S is the set of control mechanisms with s e S, Xs is the set of decision variables corresponding to control systems, and P is the set of internal parameters. Then, the mapping D is the set of external parameters.

f:(s, Xs, D, P) -> t k is the performance measure of interest. Examples of these modeling elements are given in Table 10.1. Several objectives for modeling and analyzing production systems can then be distinguished:

1. *Comparison*: Given D, P, and f, compare f(s', XS', D, P) and f(s", Xs", D, P). For example, compare a push and pull system given a specific set of external and internal parameters and a performance metric.
2. *Evaluation*: Given s, Xs, and f, evaluate f(s, Xs, D, P) as a function of D and P. For example, determine how sensitive f is to internal and external factors given a pull system and a performance metric.
3. *Static Optimization*: Given s, D, P, and f, optimize f(s, Xs, D, P) as a function of Xs. For example, calculate how many kanbans should be assigned to each workcenter in a pull system in order to optimize throughput.
4. *Dynamic Optimization*: Given s(0), Xs(0), D(t), P(t), and f, optimize g(t) f(s(t), xs(t), D(t), P(t)) dt as a function of s(t) and Xs(t), where g(t) is a discount factor. For example, determine the least expensive way to switch from a push to a pull control system.

Table 10.1 Components of categorized items

S – Control system arrangements	Uproot
	Straining
	Combination
Xs – Choice variables	Storage limit
	Batch size
	Production time
	Kanban nos.
	Dimensions
D – External influences	Impact of customer
	Distributor effects
	Product volume
	Standardization of parts
P – Set of internal parameters	Settings of internal configuration
	Idle time
	Duration of the tasks time
	Failure of the mechanical system
	Stages involved in process
f – measurement standards	All expenses of production
	Delaying of process
	Production is in process

10.11 REVIEW OF MODELS

Deterministic Model: A deterministic model is a structured mathematical or computational framework designed to produce outcomes that are entirely predictable and precisely determined by a specific set of inputs and predefined relationships, without any element of randomness or uncertainty. Unlike stochastic models, which incorporate probabilistic elements to account for variability, deterministic models operate on the principle that the same inputs will always yield the same outputs, ensuring consistency and repeatability. Synonyms for this concept include fixed-input model, which underscores the reliance on unchanging input values to generate exact results, and predictable model, which highlights the model's ability to deliver consistent outcomes under identical conditions. The term non-random model further emphasizes the absence of probabilistic factors, ensuring that all results are fully determined by the input parameters. Similarly, the exact model and definitive model suggest that the model produces precise, unambiguous outputs, free from variability or ambiguity. Other related terms, such as causal model, focus on the cause-and-effect relationships within the system, where inputs directly and logically determine outputs. The phrase static model refers to a system where inputs and relationships remain constant over time, leading to stable and unchanging results. Additionally, a closed-form model indicates that the model's solutions are derived from explicit mathematical equations without any randomness, while a determined system is a broader term often used in physics or engineering to describe systems with predictable and well-defined behavior. Finally, a rule-based model highlights the use of fixed rules or algorithms to compute outputs from inputs, ensuring a structured and logical approach to problem-solving. These terms are widely used in fields such as engineering, economics, operations research, and computer science, where deterministic models are employed to simulate systems or processes with well-defined, predictable behavior. By eliminating randomness, deterministic models provide exact and reliable results, making them ideal for scenarios where precision and consistency are critical, and uncertainty is either minimal or irrelevant.

10.12 CONCLUSION

JIT is still a very important and widely used strategy in today's manufacturing, supply chain, and business operations. It helps companies save money by reducing excess inventory, cutting down waste, and making processes more efficient by ensuring materials and products are available exactly when they are needed. Thanks to modern technology like real-time data tracking, the Internet of Things (IoT), and AI-powered demand forecasting, JIT systems have become even more accurate and flexible. This allows businesses to quickly adapt to changes in the market or customer needs. However, JIT isn't without its challenges. For example, global supply chain disruptions, like those seen during the COVID-19 pandemic, have shown that relying too much on lean inventory systems can be risky. To address this, many companies are now combining JIT with other approaches, keeping some extra stock as a backup while still using JIT principles to stay efficient. In short, JIT remains a key tool for improving efficiency and reducing costs, but businesses are now also focusing on building more resilience into their systems to handle the uncertainties of today's global economy.

10.13 FURTHER SCOPE

The future of JIT manufacturing will focus on becoming smarter, stronger, and more eco-friendly by blending new technologies and sustainable practices. Tools like artificial intelligence (AI), internet-connected sensors, and secure digital records (blockchain) will help companies make faster decisions, predict customer needs accurately, and track materials transparently across global supply chains. To protect the planet, JIT systems will cut waste, use renewable energy, and adopt greener transportation and packaging, aligning with eco-friendly goals like recycling and reusing resources. After recent disruptions like the COVID-19 pandemic, businesses will also build flexibility into JIT by working with multiple suppliers, keeping small emergency stocks of critical parts, and using data tools to foresee and avoid delays. While JIT started in car manufacturing, it will spread to industries like electronics (to manage fast-changing tech), healthcare (for precise medical equipment), and food production (to reduce spoilage). Advanced digital systems (Industry 4.0) will connect factories, suppliers, and distributors in real time, improving teamwork and efficiency. Companies will also prioritize ethical practices, using technology to ensure materials are sourced responsibly and workers are treated fairly. By balancing smart tech, environmental care, and adaptability, JIT will stay a key strategy for businesses aiming to save costs, reduce waste, and stay competitive in a world that values speed, sustainability, and reliability.

BIBLIOGRAPHY

The Design of a Factory with a Future by J.T. Black

Director of Advanced Manufacturing/Technology Center Auburn University

Juran's Quality Control Handbook by J.M. Juran

The Just-In-Time Breakthrough

Implementing the New Manufacturing Basics by Edward J. Hay

Out of the Crisis by Edward Deming

A Revolution In Manufacturing: The SMED System by Shigeo Shingo

Statistical Process Control Methods by Gary K. Griffith

World Class Manufacturing: The Lessons Of Simplicity Applied by Richard J. Schonberger

World-Class Manufacturing Casebook: Implementing JIT and TQC

Statistical Process Control Methods by Gary K. Griffith

World Class Manufacturing: The Lessons Of Simplicity Applied by Richard J. Schonberger

World-Class Manufacturing Casebook: Implementing JIT and TQC by Richard J. Schonberger

Berkley, B.(1992) "A Decomposition Approximation for Periodic Kanban-Controlled Flow Shops." *Decision Sciences*. 23.2, 291–311.

Chu, C.-S. and W.-L. Shih (1992) "Simulation Studies in JIT Production." *International Journal of Production Research*. 30.11, 2573–2586.

Glasserman, P. (1991) *Gradient Estimation via Perturbation Analysis*. Kluwer Academic Publishers. Dordrecht, The Netherlands.

Huang, P.Y. and B.L.W. Houck (1985) "Cellular Manufacturing: An Overview and Bibliography." *Production and Inventory Management*. 26, 83–93.

Jacobson, S.H. (1988) "Discrete-Event Computer Simulation Response Optimization in the Frequency Domain." Unpublished PhD Dissertation. School of Operations Research and Industrial Engineering, Cornell University. Ithaca, NY.

Kim, T.M. (1985) "JIT Manufacturing System: A Peiodic Pull System." *International Journal of Production Research*. 23, 553–562.

Kimura, O. and H. Terada (1981) "Design and Analysis of Pull Systems: A Method of Multi-Stage Production Control." *International Journal of Production Research*. 19.3, 241–253.

Krajewski, L.J., B.E. King, L.P. Ritman, and D.S. Wong (1987) "Kanban, MRP, and Shaping the Manufacturing Environment." *Management Science*. 33.1, 39–57.

Mitra, D. and I. Mitrani (1990) "Analysis of a Kanban Discipline For Cell Coordination in Production Lines. I" *Management Science*. 36.12, 1548–1566.

Mitra, D. and I. Mitrani (1991) "Analysis of a Kanban Discipline for Cell Coordinaion in Production Lines, II: Stochastic Demands." *Operations Research*.39.5, 807–823.

Schruben, L. and V.J. Cogliano (1987) "An Experimental Procedure for Simulation Response Surface Model Identificaion." *Communications of the ACM*. 30.8, 716–730.

Chapter 11

Modeling and simulation of system for power factor enhancement in commercial utilities

Bheemrao Mazumdar, Prateek Nigam, Krishnakant Gautam, Vikas Solanki, Neeraj Kumar Gupta, and Kishor Thakre

11.1 INTRODUCTION

The power factor problem is not a new phenomenon. It is one of the consequences of the harmonics present in AC supply. The effort to improve it to unity has been a major issue for electrical innovators from the initial stages (Lo et al. 2000). There has been identified research on power factor (Daut et al. 2006; Basau & Bollen 2005; Lo et al. 2005; Ali et al. 2018; Uddin et al. 2019; Mane et al. 2020; Holmes & Lipo 2003; Callies et al. 2022); in particular, Anand & Sajina (2020) and Rahman et al. (2022) have established the power factor problem by mathematical analysis for a DC drive conducting a continuous armature current. The DC/DC converter for the maximum power tracking also serves as a loader to maintain the energy of the storage battery group. The bilateral inverter, connected to the network, is used for both charging and discharging. This allows the vehicle to act as a temporary energy storage unit, providing energy to the grid during the day and charging from it at night. This also ensures that the battery's minimum charge level is maintained when the solar (PV) power is insufficient Also, a parallel function of the energy factor function (PPFc) can be completed through the checkpoint field of the inverter. The PV System submitted has simplified exercise and lower costs while offering as many features as possible. A power management algorithm is simulated in MATLAB/Simulink software with a DSP controller to manipulate power relationships between the PV system, battery group, and use.

In the present study, a simulation model of the single-phase AC–DC is undertaken to establish the poor power factor of the drive. The choice of this system has been affected by the fact that it presents the worst form of harmonics, it has a wide range of applications, and it is used in low power-drive systems. The simulation presents results for the input current, ripples, and total harmonic distortion (THD). Also, from the simulation, the behavior factors can be derived. Experimental results obtained validate the simulation results.

 DOI: 10.1201/9781003642152-11

11.2 DRIVE ANALYSIS

The circuit of a DC drive conducting in a continuous armature current conduction mode is depicted in Figure 11.1, and its associated voltage and current waveforms are in Figure 11.2

A mathematical analysis of the asymmetric single-phase drive can be found in Rahman et al. (2022) and Mishra et al. (2024). For,

$$P = mV_sI_{s_1}\cos\varphi_{s_1} \tag{11.1}$$

$$S = mV_sI_s \tag{11.2}$$

The behavior factors are summarized as.

Input Power Factor $\left(\mathrm{PF_{ac}}\right)$:

$$PF_{ac} = \frac{\sqrt{2}\left[1+\cos\alpha\right]}{\left[\pi\left(\pi-\alpha\right)\right]^{\frac{1}{2}}} \tag{11.3}$$

Harmonic Factor (HF):

$$HF = \left[\frac{\pi\left(\pi-\alpha\right)}{4\left(1+\cos\alpha\right)} - 1\right]^{\frac{1}{2}} \tag{11.4}$$

Equations (11.3) to (11.4) were simulated in MATLAB/Simulink software, and the simulation results are depicted in Figure 11.3. It shows a graphical relationship between the behavior factors and the firing angle of the converter.

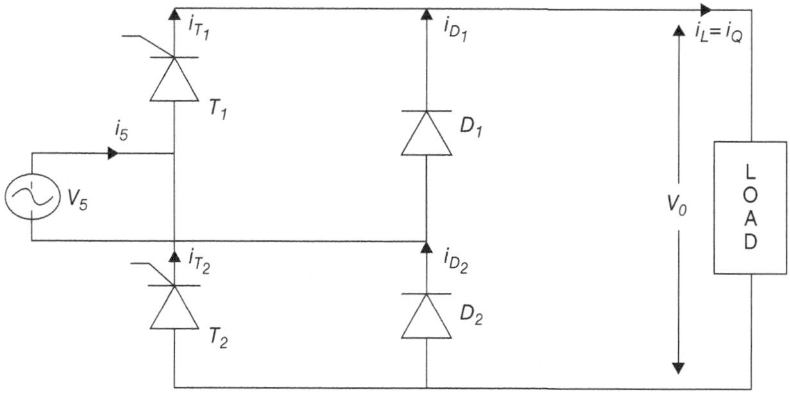

Figure 11.1 Circuit configuration of a single-phase AC–DC drive.

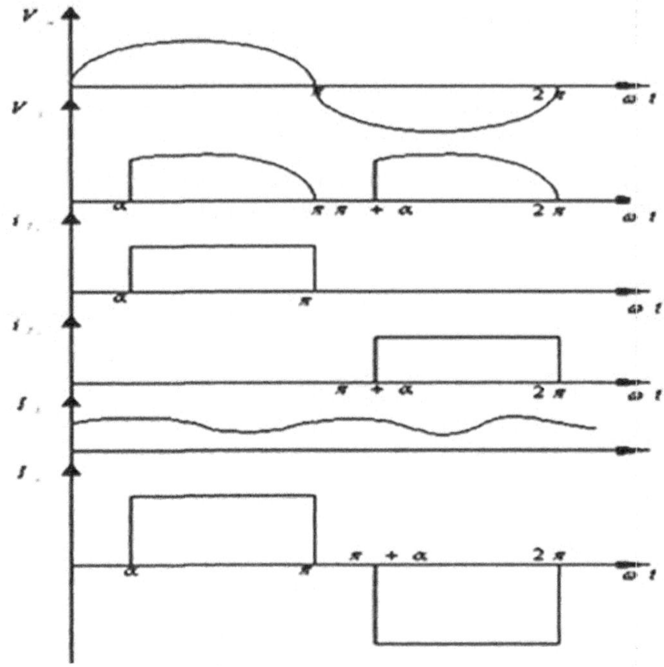

Figure 11.2 Output waveform for voltage and load current for phase angle control.

11.3 SIMULATION STUDIES

The circuit diagram of Figure 11.1 was simulated on MATLAB/Simulink software with the following machine parameters at 230 Vrms and a switching frequency of 50 Hz.

Load Resistance $R_s = 0.08\Omega$
Load Inductance $L_s = 11\mu H$
Motor parameters: $K_t = 0.56$
Converter gain $K_e = 0.055$
Rms supply current $I = 45A$
Moment of Inertia $J = 16m^4$

The simulation results show the variation of harmonic content of AC supply, AC input current and voltages, and the armature speed.

Table 11.1 was constructed using the simulation results and the equation for power factor and displacement factor of equations (11.3) and (11.4), respectively.

11.4 SIMULATION ANALYSIS

A computer program was written in MATLAB using the results of Table 11.1 to simulate the behavior factors. They are displayed in Figures 11.4 and 11.5.

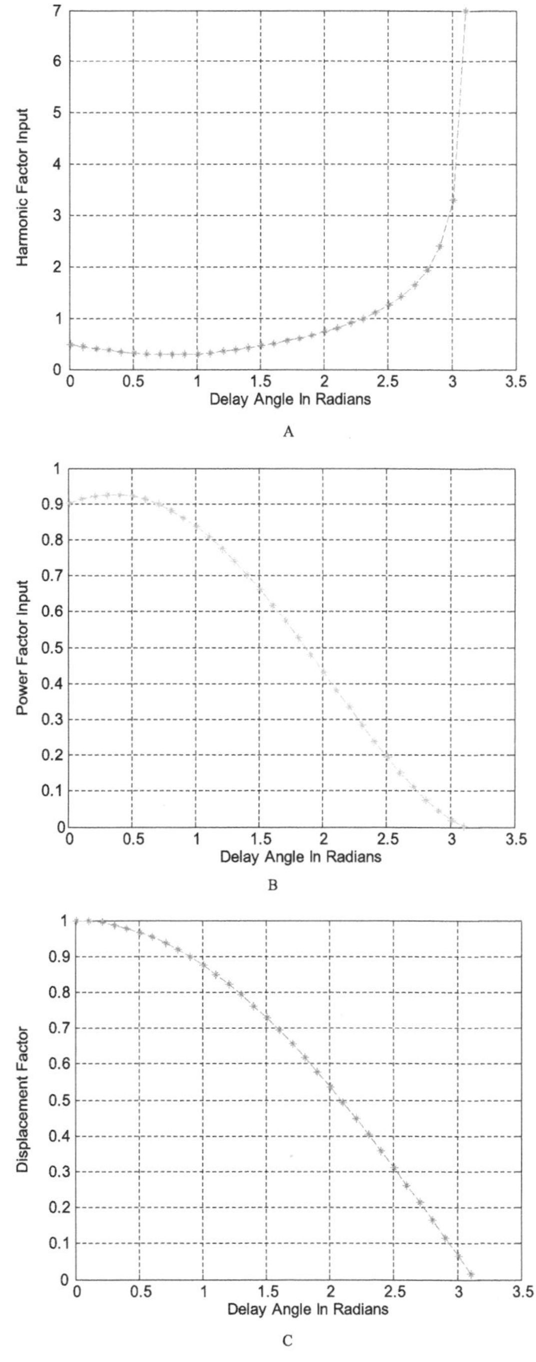

Figure 11.3 Bridge converter performance with different parameters.

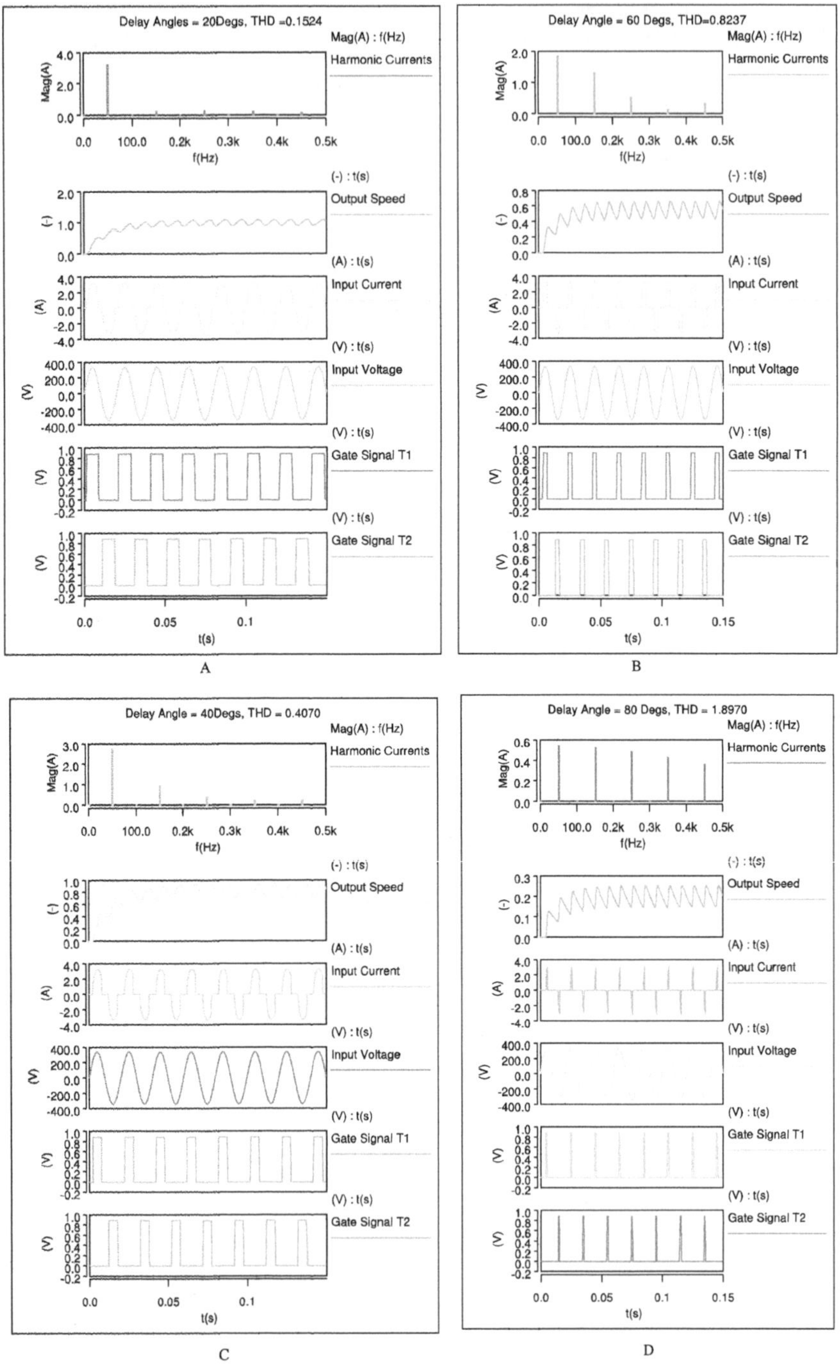

Figure 11.4 Simulation results of the single-phase asymmetric bridge drive.

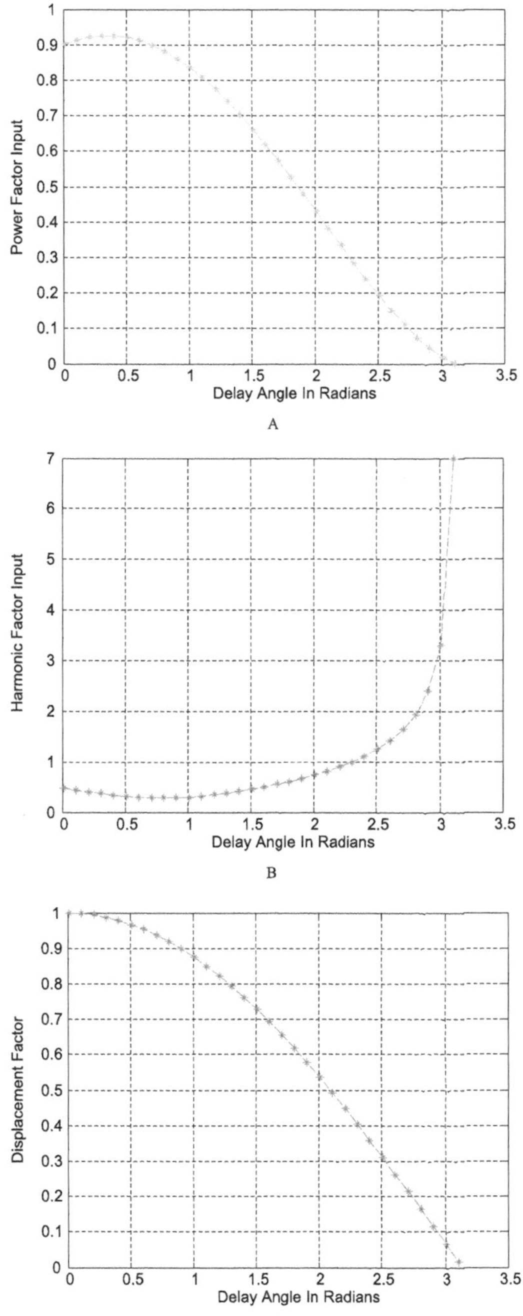

Figure 11.5 Graphical representation of simulated input power factor, total harmonic distortion, and displacement factor.

Table 11.1 Variation of THD, PF, and DF with delay angles

Firing Angle "α" in (Degs.)	*Firing Angle "α" in (Radians)*	*Input (THD)*	*Input (PF)*	*Displacement Factor (DF)*
5^0	0.08727	0.03861	0.9997	0.9999
10^0	0.17453	0.05931	0.9942	0.9960
15^0	0.26180	0.10160	0.9860	0.9910
20^0	0.34907	0.15240	0.9730	0.9848
30^0	0.52360	0.25520	0.9360	0.9660
40^0	0.69813	0.40700	0.8706	0.9400
50^0	0.87267	0.58480	0.7855	0.9100
60^0	1.04720	0.82370	0.6684	0.8660
70^0	1.22173	1.20400	0.5133	0.8190
80^0	1.39626	1.89700	0.3572	0.7660
90^0	1.57079	4.443600.	0.1552	0.7070

11.5 CONCLUSION

It is evident from the characterization that the plots of the behavioral factors obtained from simulation are in complete agreement with those obtained from mathematical analysis. Obviously, the input power factor issue exists for the asymmetrical source configuration, and it decreases with an increase in the angle of pulse for the converter. The worst PF is seen in the input current waveform as a result of THD in the AC supply, which makes it a non-sinusoidal waveform and out of phase with the input voltage waveforms. One point to note is that a decrease in power factor leads to an overdamped output speed oscillation, which results in instability of its operating system. Also, increased power factor reduces reactive load energy efficiency and conservation.

REFERENCES

Ali, W., H. Farooq, M. Jamil, A. U. Rehman, R. Taimoor and M. Ahmad, (2018) "Automatic Power Factor Correction for Single Phase Domestic Loads by Means of Arduino Based TRIAC Control of Capacitor Banks", *2018 2nd International Conference on Energy Conservation and Efficiency (ICECE)*, pp. 72–76.

Anand, E. P. and M. K. Sajina, (2020) "Single Phase Automatic Power factor Corrector with Microcontroller Based Capacitor Bank Switching", *2020 International Conference on Futuristic Technologies in Control Systems & Renewable Energy (ICFCR)*, pp. 1–4, 2020.

Basau, S. and Bollen M.H.J. (2005) "A novel common power factor correction scheme for homes and offices". *IEE Trans. Power Delivery*, pp. 2257–2263.

Callies, M., R. Ulmer, J. McHugh and Z. Reed, (2022) "Case Studies: New Options for Power Factor Correction," *2022 IEEE Rural Electric Power Conference (REPC)*, Savannah, GA, USA, 2022, pp. 60–64.

Daut, Ismail, Rosnazri Ali and Soib Taib. (2006) "Design of a Single-Phase Rectifier with Improved Power Factor and Low THD using Boost Converter Technique". *American Journal of Applied Sciences* 3(7), pp. 1902–1904.

Holmes, G.D. and T.A. Lipo, (2003) *Pulse Width Modulation for Power Converters. Principles and Practice*, IEEE Press.

Lo, Yu-Kang, Huang-Jen Chiu, Sheng-Yuan Ou, (2000) "Criteria for Setting the Reference DC Voltage of an Active Power Line Conditioner," *Proceedings of the 9th International IEEE Conference on Harmonics and Quality of Power, ICHQP IX*. pp. 658–660.

Lo, Yu-Kang, Jin-Yuan Lin and Tin-Yuan Wu, (2005) "Grid-Connection Technique for a Photovoltaic System with Power Factor Correction," *2005 International Conference on Power Electronics and Drives Systems*, Kuala Lumpur, Malaysia, pp. 522–525.

Mane, S., R. Sapat, P. Kor, J. Shelar, R. D. Kulkarni and J. Mundkar, (2020) "Microcontroller based Automatic Power Factor Correction System for Power Quality Improvement", *2020 International Conference for Emerging Technology (INCET)*, pp. 1–6.

Mishra, D. P., S. P. Padhi, B. Pati and H. Mohapatra, (2024) "Power Factor Improvement Using Noble Techniques," *2024 3rd Odisha International Conference on Electrical Power Engineering, Communication and Computing Technology (ODICON)*, Bhubaneswar, India, pp. 1–6.

Rahman, M. S., A. Memy, M. A. Mahmud and S. Siddique, (2022) "Automatic Power Factor Measurement And Improvement Using Capacitor Bank," *2022 IEEE International Power and Renewable Energy Conference (IPRECON)*, Kollam, India, pp. 1–6.

Uddin, M. M., A. Al Mahmud and N. Islam, (2019) "Design & Implementation of a Microcontroller Based Automatic Power Factor Rectification System for Different Loads", *2019 1 st International Conference on Advances in Science Engineering and Robotics Technology (ICASERT)*, pp. 1–6.

Chapter 12

Feasibility study on the implementation of Industry 4.0-based MHS for pipe manufacturing industry

Abhishek Bhandari, T. Ravi Kiran, Tejkaran Narolia, and K. Viswanath Allamraju

12.1 INTRODUCTION

Industry 4.0 is the transition from conventional machine-driven manufacturing to digital manufacturing, where machines are seamlessly connected utilizing ICT [1, 2]. Technological developments have raised consumer expectations for openness in the production process and the availability of information [3–5]. Real-time data exchange is being made possible by cloud-based systems and ontology-based data standards, which are replacing traditional file-based techniques in the data interchange space [5]. Data integration is a critical first step toward the effective deployment of Industry 4.0 technology [6]. The term "Industry 4.0" has been studied in great detail, but because it is used in a variety of contexts and encompasses concepts like competitiveness, sustainability, and smartness, it has proven difficult to define consistently and assess its wide range of applications [7, 8]. Technology-driven ways to increase productivity are becoming increasingly appealing [9, 10]. Material handling accounts for 20–35% of the overall cost of a manufactured good and makes up a sizable portion of an organization's internal activity [11]. Additionally, material handling typically requires 80% of manufacturing time and 55% of all industrial area [12]. More effective material management inside the company can lead to more efficient manufacturing and distribution processes. [13].

12.2 LITERATURE REVIEW

The rising volume of literature on the advantages of Industry 4.0 further supports the importance of these technologies to industry performance, as does the government's increased attention on technological adoption and upgrades for businesses and industries [14–17]. Two domains that were relevant to this study project were the subjects of the literature review. The data acquired from issues related to Material Handling Systems and Industry 4.0.

 DOI: 10.1201/9781003642152-12

It contributed to the development of the research's theoretical foundation. The initial goal of the literature review was to produce preliminary ideas for the material handling system in relation to Industry 4.0. It has been discovered that the manufacturing sector is currently facing several significant challenges during Industry 4.0 adoption, including a lack of funding uncertainties [18], inadequate technological infrastructure, cybersecurity risks, a lack of standards, and poor data quality [19–22]. With this paradigm shift brought about by Industry 4.0, there's a good chance that new hazards pertaining to technology, organizations, the environment, and society politics may surface.

The primary IoT technologies connected to automated support of logistics business activities were examined by Ferreira et al. [23]. They present RFID- and WSN-enabled smart products. After that, they concentrate on IoT-based runtime and design support for modifications that take dynamic shifts in business processes into account. IoT from wireless communication technologies and smart logistics information processing (e.g., business operations) are the only topics included in this survey. It disregards further relevant uses, issues, and the potential paths of IoT in intelligent logistics. The potential and emerging requirements for cyber-physical logistics systems based on the IoT were discussed by Prasse et al. [24]. This article describes the concept of CPLS, lists the basic technology layers, and illustrates how CPLS is developed using a distribution center as a real-world example.

Problems are presented in this chapter to highlight the gap and provide evidence for the need to find solutions. First, a consideration of the issues surrounding the challenges associated with material handling in Industry 3.0. Material handling experienced a number of difficulties in Industry 3.0, which is defined by the use of automation and computerization in manufacturing. These issues were resolved with the introduction of Industry 4.0.

12.3 RESEARCH METHODOLOGY

The development of a material handling system framework, which can be used as a guide to choose the optimal tactics, is necessary to achieve these study goals. To create such a framework, researchers often use one of two approaches: top-down or bottom-up. While the bottom-up technique starts with an existing plant system, the top-down approach starts with determining the needs. In this study, a bottom-up methodology is used. A material handling framework is created with the purpose of achieving this research goal. Each layer of the Industry 4.0 paradigm is mapped to a distinct area within the material handling activity. Every component of the material handling framework has to be able to communicate with the layers next to it. Furthermore, different approaches can be used for each layer depending on the needs of the application. A complicated mapping process may be

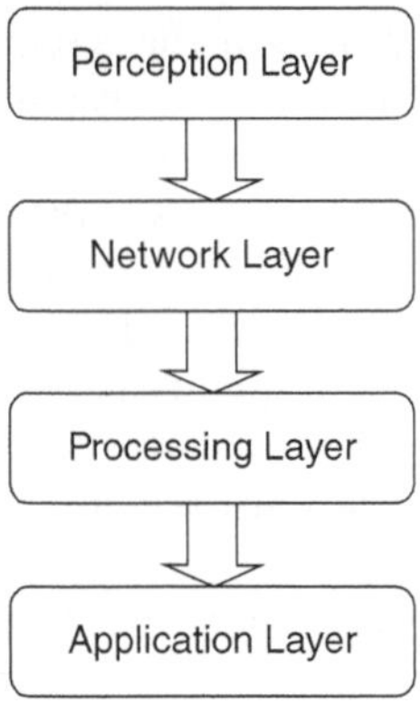

Figure 12.1 Flow chart of research methodology.

produced by combining various material handling techniques with Industry 4.0 layers (Figure 12.1).

12.3.1 Data collection

The stages listed in Table 12.1 form the basis of the data extraction approach. Every one of these processes is targeted at a certain kind of data.

When trucks carrying raw materials from suppliers arrive at RPVC pipe manufacturing enterprises in India, the procedure of receiving items begins. The chart in Figure 12.2 shows the number of trucks that arrived per shift.

A total of 8–10 laborers unload and handle the received material. PVC resin bags are stored in the godown. Goods receiving is done manually from bulk transporter to storage.

Handling is required in other processes, like master batch preparation. The chart in Figure 12.3 shows the frequency of master batch preparation.

The next chart in Figure 12.4 shows the production, wastage, and number of machines run during a day.

Table 12.1 Data collection techniques

S. No.	*Data Collection Techniques*
1	Layout of the plant
2	Operation Process Chart
3	Flow Process Chart
4	Production Per Day
5	Material Handling Checklist
6	Raw material received per shift and day

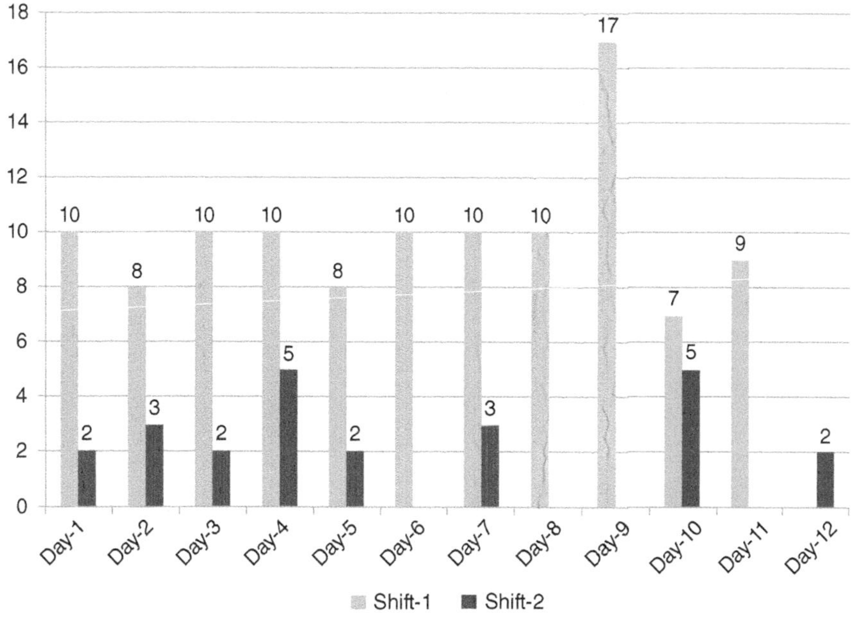

Figure 12.2 Flow of incoming trucks per shift.

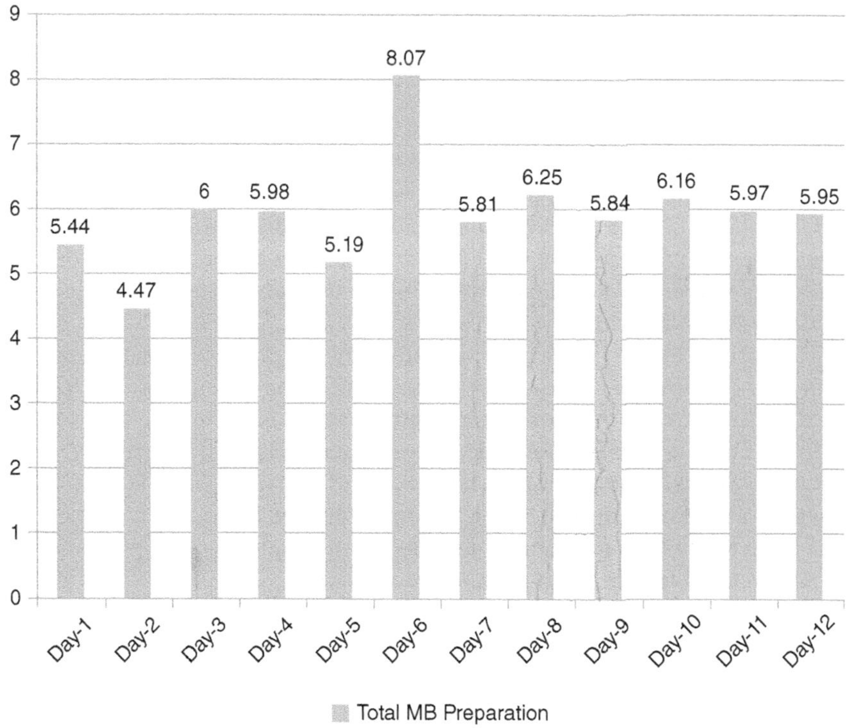

Figure 12.3 Master batch preparation per day.

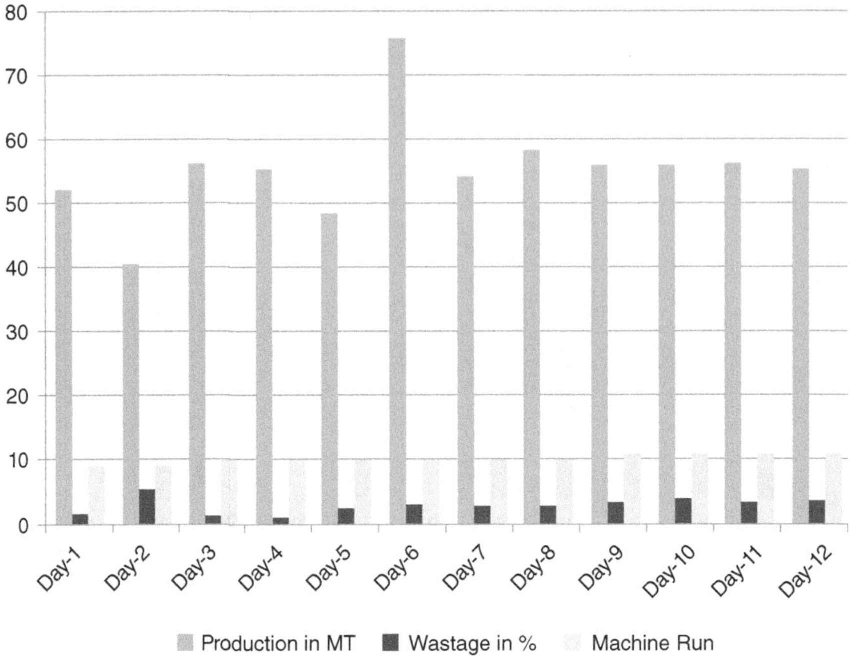

Figure 12.4 Productions per day.

12.4 ANALYSIS OF DATA COLLECTION

First of all, the extraction of relevant information from multiple data sources is identified as the research work that needs to be tackled. A reliable and accurate material handling framework is analyzed. Second, in this research, MHS data is collected with the specified number of architecture layers based on research requirements. Each layer includes different techniques for achieving the purpose of the corresponding layer. Then, each technique can be deployed on the cloud.

The current material flow is drawn after the necessary information is collected. The following charts, like outline process charts and flow charts, are used to illustrate the current material flow (Figure 12.5).

In the current material flow, the items chosen include all existing ways of goods receiving, storage, manufacturing, and delivery. For the first layer, physical assessment and the critical component of MHS are identified. In this layer, the bottleneck condition of the corresponding component and the sensors related to that equipment are identified.

The IoT is utilized in this study to link various gadgets, which may be connected to one another over any network. Certain areas may be home to the linked gadgets. A vast quantity of data that is processed and stored in a centralized location, like the cloud, may be produced and shared by the linked devices. The linked gadgets generate a vast quantity of data. These gadgets generate data that must be managed methodically.

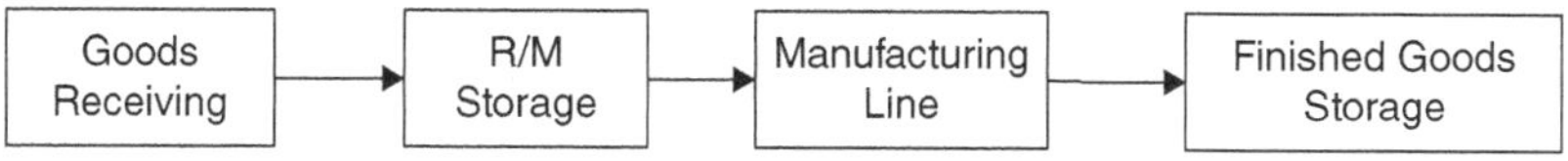

Figure 12.5 Current material flow at RPVC pipe manufacturing industries in India.

12.4.1 Step 1 Perception layer

The perception layer is the lowest layer in the IoT architecture. The perception layer's primary duty is to gather information from various devices, process it, and digitize it. Additionally, it moves the processed data into the IIoT architecture's higher tiers. Various kinds of sensors are affixed to things to gather data.

12.4.2 Step 2 Transport layer

End-to-end network communication is handled by the transport layer. For smooth data packet delivery, it also incorporates characteristics like data integrity, flow management, traffic control, connection-oriented delivery, same-order delivery, dependability, congestion detection and avoidance, multiplexing, and byte orientation. This layer in the IIoT is in charge of tying networks, smart objects, and network devices together, which makes it extremely vulnerable to security concerns pertaining to the authenticity and integrity of data being sent over the network.

12.4.3 Step 3 Processing layer

Together with hardware platforms and clever algorithms, like the cloud platform, the processing layer offers a facility for data access, storage, and processing.

12.4.4 Step 4 Application layer

IIoT's primary goal is to guarantee efficient communication between items and create a long-lasting connection between them through a variety of applications. Choosing a selection of protocols and ensuring reliable communication between various apps are the responsibilities of the application layer.

12.5 RESULTS AND DISCUSSIONS

When the same materials are received in the same cycle time, the handling may be both easier and faster. The advantages of the proposed solution are less transportation and the requirement of less effort. Laborers can put the raw material bags one by one in the storage area.

By fully using information from material handling activities, IoT demonstrates its enormous potential to enhance the material handling system. IoT helps to identify the accurate tracking of material handling equipment, delay and non-value-added activity elimination, cost reduction, and monitoring inventory.

IoT helps to maintain accurate tracking of material handling equipment in the industry. Depending on the kind of product, the material handling system uses sophisticated sensors and tracking devices to always monitor its movements. They can obtain data on whether the cars are following the suggested path. This then makes it easier for them to respond to the operator about delivery delays, wherever they may be. Using IoT services, a fleet vehicle's driver may notify other vehicles on the same route if their vehicle becomes stuck. It also reduces unnecessary delay elimination. Product delivery delays are the main issue with any material handling system. It is simpler to share traffic and weather information in real time when material handling systems are connected via IoT-enabled devices. It helps to reduce production costs for the companies. Despite the implementation of technology in warehouses to keep track of inventory stock, material handling systems still rely heavily on human labor and observation. Inventory management is enhanced by the use of RFID and CCTV cameras connected by IoT-enabled devices, which provide digital and accurate surveillance around the clock. Businesses may always be informed about the inventory levels in warehouses. This enables them to more effectively satisfy changing client requests. The business may accumulate more inventory before a product runs out in the warehouse, maintaining a balance between supply and demand from customers.

12.6 CONCLUSION

The present study is carried out for a critical study of material handling analysis at RPVC Pipe Manufacturing Companies in India based on material handling principles. During the study of industry, a material handling checklist, flow process chart, outline process chart, plant layout, and daily production chart were constructed. The gathered information is used to compare the different plants with each other. It is found that the system used at RPVC pipe manufacturing companies in India is not familiar with the material handling principles. Based on material handling principles, a suggestion has been proposed. Moreover, solutions to eliminate the identified waste and balance the material handling processes are suggested. The recommended changes will help RPVC pipe manufacturing companies in India to decrease time and costs, enhancing productivity. Also, IoT has been introduced as a key subject for Industry 4.0, where its application allows for reduced downtime and a consequent improvement in productivity and reduced production cost. MHS is now changing in tandem with the industrial market; manufacturers who do not adopt innovative practices risk falling behind in the new industrial market.

12.7 FUTURE SCOPE

A similar study could be done for other companies working in a similar sector. A more detailed value stream mapping could be done in the future, including the information flow and measuring lead times. Scenario simulation could be done in order to see how changes in critical areas could affect the company.

REFERENCES

1. E. Oztemel, S. Gursev, Literature review of Industry 4.0 and related technologies, *J Intell Manuf*, 31 (1) (2020), pp. 127–182.
2. M. Budler, B.F. Quiroga, P.A. Trkman, Review of supply chain transparency research: Antecedents, technologies, types, and outcomes, *J Bus Logist*, 45 (1) (2024), Article e12368.
3. V. Harish, R. Sharma, R.K. Saini, D. Negi, Intelligent supply chain orchestration: A framework for seamless integration of Industry 4.0 technologies *2023 seventh international conference on image information processing*, ICIIP, IEEE (2023), pp. 265–270.
4. M. Igamberdiev, G. Grossmann, M. Selway, M. Stumptner, An integrated multi-level modeling approach for industrial-scale data interoperability, *Softw Syst Model*, 17 (1) (2018), pp. 269–294.
5. A. Villalonga, G. Beruvides, F. Castano, R.E. Haber, Cloud-based industrial cyber–physical system for data-driven reasoning: A review and use case on an Industry 4.0 pilot line IEEE *Trans IndInf*, 16 (9) (2020), pp. 5975–5984.
6. S. Janík, P. Szabó, M. Mĺkva, M. Mareček-Kolibiský, Effective data utilization in the context of Industry 4.0 technology integration, *Appl Sci*, 12 (20) (2022), p. 10517.
7. A. Adamik, D. Sikora-Fernandez Smart organizations as a source of competitiveness and sustainable development in the age of Industry 4.0: Integration of micro and macro perspective, *Energies*, 14 (6) (2021), p. 1572.
8. H. Ullah, M. Uzair, Z. Jan *et al.*, Integrating industry 4.0 technologies in defense manufacturing: Challenges, solutions, and potential opportunities. Array (2024).
9. K. Fukuda, Science, technology and innovation ecosystem transformation toward society 5.0 *Int. J. Prod. Econ.*, 220 (2020), Article 107460.
10. Dilupa Nakandala, The role of lean, agility and learning ambidexterity in Industry 4.0 implementations, Technological Forecasting and Social Change Volume 206, September 2024, 123533.
11. Theodore H. Allegri, Sr.; 1987 *Material Handling Principles and Practice*, CBS Publishers and Distributors, New Delhi.
12. Telsang Mertand; 2007 *Industrial Engineering and Production Management*, S. Chand & Company Ltd, New Delhi.

13. James M. Apple; 1972, "*Material Handling systems Design*"; John Wiley and Sons, New York.
14. L.S. Dalenogare, G.B. Benitez, N.F. Ayala, A.G. Frank The expected contribution of Industry 4.0 technologies for industrial performance *Int. J. Prod. Econ.*, 204 (2018), pp. 383–389.
15. A.B.L. de Sousa Jabbour, C.J.C. Jabbour, T.-M. Choi, H. Latan 'Better together': evidence on the joint adoption of circular economy and industry 4.0 technologies *Int. J. Prod. Econ.*, 252 (2022), Article 108581.
16. D. Nakandala, R. Yang, H. Lau, S. Weerabahu Industry 4.0 technology capabilities, resilience and incremental innovation in Australian manufacturing firms: a serial mediation model *Supply Chain Manag. Int. J.*, 28 (4) (2023), pp. 760–772.
17. D. Bianco, A. Bueno, M. GodinhoFilho, H. Latan, G.M.D. Ganga, A.G. Frank, C.J.C. Jabbour The role of Industry 4.0 in developing resilience for manufacturing companies during COVID-19 *Int. J. Prod. Econ.*, 256 (2023), Article 108728.
18. G. Büchi et al Smart factory performance and Industry 4.0 *Technological Forecasting and Social Change* (2020).
19. A.R. Bakhtari et al Evaluating Industry 4.0 Implementation Challenges Using Interpretive Structural Modeling and Fuzzy Analytic Hierarchy Process *Cybernetics and Systems* (2021).
20. S. Gabriel *et al.* Analyzing socio-technical risks in implementation of Industry 4.0-use cases *Procedia CIRP* (2021).
21. R. Gadekar et al Investigating the relationship among Industry 4. 0 drivers, adoption, risks reduction, and sustainable organizational performance in manufacturing industries: An empirical study *Sustainable Production and Consumption* (2022)
22. J.P. Tamvada et al Adopting new technology is a distant dream? The risks of implementing Industry 4.0 in emerging economy SMEs *Technological Forecasting and Social Change* (2022).
23. P. Ferreira, R. Martinho, and D. Domingos, "IoT-aware business processes for logistics: Limitations of current approaches," in *Proc. 16th Annu. Conf. INFORUM*, May 2010, pp. 611–622.
24. C. Prasse, A. Nettstraeter, and M. Hompel, "How IoT will change the design and operation of logistics systems," in *Proc. Int. Conf. Internet Things (IoT)*, Cambridge, MA, Oct. 2014, pp. 55–60.

Chapter 13

Fixed-bed column studies for removal of contaminants from synthetic wastewater

Amit Kharwade, Shalini Yadav, Jaicky H. Gurnani, Buddharatna Godboley, Nitesh Thikare, Mayank Gupta, and Meghna Jumde

13.1 INTRODUCTION

One of the biggest problems of the twenty-first century is pollution brought on by human and industrial activity (N. M. Y. Almahbashi et al. 2021). The term "environmental pollution" refers to a deficiency in an ecosystem's environmental processes brought on by contamination of the soil, the atmosphere, or the aquatic system (A. Ghaedi et al. 2015). But one of the most prevalent pollutants in the environment is heavy metals, especially copper and cadmium, which are hazardous and bad for the ecosystem in large quantities because they are carcinogenic and non-biodegradable (R. Muhammad et al. 2016; S. Chen et al. 2012).

Naturally occurring fluorine makes up 6% to 9% of the earth's layers on the surface. Fluoride metal is commonly found in the earth's layers in a variety of mineral compositions, including fluorite, phosphate rocks, cryolites, apatites, and mica. Of these, fluorite, apatite, and mica are the most commonly occurring forms (A. Satya et al. 2021; John et al. 2016). The WHO recommends that the maximum acceptable level of fluoride in water be 15 parts per million. Nonetheless, 0.0005 and 0.0010 g/L as limits have been established by the Bureau of Indian Standards (BIS) (Kumar Rajeev et al. 2020). The potential corrective action is a serious issue.

The amount of heavy metals in wastewater is constantly rising; thus, many methods are employed to get rid of them, including membranes, sophisticated oxidation processes, magnetic fields, adsorption with both natural and manufactured adsorbents, and electrocoagulation (EC). Selecting the most appropriate and effective methods, however, is crucial and dependent on numerous variables (F. Feizi et al. 2021). Despite the high cost of producing granular and powdered activated carbon (M. Basu et al. 2017), adsorption is an effective way to remove heavy metals. As a result, a promising method for getting rid of a variety of pollutants is using adsorbents (J. Liu et al. 2018). However, a variety of waste products have been utilized as precursors to create activated carbon, including wood, sawdust, tea leaves, nutshells (M.L. Soto et al. 2017), bagasse (M. Shanmugaprakash et al. 2018), and wood. The powdered seeds and leaves of *Moringa oleifera*

DOI: 10.1201/9781003642152-13

are useful as a powerful coagulant when treating water or wastewater in a practical, affordable, and natural manner (Eman Noori Ali 2020). Fluoride removal from aqueous solutions is being replaced by innovative adsorption techniques that use agricultural waste and natural adsorbents. Compared to *M. oleifera* leaf powder treated with alkali, the powder treated with acid demonstrated a higher ability for adsorption (Shabnam Dan et al. 2018). For extremely turbid, untreated, and diseased surface water, ground seeds from moringa trees can be used to make a soluble extract in water that can be a helpful natural clarifying agent (Chepkwony et al. 2019).

The adsorption process is typically conducted in moving beds, fluidized beds, fixed-bed columns, or batch systems. Nonetheless, a fixed-bed column is recommended in ongoing research because of its low cost, high efficiency, and practicability (N. Blagojev et al. 2019). A breakthrough curve is used in column studies to characterize the breakthrough process (C. M. Hasfalina et al. 2012). The adsorption experiment data are used to model breakthrough curves. Furthermore, additional models were developed to investigate the breakthrough profiles and clarify the dynamic adsorption behavior of heavy metal ions, such as the Adams-Bohart models, Yoon-Nelson models, and Thomas models (H. T. Van et al. 2019). This work aims to evaluate the effectiveness of activated carbon derived from natural-based media in removing fluoride ions through aqueous solutions under various operating conditions in column experiments.

13.2 MATERIAL METHODOLOGY

13.2.1 Adsorbent preparation

The adsorbent in this study was natural-based activated carbon, which had been prepared in a prior study. Analytical-grade chemicals were used to prepare stock solutions (10 ppm) of fluoride using sodium fluoride (NaF), which were then diluted to the desired concentrations (R. Muhammad et al. 2016).

13.2.2 Column study

A downflow fixed-bed column with a 6.5 cm diameter and an activated carbon pack at 6 cm and 10 cm depth was used for the column investigation. Glass wool was utilized at the column's base to keep the drainage area from becoming clogged with activated carbon leaks. Additionally, it was positioned above the adsorbent bed to keep the flow constant and evenly distribute the solution across the adsorbent surface. Fluoride solution starting concentrations were set at 10 mg/L. The pH was reduced to 5.0 using sodium hydroxide and hydrochloric acid. The column experiment was conducted under ideal batch adsorption conditions. Using a peristaltic pump, the solution was continuously poured into the column at a rate of 12 and 15 milliliters per minute in a downward flow direction. Wastewater was collected every five minutes for the first hour and then every half hour after that until the exhaust point was achieved in the effluents' metal ion concentration.

13.3 EQUILIBRIUM ISOTHERM MODELS

Adsorption isotherms show how the concentration of a material in solution at a certain temperature affects the amount of material adsorbed on a solid surface. Understanding how adsorption works and creating effective solutions to maximize the process depend on these models. The Freundlich isotherm and Langmuir models were used to examine the experimental equilibrium data. These are most commonly used in adsorption studies (Table 13.1).

13.4 RESULT AND DISCUSSION

An adsorption experiment using natural carbon made from moringa seeds at 25°C is shown in Figure 13.1 as a Freundlich isotherm plot. The logarithm of the adsorbate's equilibrium concentration, or log Ce, is plotted on the x-axis, while the logarithm of the amount adsorbed per adsorbent mass, or log qe, is plotted on the y-axis. According to the Freundlich isotherm model equation, $\log qe = \log KF + (1/n)\log Ce$, the linear trend that was seen makes sense.

Here, the slope of the line is 0.8864, corresponding to the value of $1/n$. Since this value is less than 1, it shows the adsorption process is favorable. Y-intercept is –0.4194, which corresponds to $\log KF$. Taking the antilog of this intercept yields a KF value of approximately 0.38, representing the adsorption capacity of the moringa-based carbon. The coefficient of determination (R^2) is 0.8168, which suggests a reasonably strong agreement between the experimental value and the Freundlich model. This implies that the adsorption occurs on a heterogeneous surface and likely involves multilayer adsorption.

The Langmuir isotherm plot for the adsorption of a material using natural carbon from moringa seeds at 25°C is shown in Figure 13.2. The x-axis in this plot represents the reciprocal of the adsorbate's equilibrium concentration $(1/Ce)$, and the y-axis represents the reciprocal of the amount of adsorbed adsorbent per unit weight $(1/qe)$. According to this concept, there are only a few similar active spots on a homogenous surface where adsorption occurs, each of which can only hold a single molecule, indicating monolayer adsorption. The adsorption behavior in this instance appears to be

Table 13.1 Comparison of adsorption equilibrium isotherm model [Wang and Guo, 2020]

Model	*Equation*	*Linear Form*	*Assumptions*	*Constants*	*Interpretation*
Langmuir isotherm model	$q_e = \left(q_{ax}{}^{*}b^{*}C_e\right)/\left(1+b^{*}C_e\right)$	$C_e/q_e = 1/\left(q_{ax}{}^{*}b\right) + C_e/q_{ax}$	Monolayer adsorption on a homogeneous surface	q_{ax}, b	For homogeneous surfaces, it accurately predicts the maximal adsorption capacity
Freundlich isotherm model	$q_e = K\_F^{*}C_e \wedge (1/n)$	$\log q_e = \log K\_F + (1/n)\log C_e$	Adsorption of several layers on a heterogeneous surface	K_F,1/n	Empirical indicates surface heterogeneity and adsorption intensity
Temkin isotherm model	$q_e = B\ln\left(A^{*}C_e\right)$	$q_e = B\ln C_e + B\ln A$	As surface coverage increases, the heat of adsorption falls linearly	A, B = RT/b_T	Considers adsorbate–adsorbent interactions and variation in adsorption energy
Dubinin–Radushkevich (D–R)	$q_e = q^{*}\exp\left(-B^{*}\varepsilon^2\right)$	$\ln\ q_e = \ln\ q - B^{*}\varepsilon^2$	Porous adsorption and energy-based model	q, B, ε	Distinguishes between physical and chemical adsorption mechanisms

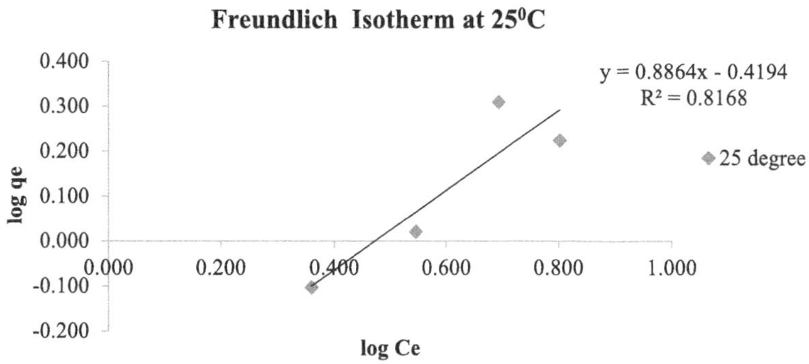

Figure 13.1 Concentration of the adsorbate (log Ce) versus adsorbent per unit weight (log qe).

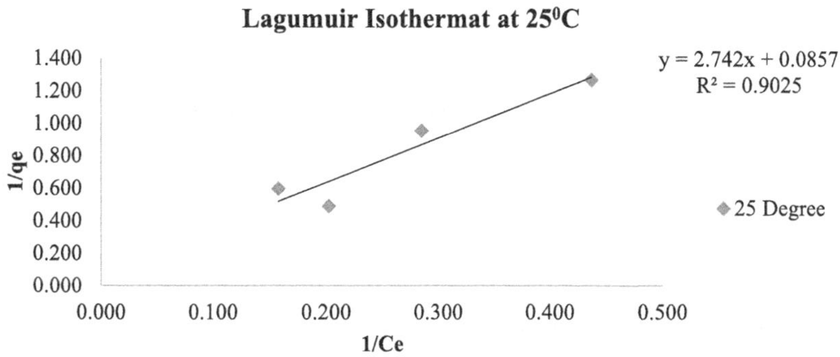

Figure 13.2 Reciprocal of the concentration of the adsorbate (1/Ce) versus adsorbed adsorbent per unit weight (1/qe).

well described by the Langmuir isotherm model, as indicated by the strong linear fit shown by the R^2 value of 0.9025. With little contact between adsorbed molecules, adsorption occurs as a monolayer on a homogenous surface, according to this finding.

Figure 13.3 shows a Temkin isotherm plot for the adsorption process using natural carbon from moringa seeds at 25°C. The x-axis of the plot shows the amount of adsorbate absorbed per gram of adsorbent (qe), while the y-axis shows the equilibrium concentration's natural logarithm (ln Ce). The Temkin isotherm model implies that the heat of adsorption progressively drops as the surface becomes increasingly covered and takes into consideration the adsorbent and adsorbate's interactions. A moderate degree of agreement between the Temkin isotherm and the experimental data is shown by the R^2 value of 0.7325. The Temkin model nonetheless provides valuable information about the energy changes and interaction behavior during

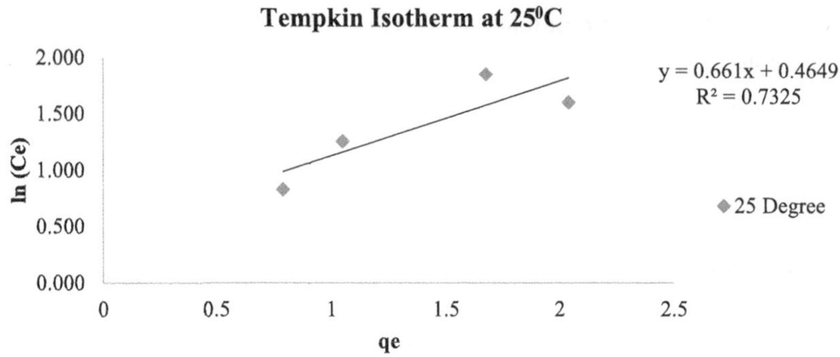

Figure 13.3 Adsorbed adsorbate per gram of adsorbent (qe) versus logarithm of the adsorbate equilibrium concentration (lnCe).

adsorption, despite the fact that it does not fit this situation as well as the Langmuir or Freundlich models.

13.4.1 Comparison of the equilibrium isotherm model

Using the Langmuir isotherm models, Freundlich isotherm models, and the Temkin isotherm models, the adsorption behavior of natural carbon from moringa seeds was examined at a temperature of 25°C. With their distinct viewpoints on the adsorption process, each model contributes to an enhanced comprehension of the nature of adsorption on the carbon surface derived from moringa.

The Langmuir isotherm assumes monolayer adsorption on a homogeneous surface with a finite number of identical adsorption sites and no interaction between adsorbed molecules. The strong linear relationship and R^2 value of 0.9025 in the Langmuir plot demonstrate how well this model fits the experimental data. Calculating the maximum adsorption capacity (q_{max}), which is approximately 11.67 mg/g, is made possible by the line's slope and intercept. This suggests that the carbon in moringa seeds has a surface that is comparatively homogeneous and that it facilitates monolayer adsorption with little interaction between adsorbed molecules. Given the circumstances, the Langmuir model is the most suitable for explaining the adsorption process, as indicated by the high R^2 value.

However, the Freundlich isotherm model is empirical and presupposes that adsorption takes place on a heterogeneous surface with a variety of adsorption sites and energies, enabling multilayer adsorption. The Freundlich model's graph, which has an R^2 value of 0.8168, likewise displays a respectably excellent match. Because favorable adsorption is typically reflected by values of $1 / n$ less than 1, the plot's slope, which corresponds to $1 / n$, is 0.8864. The antilog value of the intercept, which gives the Freundlich constant (KF), is roughly 0.38, indicating a moderate adsorption capability.

Even if the Langmuir isotherm model fits the data better, the model still offers valuable insights, especially when it comes to the surface's heterogeneity and the potential for multilayer creation.

According to the Temkin isotherm model, surface coverage causes the heat of adsorption to decrease linearly, accounting for interactions between the adsorbent and adsorbate. Compared to the Langmuir and Freundlich isotherm models, the Temkin plot's correlation coefficient $\left(R^2\right)$ is 0.7325, which is lower. This suggests a less satisfactory fit to the experimental value. The values of the slope and intercept correspond to the Temkin isotherm binding constant and the adsorption energy, respectively. The Temkin isotherm model helps us understand the energetics of adsorption, but in this instance, it does not precisely depict the system as well as the other two models.

13.4.2 Bed depth's impact on breakthrough curves

The experiment was conducted by applying variation in the bed depths (Z) by 6 and 10 cm while maintaining the flow rate constant and starting concentration constant at 15 mL per min and 10 mL, respectively. Breakthrough curves were plotted. The column's performance at each depth is shown in Figure 13.4. The findings indicated that the breakthrough point could be reached more quickly with a shallower bed depth. The breakthrough points were reached at 900 and 1140 minutes for a depth of 6 and 10 cm, respectively.

The following interpretation of this trend is possible: it takes less time to reach a breakthrough point because there are fewer adsorbent active sites at the lowest bed depth (M. L. Soto et al. 2017). Figure 13.5 shows measured curves for fluoride adsorption on the activated surface. It is evident that the breakthrough time exhibits a variety of tendencies at varying depths. It took the longest and had the highest adsorption capacity to break through at a depth of 10 cm. This might be because there is enough surface area and

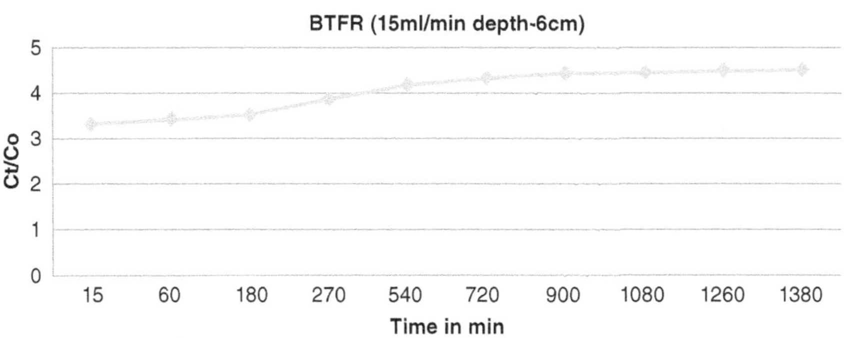

Figure 13.4 Fluoride adsorption breakthrough curves on activated carbon.

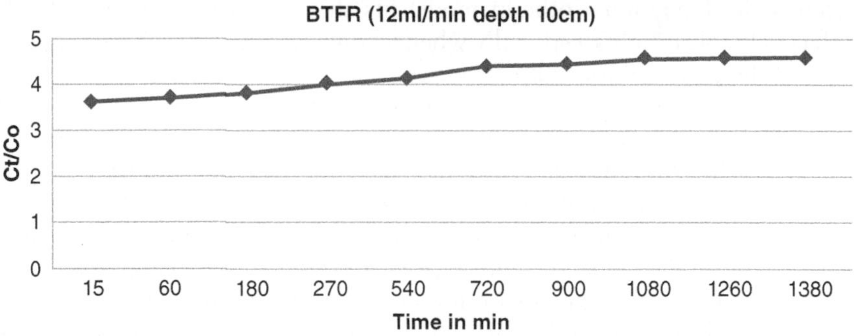

Figure 13.5 Fluoride adsorption breakthrough curves on activated carbon.

interaction time available (M. Shanmugaprakash et al. 2018). The breakthrough time increases as the bed depth increases because diffusion mass transfer becomes more dominant over axial dispersion (A. Abdolali et al. 2017). The exhaustion points for the two distinct depths are displayed in Figure 13.6. It is clear that the time it takes to exhaust increases with depth. A bed depth of 6 cm was tried at 900 minutes, and a bed depth of 10 cm was exhausted at 1140 minutes. The larger surface area and more adsorbent sites also help to explain this (C. Sukumar et al. 2017). The breakthrough had an uneven S-shape (N. Blagojev et al. 2019), and it was noted that as bed depth increased, so did breakthrough and exhaustion times (C. M. Hasfalina et al. 2012) (Table 13.2; Figure 13.7).

Fluoride adsorption performance can be improved by both raising the bed depth and decreasing the flow rate in the bed. The best configuration for

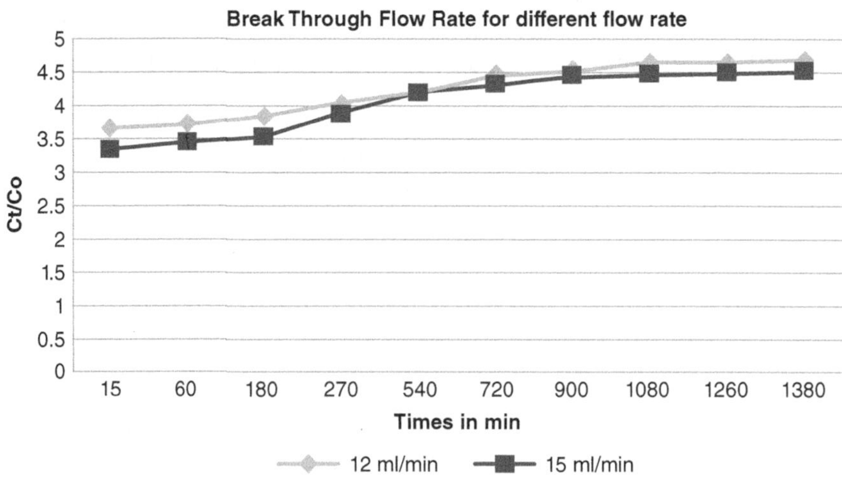

Figure 13.6 Fluoride adsorption breakthrough curves on activated carbon.

Table 13.2 Comparison of breakthrough curves

Parameter	*Figure no. 13.4*	*Figure no. 13.5*	*Figure no. 13.6*	*Figure no. 13.7*
Flow Rate in bed	15 mL/min	12 mL/min	12 vs 15 mL/min	Constant
Bed Depth	6 cm	10 cm	Constant	6 cm versus 10 cm
Breakthrough Time	~720 min	~900 min	12 mL/min breaks through later	10 cm breaks through later
Performance	Least effective	Most effective	Lower flow rate performs better	Deeper bed performs better

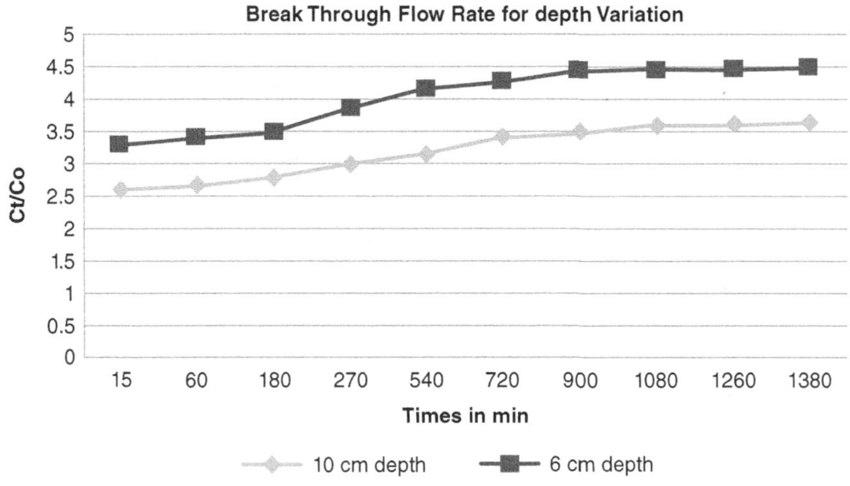

Figure 13.7 Fluoride adsorption breakthrough curves on activated carbon.

fluoride removal is when the flow rate is 12 mL/min and the bed depth is 10 cm (as illustrated in Figure 13.5). The optimization of fixed-bed column designs in water treatment applications depends on these discoveries.

13.5 CONCLUSIONS AND FUTURE SCOPE

When the three models are compared, the Langmuir isotherm model shows the best fit to the experimental value. This suggests that the solute largely adsorbs as a monolayer on a uniform surface onto moringa seed-based carbon at 25°C. The existence of some heterogeneity and the possibility of multilayer adsorption are also supported by the Freundlich model, which fits the data very well. Because it has a poorer correlation with the actual data, the Temkin model is less appropriate in this situation, even though it is helpful for evaluating energy changes during adsorption. Consequently, the Langmuir isotherm provides the best description of the adsorption mechanism for this system at 25°C.

The Langmuir isotherm model shows the highest R^2 value (0.902), indicating it is the best fit for your system. This suggests heterogeneous adsorption and multilayer formation on the natural-based activated carbon surface.

REFERENCES

Abdolali, A., H.H. Ngo, W. Guo, JL Zhou, J. Zhang, and S. Liang (2017), Application of a breakthrough biosorbent for removing heavy metals from synthetic and real wastewaters in a labscale continuous fixed-bed column, *Bioresource Technology*, vol. 229, pp. 78–87.

Almahbashi, N. M. Y., S. R. M. Kutty, M. Ayoub, A. Noor, I. U. Salihi, and A. Al-Nin (2021), Optimization of preparation conditions of sewage sludge based activated carbon, *Ain Shams Engineering Journal*, vol. 12, no. 2, pp. 1175–1182.

Basu, M., A.K. Guha, and L. Ray (2017), Adsorption of cadmium ions by cucumber peel in continuous mode, *International Journal of Environmental Science and Technology*, vol. 16, no. 1, pp. 237–248.

Blagojev, N., D. Kukic, V. Vasic, M. Sciban, J. Prodanovic, and O. Bera (2019), A new approach for modelling and optimization of Cu(II) biosorption from aqueous solutions using sugar beet shreds in a fixed-bed column, *Journal of Hazardous Materials*, vol. 363, pp. 366–375.

Chen, S., Q. Yue, B. Gao, Q Li, X Xu, and K Fu (2012), Adsorption of hexavalent chromium from aqueous solution by modified corn stalk: a fixed-bed column study, *Bioresource Technology*, vol. 113, pp. 114–120.

Chepkwony, G., Mwamburi, E. and Makumba, B. (2019), Effect of charcoal filters integrate with Moringaoleifera seed extracts on microbial population in water from unprotected sources of Kapseret division, Kenya, *Journal of Advanced in Microbiology*, 1–8 https://doi.org/10.9734/JAMB/2019/44978

Dan, Shabnam, and Amit Chattree (2018), Sorption of fluoride using chemically modified Moringa oleiferaleaves, *Applied Water Science*, vol. 8, Article Number 76.

Feizi, F., A.K. Sarmah, and R. Rangsivek (2021), Adsorption of pharmaceuticals in a fixed-bed column using tyre-based activated carbon: experimental investigations and numerical modelling, *Journal of Hazardous Materials*, vol. 417, 126010.

Ghaedi, A., M. Ghaedi, A. Vafaei, N. Iravani, M. Keshavarz, and M. Rad (2015), Adsorption of copper (II) using modified activated carbon prepared from Pomegranate wood: optimization by bee algorithm and response surface methodology, *Journal of Molecular Liquids*, vol. 206, pp. 195–206.

Hasfalina, C.M., R.Z. Maryam, C.A. Luqman, and M. Rashid (2012), Adsorption of copper (II) from aqueous medium in fixedbed column by kenaffibers, *APCBEE Procedia*, vol. 3, pp. 255–263.

John, Bentiland William Kwame Buah (2016), Removal of fluoride in water using activated carbons prepared from selected agricultural waste materials, *Journal of Environment and Earth Science*, vol. 6, no.4, pp. 146–152.

Liu, J., C. Hu, and Q. Huang (2018), Adsorption of Cu2+, Pb2+, and Cd2+ onto oiltea shell from water, *Bioresource Technology*, vol. 271, pp. 487–491.

Muhammad, R., S. Ali, T. Abbas, M. Zia-Ur-Rehman, F. Hannan, and C. Keller (2016), Cadmium minimization in wheat: a critical review, *Ecotoxicology and Environmental Safety*, vol. 130, pp. 43–53.

Noori Ali, Eman (2020), Removal of heavy metal from water and wastewater by using moringa oleifera, *Trace Metals in the Environment - New Approaches and Recent Advances*, https://doi.org/10.5772/intechopen.89769

Rajeev, Kumar and Chawla Jyoti (2020),Carbon-based materials for de-fluoridation of water: current status and challenges, Carbon-Based Material for Environmental Protection and Remediation, chapter 2, pp. 29–44, http://doi.org/10.5772/intechopen.90879

Satya, A., A. Harimawan, G. Sri Haryani, M. A. H. Johir, L. N. Nguyen, and L. D. Nghiem (2021), Fixed-bed adsorption performance and empirical modeling of cadmium removal using adsorbent prepared from the cyanobacterium Aphanothecesp cultivar, *Environmental Technology & Innovation*, vol. 21, Article ID 101194.

Shanmugaprakash, M., S. Venkatachalam, K. Rajendran, and A. Pugazhendhi(2018), Biosorptive removal of Zn(II) ions by Pongamia oil cake (Pongamiapinnata) in batch and fixed-bed column studies using response surface methodology and artificial neural network, *Journal of Environmental Management*, vol. 227, pp. 216–228, https://doi.org/10.1016/j.jenvman.2018.08.088

Soto, M.L., A. Moure, H. Dominguez, and J.C. Parajo (2017), Batch and ´ fixed bed column studies on phenolic adsorption from wine vinasses by polymeric resins, *Journal of Food Engineering*, vol. 209, pp. 52–60.

Sukumar, C., V. Janaki, K. Vijayaraghavan, S. Kamala-Kannan, and K Shanthi (2017), Removal of Cr(VI) using co-immobilized activated carbon and Bacillus subtilis: fixed-bed column study, *Clean Technologies and Environmental Policy*, vol. 19, no. 1, pp. 251–258.

Van, H.T., L.H. Nguyen, V.D. Nguyen, X.H. Nguyen, T.H. Nguyen, and T.V. Nguyen (2019), Characteristics and mechanisms of cadmium adsorption onto biogenic aragonite shells-derived biosorbent: batch and column studies, *Journal of Environmental Management*, vol. 241, pp. 535–548.

Wang, Jianlong and Xuan Guo (2020), Adsorption isotherm models: classification, physical meaning, application and solving method, *Chemosphere*, vol. 258, 127279, https://doi.org/10.1016/j.chemosphere.2020.127279

Chapter 14

Mapping the research landscape of genetically modified crops and food security

A bibliometric perspective

Hrishikesh Kokate, H. D. Verma, Balkrishna Namdeo, Mehul Barai, Aishwarya Maharshi, and Abid Salati

14.1 INTRODUCTION

Genetically modified (GM) crops have been a subject of extensive research due to their potential to enhance food security, improve agricultural productivity, and address challenges posed by climate change and resource limitations (Mandal et al., 2020). As global populations rise and food demand increases, scientific literature on GM crops has expanded significantly, necessitating a systematic evaluation of research trends, key themes, and collaboration networks. This study employs bibliometric analysis to map the intellectual structure of GM crop research and its role in food security, highlighting major research clusters, evolving thematic trends, and influential contributors in the field.

Using advanced bibliometric techniques such as dendrogram analysis, factorial mapping, word cloud visualization, thematic mapping, and network analysis, this study identifies dominant research themes, emerging areas of interest, and the interconnectedness of different subfields (Rani et al., 2024). By analyzing co-occurrence patterns, keyword distributions, and international collaborations, the study provides a comprehensive overview of the scholarly landscape in GM crops and food security.

The findings reveal how research has evolved from early discussions on transgenic crop development and regulatory policies to broader themes such as sustainability, climate resilience, and socioeconomic impacts. Country collaboration networks highlight the leading contributors in GM crop research, while reference spectroscopy sheds light on highly influential studies shaping the discourse.

 DOI: 10.1201/9781003642152-14

Through this bibliometric review, scholars, policymakers, and industry stakeholders can gain insights into the trajectory of GM crop research, enabling evidence-based decision-making for future agricultural policies, innovation strategies, and global food security initiatives. This study serves as a guide for researchers aiming to make impactful contributions to the field.

14.2 METHODOLOGY

Data for this analysis were collected from a relevant database, processed using bibliometric software, and visualized through different mapping techniques. The study employs clustering algorithms for hierarchical representation (dendrograms), principal component analysis (factorial maps), keyword frequency (word clouds), thematic evolution (thematic maps), and network visualization (co-occurrence networks and country collaboration maps) (Figure 14.1).

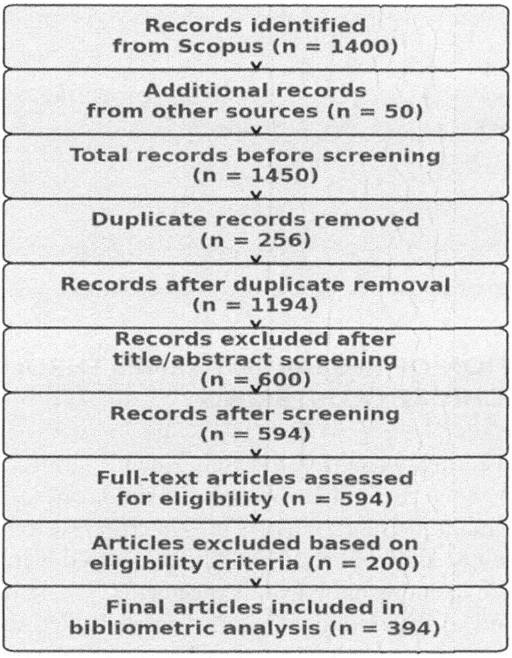

Figure 14.1 Article screening process.

14.3 RESULTS AND DISCUSSION

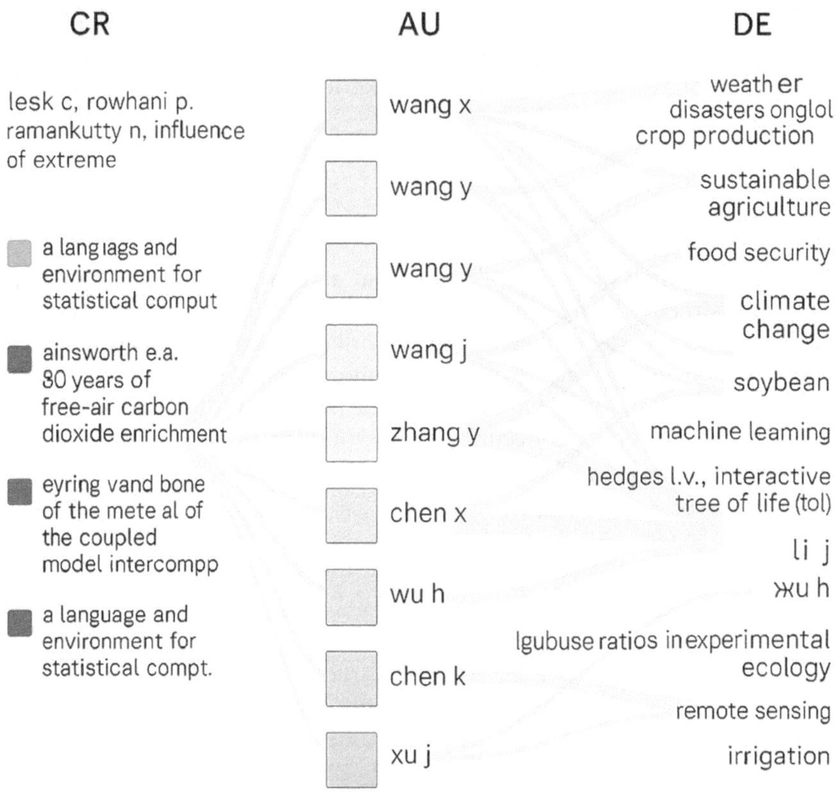

Figure 14.2 Hierarchical clustering and thematic evolution.

14.4 EVOLUTION OF RESEARCH THEMES THROUGH HIERARCHICAL CLUSTERING

Hierarchical clustering, visualized through a dendrogram, groups research topics based on semantic, and co-occurrence similarities, allowing for the identification of major thematic clusters in GM crop research and food security (Chandra Sen & Kushwaha, 2017). The structured hierarchy highlights key domains such as biotechnology advancements, regulatory frameworks, sustainability, and socioeconomic impacts. Closely linked subdomains indicate a well-integrated research field, where interdisciplinary studies bridge various aspects of GM crop development, policy, and environmental concerns. This clustering approach reveals how research themes have evolved, with older, well-established topics providing a foundation for emerging areas such as climate-resilient GM crops and precision agriculture innovations (Figure 14.3).

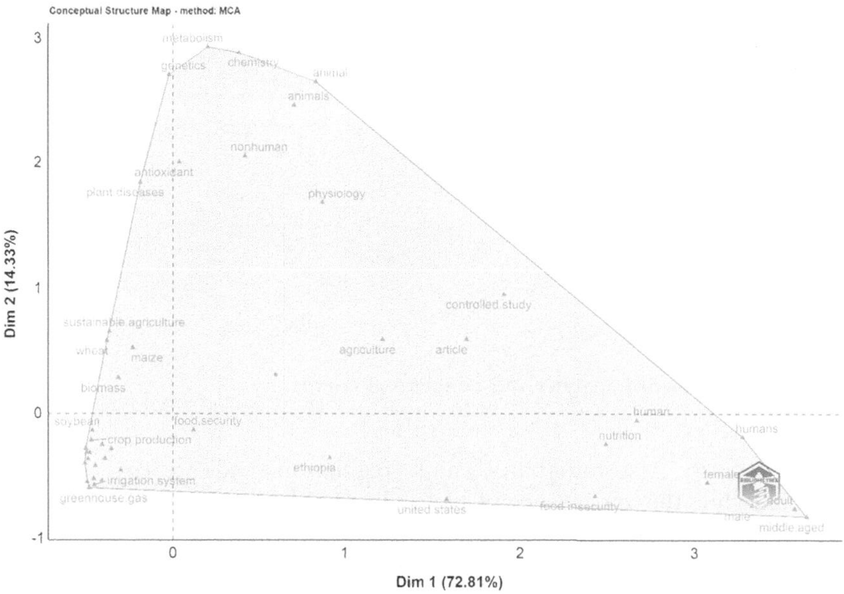

Figure 14.3 Factorial analysis and knowledge structure.

14.5 UNVEILING THE KNOWLEDGE STRUCTURE VIA FACTORIAL ANALYSIS

Factorial analysis, represented through a factorial map, uncovers the principal components defining the knowledge structure of GM crop research. The mapping positions research themes along two primary axes—one reflecting fundamental topics such as genetic engineering and food production, and the other indicating evolving areas like consumer acceptance, trade policies, and sustainability (Deprá et al., 2022). This clear thematic distinction enables the identification of research trajectories, showing shifts from laboratory-based genetic advancements to broader discussions on policy implications and societal concerns. Factorial mapping helps researchers track conceptual progress, locate research gaps, and explore interdisciplinary intersections critical to addressing food security challenges through GM crop innovations (Azmi et al., 2023) (Figure 14.4).

14.6 CONCEPTUAL TRENDS IDENTIFIED THROUGH KEYWORD ANALYSIS

Keyword analysis depicted through a word cloud provides a visual representation of dominant and emerging concepts in GM crop research (Ji et al., 2025). High-frequency terms such as "genetically modified crops,"

Figure 14.4 Keyword analysis and conceptual trends.

"food security," "sustainability," and "regulation" indicate core research themes, while the appearance of newer keywords like "climate resilience," "CRISPR," and "public perception" reflects evolving scientific and policy interests. This technique helps in tracking shifts in academic focus, revealing growing attention toward advanced genetic modification techniques and socioeconomic considerations (Hamdan & Tan 2024). By analyzing keyword patterns, researchers can identify trending topics, forecast future directions, and align their studies with the most relevant and impactful areas of the field (Figure 14.5).

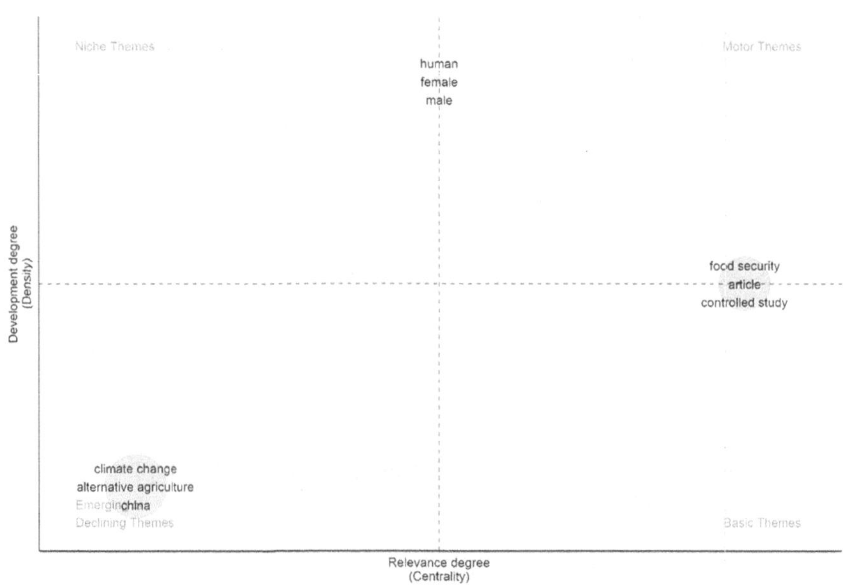

Figure 14.5 Thematic mapping and research dynamics.

14.7 DYNAMICS OF RESEARCH EXPLORED THROUGH THEMATIC MAPPING

The thematic map categorizes research themes in GM crop studies and food security into four quadrants, offering insights into the field's intellectual structure (Qaim & Kouser, 2013). Motor themes, such as biotechnology advancements and regulatory policies, drive major research discussions, while niche themes like ethical concerns and consumer attitudes are highly specialized but less widespread. Emerging or declining themes include outdated methodologies or emerging innovations like CRISPR technology (Ahmad, 2023). Basic themes, such as food security and sustainability, remain foundational but require interdisciplinary exploration. This classification helps researchers identify gaps, refine focus areas, and align their studies with the most impactful research trajectories (Figure 14.6).

14.8 MAPPING CONCEPTUAL RELATIONSHIPS THROUGH CO-OCCURRENCE NETWORKS

The co-occurrence network illustrates the conceptual structure of GM crop research by mapping relationships between key terms (Xiao et al., 2024). Strongly linked concepts, such as "genetic engineering," "crop yield," and "regulation," indicate well-established research domains, whereas weaker

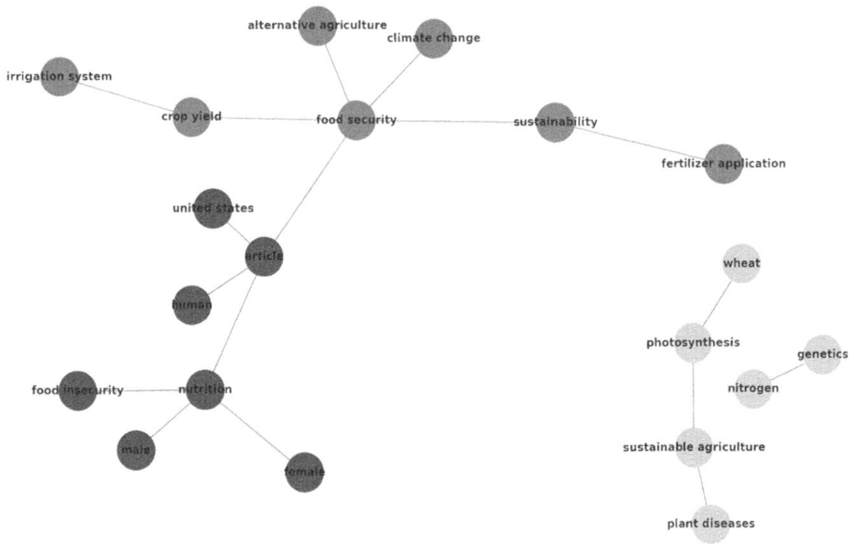

Figure 14.6 Co-occurrence network and conceptual linkages.

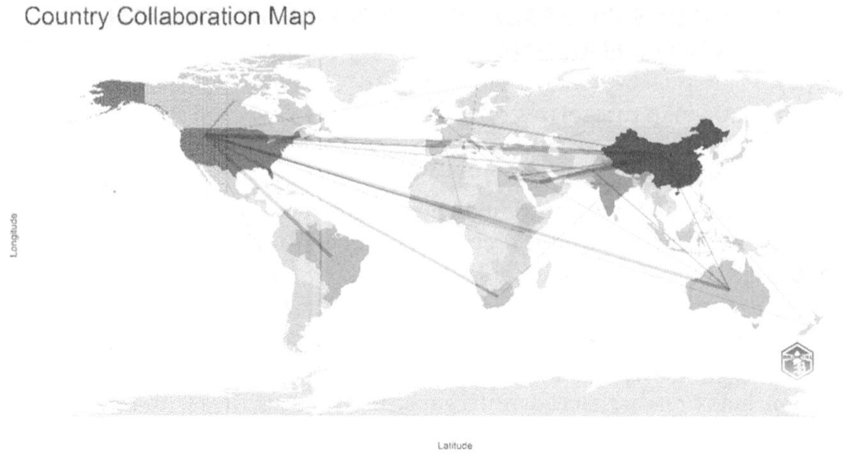

Figure 14.7 Country collaboration and global research landscape.

links suggest underexplored or emerging areas like "climate resilience" and "consumer perception." This network visualization aids in understanding knowledge integration across disciplines, showing how different aspects of GM crops—scientific, economic, and policy-related—interact. Identifying weakly connected nodes highlights research gaps, encouraging interdisciplinary studies to bridge these areas (Bergsten et al., 2019). Strengthening conceptual linkages ensures a more holistic approach to addressing global food security through GM technology (Figure 14.7).

14.9 GLOBAL RESEARCH COLLABORATIONS IN THE CONTEXT OF GM CROPS AND FOOD SECURITY

The country collaboration map provides a global perspective on research partnerships in GM crops and food security (R. et al., 2022). Leading contributors, such as the United States, China, and European nations, exhibit strong collaboration networks, reflecting active knowledge exchange and joint innovation efforts. International partnerships foster advancements in biotechnology, policy harmonization, and sustainable agricultural practices. However, isolated clusters suggest limited integration in certain regions, highlighting opportunities for increased global research cooperation. Strengthening cross-border collaborations can enhance technology transfer, improve regulatory frameworks, and address region-specific challenges, ultimately fostering a more inclusive and comprehensive approach to solving food security concerns through GM crops (Apeh & Nwulu, 2025) (Figure 14.8).

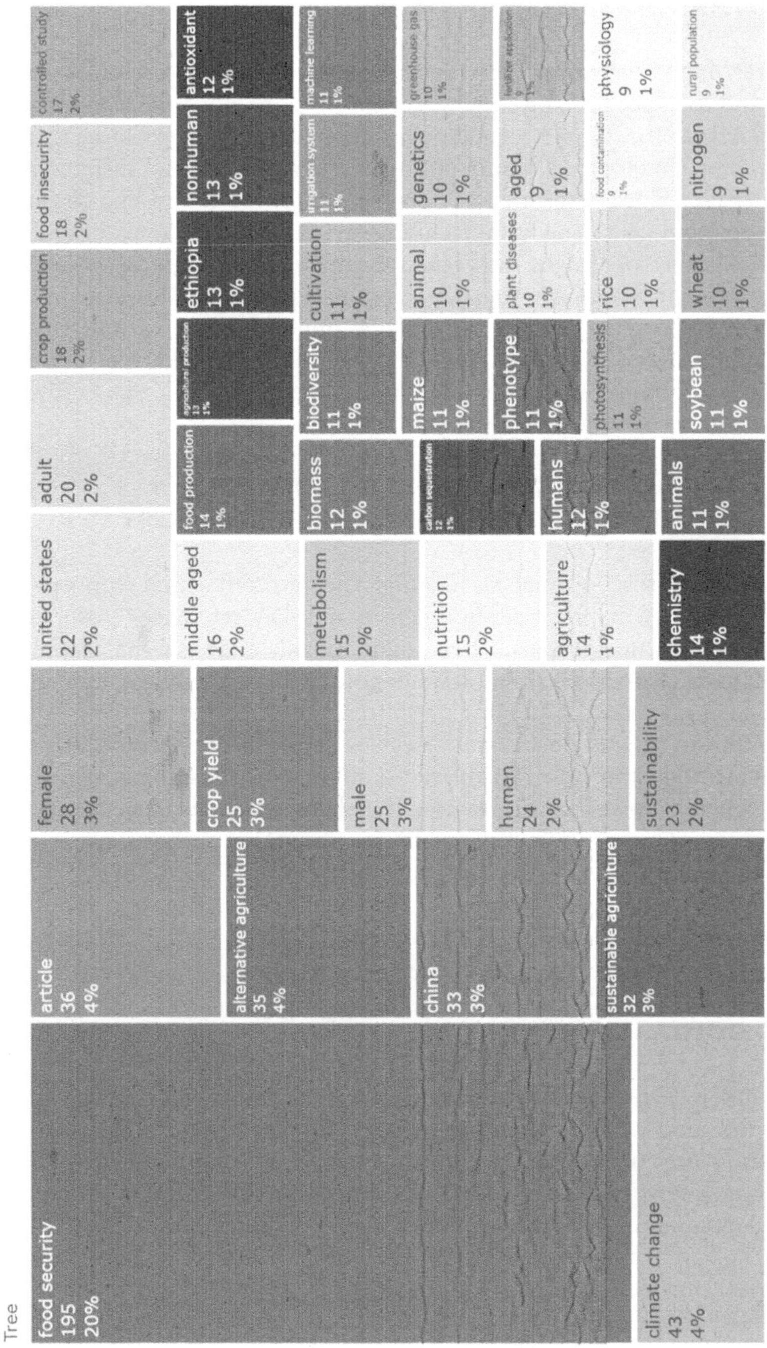

Figure 14.8 TreeMap and field-specific focus.

14.10 FIELD-SPECIFIC EMPHASIS ILLUSTRATED THROUGH TREEMAP VISUALIZATION

The TreeMap visualization hierarchically categorizes research topics in GM crops and food security, offering a comparative analysis of key focus areas (Abdul Aziz et al., 2022). Larger blocks represent dominant research themes, such as biotechnology advancements, regulatory policies, and sustainability, while smaller blocks indicate emerging or specialized topics like consumer perception and genome-editing techniques. The color gradients provide an additional layer of insight into research intensity, highlighting areas with significant contributions and growing scholarly interest. This visualization helps researchers understand the relative importance of various themes, guiding future studies toward underexplored yet impactful areas within the GM crop and food security domain (Figure 14.9).

14.11 CITATION PATTERNS AND SCHOLARLY INFLUENCE VIA REFERENCE SPECTROSCOPY

Reference spectroscopy analysis identifies the most influential works in GM crop research by mapping citation density and key references (Joshi et al., 2023). Highly cited studies, often foundational, establish the theoretical and technical frameworks that drive ongoing research, while emerging references indicate new paradigms or shifting scientific perspectives. A high concentration of citations around specific topics, such as genetic engineering innovations and policy regulations, suggests well-established research areas, whereas lower-density regions highlight potential gaps for further exploration. This analysis helps scholars recognize authoritative works, build upon critical findings, and align their research with impactful contributions, thereby strengthening the academic discourse on GM crops and food security.

14.12 CONCLUSION

This bibliometric review systematically maps the research landscape of GM crops and food security, providing insights into thematic progressions, collaboration networks, and research trajectories. By analyzing hierarchical clustering, keyword co-occurrence, and citation networks, this study highlights dominant research themes, emerging trends, and interdisciplinary linkages. The findings reveal a well-integrated field with ongoing developments in genetic engineering, regulatory frameworks, and sustainability considerations. By leveraging bibliometric tools, scholars can navigate the expanding literature, identify knowledge gaps, and contribute to innovative research avenues. Strengthening international collaboration and interdisciplinary

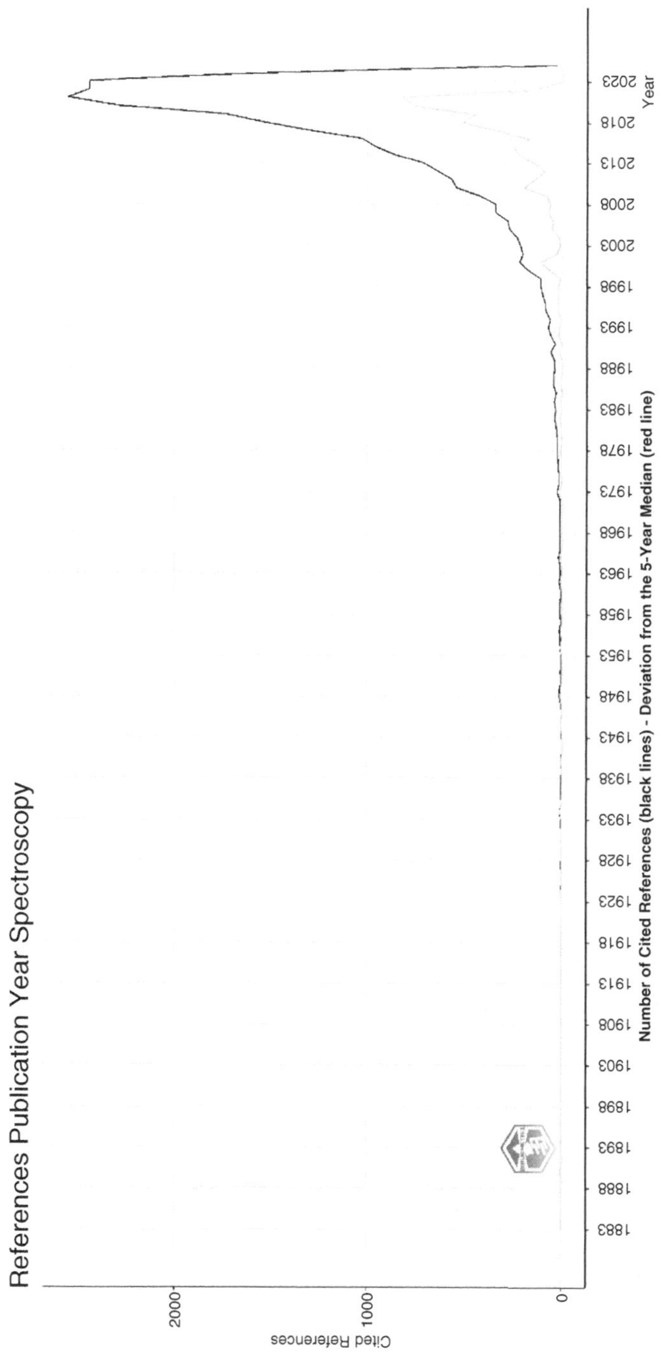

Figure 14.9 Reference spectroscopy and citation impact.

approaches will be crucial for advancing GM crop research and addressing global food security challenges.

REFERENCES

Abdul Aziz, M., Brini, F., Rouached, H., & Masmoudi, K. (2022). Genetically engineered crops for sustainably enhanced food production systems. *Frontiers in Plant Science*, *13*, 1027828. https://doi.org/10.3389/fpls.2022.1027828

Ahmad, M. (2023). Plant breeding advancements with "CRISPR-Cas" genome editing technologies will assist future food security. *Frontiers in Plant Science*, *14*, 1133036. https://doi.org/10.3389/fpls.2023.1133036

Apeh, O. O., & Nwulu, N. I. (2025). Improving traceability and sustainability in the agri-food industry through blockchain technology: A bibliometric approach, benefits and challenges. *Energy Nexus*, *17*, 100388. https://doi.org/10.1016/j.nexus.2025.100388

Azmi, F. R., Zailani, S., & Roni, M. (2023). A Review of the Critical Gaps in the Food Security Literature: Addressing Key Issues for Sustainable Development. *Information Management and Business Review*, *15*(2(I)), 35–46. https://doi.org/10.22610/imbr.v15i2(I).3440

Bergsten, A., Jiren, T. S., Leventon, J., Dorresteijn, I., Schultner, J., & Fischer, J. (2019). Identifying governance gaps among interlinked sustainability challenges. *Environmental Science & Policy*, *91*, 27–38. https://doi.org/10.1016/j.envsci.2018.10.007

Chandra Sen, L. K. M., & Kushwaha, S. (2017). Cluster analysis to form similarity for major selected crops in Rajasthan, India. *International Journal of Current Microbiology and Applied Sciences*, *6*(4), 2673–2682. https://doi.org/10.20546/ijcmas.2017.604.311

Deprá, M. C., Dias, R. R., Zepka, L. Q., & Jacob-Lopes, E. (2022). Building cleaner production: How to anchor sustainability in the food production chain? *Environmental Advances*, *9*, 100295. https://doi.org/10.1016/j.envadv.2022.100295

Hamdan, M. F., & Tan, B. C. (2024). Genetic modification techniques in plant breeding: A comparative review of CRISPR/Cas and GM technologies. *Horticultural Plant Journal*, S2468014124001675. https://doi.org/10.1016/j.hpj.2024.02.012

Ji, X., Wu, X., Deng, R., Yang, Y., Wang, A., & Zhu, Y. (2025). Utilizing large language models for identifying future research opportunities in environmental science. *Journal of Environmental Management*, *373*, 123667. https://doi.org/10.1016/j.jenvman.2024.123667

Joshi, A., Pradhan, B., Gite, S., & Chakraborty, S. (2023). Remote-sensing data and deep-learning techniques in crop mapping and yield prediction: A systematic review. *Remote Sensing*, *15*(8), 2014. https://doi.org/10.3390/rs15082014

Mandal, A., Sarkar, B., Owens, G., Thakur, J. K., Manna, M. C., Niazi, N. K., Jayaraman, S., & Patra, A. K. (2020). Impact of genetically modified crops on rhizosphere microorganisms and processes: A review focusing on Bt cotton. *Applied Soil Ecology*, *148*, 103492. https://doi.org/10.1016/j.apsoil.2019.103492

Qaim, M., & Kouser, S. (2013). Genetically Modified Crops and Food Security. *PLoS ONE*, *8*(6), e64879. https://doi.org/10.1371/journal.pone.0064879

Rani, P., Hussain, A., Shaikh, Md. K., & Babu, M. S. (2024). Mapping the Blockchain technology research: A bibliometrics and visualization approach. *International Journal of Research in Library Science*, *10*(1), 201–217. https://doi.org/10.26761/ijrls.10.1.2024.1746

Sendhil, R., Nyika, J., Yadav, S., Mackolil, J., Prashat, G., Workie, E., Ragupathy, R., Ramasundaram, P. (2022). Genetically modified foods: Bibliometric analysis on consumer perception and preference. *GM Crops & Food*, *13*(1), 65–85. https://doi.org/10.1080/21645698.2022.2038525

Xiao, Q., Wang, B., Li, Z., Zhang, Z., Xie, K., Zhou, J., Lin, K., Geng, X., Li, X., & Chen, J. (2024). The assembly process and co-occurrence network of soil microbial community driven by cadmium in volcanic ecosystem. *Resources, Environment and Sustainability*, *17*, 100164. https://doi.org/10.1016/j.resenv.2024.100164

Chapter 15

Characteristic comparative study of metallic toxicants by using thin layer chromatography and atomic adsorption technique

Meghna H. Jumde, Jaicky H. Gurnani, Ravindra Gandhe, Jitendra Bhaiswar, Shweta Tidake, and Amit Kharwade

SUMMARIZATION

TLC	Thin layer chromatography
AAS	Atomic adsorption spectroscopy
HA	Humic acid
DMSO	Dimethyl sulfoxide
DMF	N,N-Dimethyl methanamide

15.1 INTRODUCTION

Contamination of metallic toxicants causes well-being issues and is hard to treat. This is the most flexible issue in climate (Shen et al., 2022). Indeed, metallic toxicants are still unsafe because of their solid restricting affinity toward nitrogen, oxygen, and sulfur ions of biomolecules, which hinders natural capabilities and hinders physiological cycles at very low concentrations (Buhari et al., 2019). Heavy metal harming, for example, occurs when Hg^{2+} or Cu^{2+} restricts, which significantly impairs the reactant movement of some compound. Amyotrophic lateral sclerosis, Parkinson's disease, and Alzheimer's disease are among the most severe illnesses that are brought on by an excess of metallic toxicants (Chen et al., 2017; Whitaker et al., 2022). In addition, compared to natural compounds, heavy metals exhibit high reliability under a variety of circumstances (Zhao et al., 2021). The outcome of this, the recognition with expulsion of naturally occurring toxic metal ions, is influential for biological security. In summary, metallic components that are poisonous metallic toxicants typically have higher layers than water (Fergusson, 1990). Recent evidence indicates that the modern and natural contamination of these metals is expanding the worldwide general well-being and biological concern. Likewise, it shows that human openness has increased emphatically because of a dramatic increase in their utilization in a few production lines, farming, homemade, and mechanical applications. The defilement of heavy metals in industry incorporates different sources. For instance, activities such as consumption, management, processing

 DOI: 10.1201/9781003642152-15

facilities, combustion of fuel, thermal power plants, stress on materials, and safety measures are involved in operations (Arruti et al., 2010; Strater et al., 2010; Pacyna, 2010). Toward this objective, thin-layer chromatography is an exceptionally advantageous and compelling method for detachment and recognizable proof of inorganic heavy metal particles. It permits basic location, specific detachments, and simple handling of the portable stage. The beginning of the thin layer chromatography procedure can be followed by Izmailov and Shraiber, which is based on drop chromatography (Berezkin et al., 2008). It is an extremely effective and useful technique for separating and identifying inorganic ions in thin-layer chromatography. It enables simple detection, selective separations, and mobile phase handling. Selectivity, resolution, speed, and repeatability are the main goals of the enormous variety of sorbents and mobile phases that have been created to enhance chromatographic performance (Jumde and Gurnule, 2015a, 2016).

The strong stage in TLC, the glass plate (thin), is covered with aluminum oxide or silica gel G. Silica gel G is a crucial substrate for the manufacturing of chemically bound sorbents and the main stationary phase in TLC. This work uses humic acid, aqueous humic acid mixed with DMF (Jumde and Gurnule, 2015b), and aqueous humic acid with DMSO (Jumde and Gurnule, 2016) as a versatile stage to investigate the effects of pH and concentration on the R_f values obtained for different metallic toxicants. Humic acid is the main ingredient in this study since it is one of the main organic components of soil (humus), that is, humic acid (Stevenson, 1994), which includes peat, coal, natural streams, ocean water, and dystrophic lakes.

15.2 RESOURCES AND PLANS

Equipment:

- 4 x 20 cm glass plates
- Glass containers
- A glass sprayer for applying reagents
- pH meter

Substance:

- Silica gel G
- Humic acid
- L-glycine
- Dimethylformamide
- Dimethyl sulfoxide
- HCl
- NaOH

Metallic Toxicants:

Cr + 6, Cr + 3, Fe + 3, Cu + 2Co + 2, Zn + 2, Mo + 6, Ni + 2

Standard solution:
1% concentration of standard stock solutions was prepared in 0.1 M HCl as follows.

1. Potassium chromate (Cr+6)
2. Chlorides of chromium (+3), nickel (+2), zinc (+2), and iron (+3)
3. Copper(II) sulfate
4. Molybdenum (VI) trioxide
5. Cobalt (II) nitrate

Double-distilled water was used to prepare the moving phase.

Spotting reagents:
The reagents listed below were utilized to detect different cations:

1. Dithizone in carbon tetrachloride at 0.05%.
2. $AgNO_3$ that is saturated in alcohol.
3. Alizarin crimson that is saturated with alcohol.
4. A solution of DMG (dimethylglyoxime) in 1% alcohol.
5. A potassium ferrocyanide solution containing 1% alcohol.

Fixed adsorbent:

- Silica gel G

Eluent:

- Aqueous humic acid + L-glycine
- Aqueous humic acid + DMF
- Aqueous humic acid + DMSO

15.3 METHOD: THIN-LAYER CHROMATOGRAPHY

1. **Plate Preparation:** The slurry of silica gel with water in a 1:2 proportion is used to coat the TLC glass plate by using the dipping method. Apply the slurry on the plate and then allow it to dry well overnight.
2. **TLC Plate operation:** On the well-dried glass plate, the sample solutions were spotted by using fine glass capillaries and then dried. The chosen eluent with dissimilar concentrations was maintained at the optimum pH using NaOH and HCl solutions. The TLC plates, which are coated with a silica gel layer, are placed in glass containers containing 15 ml of the mobile phase (solvent or solvent mixture). About 2–3 ml of eluent is needed to run the sample (Jumde and Gurnule, 2015a).
3. **TLC plate development:** To identify or differentiate specific metal particles on the TLC plate, various spot test reagents are sprayed onto the plate. These included soaking alcoholic alizarin red, dithizone in

carbon tetrachloride, dimethylglyoxime, and potassium ferrocyanide for **Mo^{+6}, Zn^{+2}, Co^{+2}, Ni^{+2}, Cr^{+3}, Cr^{+6}, Fe^{+3}**, and **Cu^{+2}**, all such metallic toxicants independently. Every experiment was carried out at room temperature. For every set of conclusions, the R_f values were estimated three times. The necessary conditions for the R_f value measurement of metallic toxicants are as below.

- The concentration of the mobile phase is between **0.005 and 0.1**.
- The **pH** of the mobile phase is between **1.0 and 7.0**.
- The development time of the TLC plates is adjusted between **5 and 20 minutes**.

15.4 RESULTS AND CONVERSATION

15.4.1 Analysis of the effect of pH

By conducting a few sets of tests, this section provides a detailed analysis of the effect of pH on the R_f values of distinct metallic particles based on a comparative study of pH. Close runs of the plates were made up to 10 cm above the baseline. Table 15.1 presents the findings, indicating variation in the R_f values with pH of different humic acid configurations. The pH range of 2.0 to 7.0 at a strength of 0.05 M was used for the determination of the R_f value of different metal particles. The R_f value of 3.5 for aqueous humic acid + L-glycine metal particles were found to exhibit minimal contrast, but when the pH was raised to 4, there was the biggest level of separation, so only at a pH of 4.0 was the distinct partition visible. As a result, partition estimation has been carried out at pH 4.0 for all 3 systems. At pH 4.0, Ni

Table 15.1 pH impact on metallic toxicants (T: Tailing and Spr: Spreading)

Metallic Toxicants	*M-1*		*M-2*		*M-3*	
	pH = 3	*pH = 4*	*pH = 3*	*pH = 4*	*pH = 3*	*pH = 4*
Cr^{+6}	0.96	0.97	0.97	0.98 T	0.95	0.97
Cr^{+3}	0.71	0.74	0.54	0.53	0.61	0.41
Co^{+2}	0.78	0.58	0.82	0.62	0.48	0.44
Ni^{+2}	0.80	0.69 Spr	0.97	0.99 Spr	0.59	0.58 Spr
Zn^{+2}	0.51	0.18 T	0.48	0.17 T	0.41	0.41 T
Cu^{+2}	0.45	0.17	0.38	0.21	0.25	0.18
Fe^{+3}	0.49	0.17	0.42	0.16	0.14	0.11
Mo^{+6}	0.48	0.11	0.57	0.44 T	0.82	0.48

(II) exhibits minimal spreading in all three situations (Figures 15.1–15.3). The mobile phase was used for comparisons as depicted.

1. M1: Humic acid dissolved in L-glycine solution.
2. M2: Humic acid dissolved in dimethylformamide (DMF) solution.
3. M3: Humic acid dissolved in dimethyl sulfoxide (DMSO) solution.

15.4.2 Analysis of the concentration effect

The outcomes of controlling the influence of various portable stages, such as aqueous HA in L-glycine, HA in DMF, and HA in DMSO on the R_f values of distinct metallic toxicants, including **Mo^{+6}**, **Fe^{+3}**, **Zn^{+2}**, **Co^{+2}**, **Cr^{+3}**, **Ni^{+2}**, **Cu^{+2}**, and **Cr^{+6}**, are shown visually in Figure 15.4 and summarized in Table 15.2, and graphically represented in Figure 15.4. R_f values were read up for the

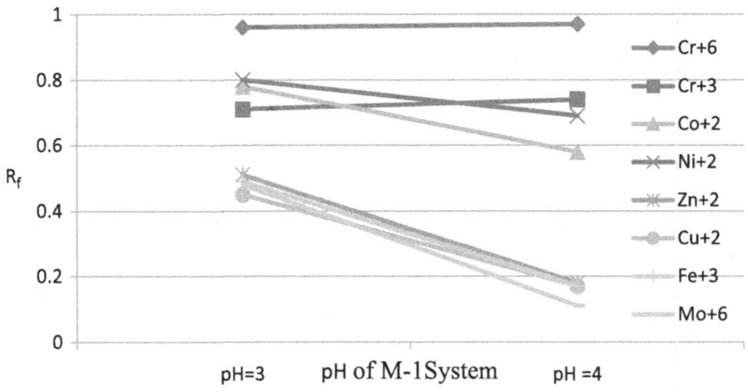

Figure 15.1 pH of M-1 system.

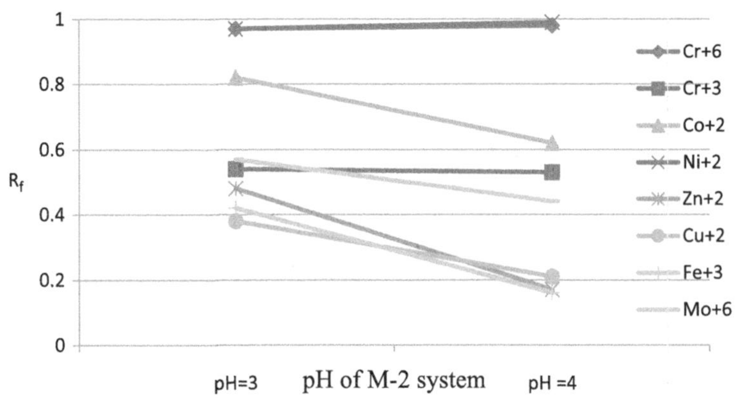

Figure 15.2 pH of M-2 system.

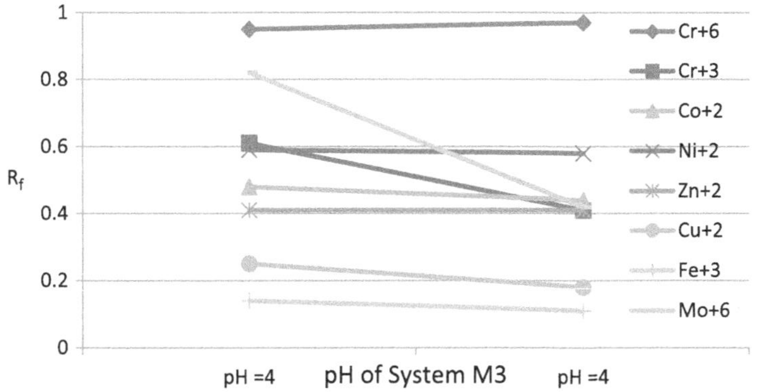

Figure 15.3 pH of M-3 system.

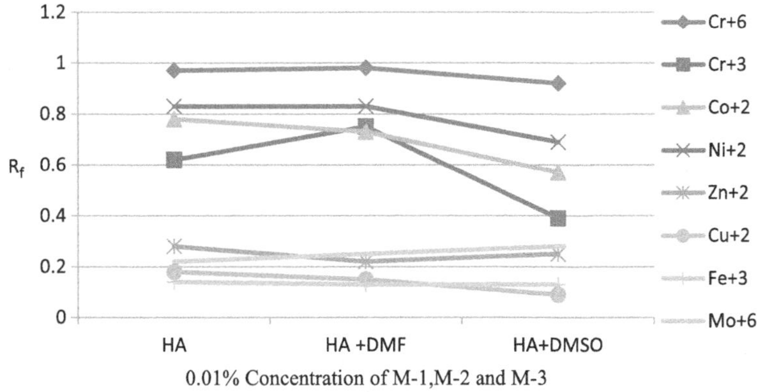

Figure 15.4 Concentration of M-1, M-2, and M-3.

Table 15.2 Concentration impact on metallic toxicants

Metallic toxicants	*M-1*	*M-2*	*M-3*
Cr^{+6}	0.97	0.98	0.92
Cr^{+3}	0.62	0.75	0.39
Co^{+2}	0.78	0.73	0.52
Ni^{+2}	0.83	0.83	0.69
Zn^{+2}	0.28	0.22	0.25
Cu^{+2}	0.18	0.15	0.09
Fe^{+3}	0.14	0.13	0.13
Mo^{+6}	0.22	0.25	0.28

selected versatile stage at their specific optimal pH and development time, with an emphasis on the range of 0.005% to 0.1%. Better separation with distinct and obvious spots was achieved, though, as the concentration of the M-1, M-2, and M-3 system was increased to 0.01%. Mo(IV) shows spreading, and Zn(II) displays two distinct spots, while Cr and Ni(II) do not exhibit any recognizable spots when the 0.1% aqueous phase of all systems was used.

Furthermore, it was observed that when fixation increases, so do R_f levels. Thus, at pH 4.0, 0.01% fixation concentration for all three systems was selected as the optimal target for further separation of the mixture.

15.4.3 Ideal separation conditions

Ideal separation conditions for systems M-1, M-2, and M-3 are as follows:

1. Mobile phase: 0.01% aqueous phase of the system
2. Adsorbent: Silica gel–G
3. Developing Time =15 min
4. pH = 4

After establishing the ideal condition of pH, concentration, development time, and proper adsorbent, the binary separation of all systems M-1, M-2, and M-3 took place, and it was tabulated in Tables 15.3, 15.4, and 15.5 (Figures 15.5, 15.6, and 15.7).

Table 15.3 Binary separations of system M-1

S. No	*Constituents of a binary mixture*	*Metallic toxicants with the R_f values*
a	Co^{+2}; Cu^{+2}	Cr^{+6}–0.94; Co^{+2}–0.71
b	Ni^{+2}; Zn^{+2}	Ni^{+2}–0.82; Zn^{+2}–0.20
c	Ni^{+2}; Mo^{+6}	Ni^{+2}–0.85; Mo^{+6}–0.15
d	Cr^{+3}; Ni^{+2}	Cr^{+3}–0.29; Ni^{+2}–0.82
e	Co^{+2}; Zn^{+2}	Co^{+2}–0.70; Zn^{+2}–0.16
f	Cr^{+6}; Cr^{+3}	Cr^{+6}–0.95; Cr^{+3}–0.40
g	Ni^{+2}; Co^{+2}	Ni^{+2}–0.82; Co^{+2}–0.70
h	Ni^{+2}; Cu^{+2}	Ni^{+2}–0.83; Cu^{+2}–0.20
i	Ni^{+2}; Zn^{+2}	Ni^{+2}–0.80; Zn^{+2}–0.16

Table 15.4 Binary separations of system M-2

S. No.	*Constituents of a binary mixture*	*Metallic toxicants with the Rf values*
a	Ni^{+2};Co^{+2}	Ni^{+2}–0.75; Co^{+2}–0.62
b	Ni^{+2};Zn^{+2}	Ni^{+2}–0.76; Zn^{+2}–0.22
c	Ni^{+2}; Cr^{+6}	Ni^{+2}–0.74; Cr^{+6}–0.95
d	Cr^{+6};Co^{+2}	Cr^{+6}–0.94; Co^{+2}–0.67
e	Cr^{+3}; Zn^{+2}	Cr^{+3}–0.95; Zn^{+2}–0.20
f	Cr^{+6}; Cr^{+3}	Cr^{+6}–0.94; Cr^{+3}–0.23
g	Ni^{+2}; Cu^{+2}	Ni^{+2}–0.73; Cu^{+2}–0.11

Table 15.5 Binary separations of system M-3

S. No.	*Constituents of a binary mixture*	*Metallic toxicants with the Rf values*
a	Cr^{+6}; Co^{+2}	Cr^{+6}–0.95; Co^{+2}–0.44
b	Ni^{+2}; Zn^{+2}	Ni^{+2}–0.60; Zn^{+2}–0.11
c	Ni^{+2}; Mo^{+6}	Ni^{+2}–0.71; Mo^{+6}–0.17
d	Cr^{+3}; Ni^{+2}	Cr^{+3}–0.19; Ni^{+2}–0.54
e	Co(II); Zn^{+2}	Co^{+2}–0.67; Zn^{+2}–0.13
f	Cr^{+6}; Cr^{+3}	Cr^{+6}–0.92; Cr^{+3}–0.20
g	Ni^{+2}; Co^{+2}	Ni^{+2}–0.68; Co^{+2}–0.60
h	Ni^{+2}; Cu^{+2}	Ni^{+2}–0.65; Cu^{+2}–0.11
i	Ni^{+2}; Zn^{+2}	Ni(II)–0.64; Zn^{+2}–0.13
j	Cr^{+6}; Ni^{+2}	Cr^{+6}–093; Ni^{+2}–0.70

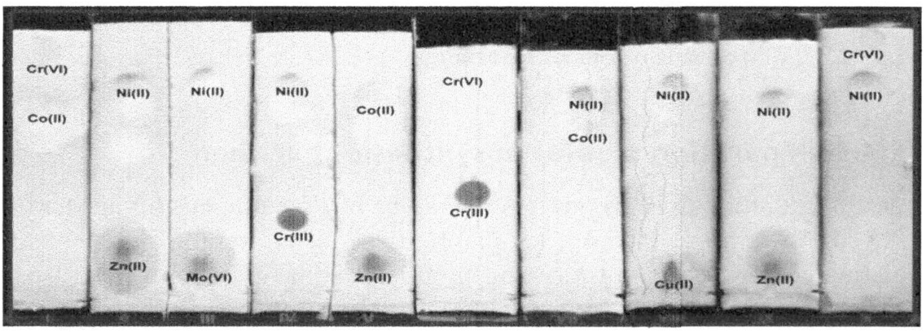

Figure 15.5 Binary separations of system M-1.

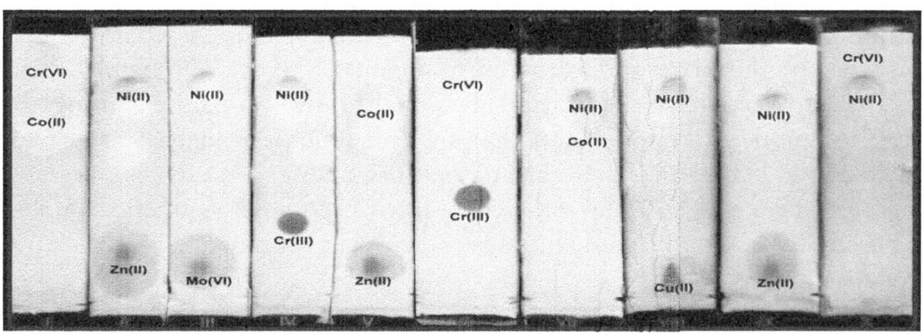

Figure 15.6 Binary separations of system M-3.

15.4.4 Separation studies in mixture

After establishing optimum conditions, quantitative separation of ten binary mixtures of the entire three systems was carried out in synthetic samples, and their images have been depicted in images of plates 3.1, 3.2, and 3.3 and tabulated in Tables 15.3, 15.4, and 15.5, respectively.

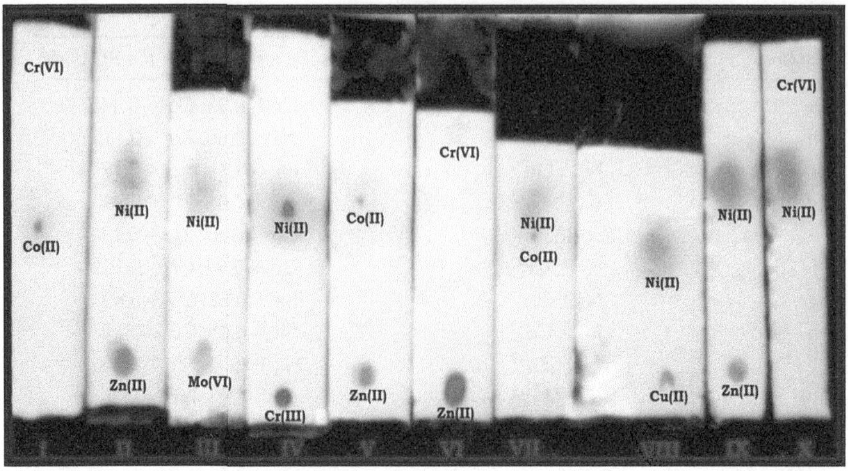

Figure 15.7 Binary separations of system M-3.

15.4.5 Separation studies in synthetic specimen

The current study used the system M-1, which is an aqueous humic acid + L-glycine separation method, to quantitatively separate and detect metal ions in synthetic chromate ore and industrial effluent samples from Ispat Industry Pvt. Ltd. in the industrial area of the Hingna region at Nagpur. Using a micropipette, a predetermined volume of aliquot (10 mL) was spotted on the TLC plate to perform a quantitative examination of the metal ions contained in the specimen. To perform the chromatogram, ideal separation conditions were used. The metal ion that had been separated was scraped off the plate and gathered in various beakers. Ultimately, those metallic toxicants were extracted and quantitatively identified using the AAS technique. For comparison analysis, metal ion AAS analyses were conducted both before and after chromatogram running.

Tables 15.6 and 15.7 show the comparative results of synthetic samples, a chromate ore sample, and an industrial effluent specimen.

Table 15.6 Separations of metallic toxicants in synthetic specimens in PPM

Components	*Metallic toxicants*	*Metallic toxicants before chromatogram*	*Metallic toxicants by AAS after chromatogram*
Cr^{+6}, Ni^{+2}	Cr^{+6}	Cr^{+6}–11.07	Cr^{+6}–10.23
Co^{+2}, Cr^{+3}	NI^{+2}	Ni^{+2}–8.79	Ni^{+2}–8.64
	Co^{+2}	Co^{+2}–7.01	Co^{+2}–6.88
	Cr^{+3}	Cr^{+3}–18.83	Cr^{+3}–17.98

Table 15.7 Separations of metal ions in chromate ore and industrial effluent specimens in PPM

Name of sample	*Metallic toxicants*	*Metallic toxicants before chromatogram (ppm)*	*Metallic toxicants after chromatogram (ppm)*
Chromate ore	Cr^{+6}, Cr^{+3}	Cr^{+6}– 18.75	Cr^{+6}– 18.59
		Cr^{+3}– 29.53	Cr^{+3} – 29.39
Industrial Effluent	Cr^{+6}	Cr^{+6}– 0.019	Cr^{+6} – 0.017
	Cr^{+3}	Cr^{+3}– 0.013	Cr^{+3}– 0.012
	Ni^{+2}	Ni^{+2}– 0.035	Ni^{+2}– 0.032
	Co^{+2}	Co^{+2}– 0.020	Co^{+2}– 0.018
	Zn^{+2}	Zn^{+2}– 0.013	Zn^{+2}– 0.011
	Mo^{+6}	Mo^{+6}– 0.023	Mo^{+6}– 0.018
	Fe^{+3}	Fe^{+3}– 0.18	Fe^{+3}– 0.17

15.5 CONCLUSIONS AND FUTURE SCOPE

According to the above result, the hazardous heavy metal particles were identified and separated by using aqueous phases of system M-1, M-2, and M-3 at their optimal pH, concentration, and adsorbent. The optimal conditions for the separation of a mixture are a 4.0 pH level and 0.01% concentration of mobile phase at 15 min of separating time with the proper silica gel as an adsorbent. By choosing such a condition of pH, adsorbent, and concentration, the double, ternary, and quaternary combinations of metal particles have been subjectively separated, isolated, and quantitatively determined by AAS, and their results were arranged in Tables 3.1–3.3, with their pictures displayed in Figures 3.1–3.3. The R_f value of metal particles shows up on the chromatographic plate. Metal ions are quantitatively separated and detected in the synthetic material, chromate ore specimen, and industrial effluent specimen from Ispat Industry Pvt. Ltd., Nagpur, employing all ideal pH, concentration, and adsorbent conditions. It was clearly demonstrated that the chromatographic separation method is not only effective for synthetic mixtures but also **highly applicable to natural and industrial samples**:

- **Chromate ore** shows successful separation of Cr^{6+} and Cr^{3+} with minimal loss.
- **Industrial effluent**, despite its complex composition and trace metal concentrations, still yields high recovery post-chromatography.

Thus, the technique is **robust, sensitive, and suitable for environmental monitoring and industrial waste analysis** in the future.

REFERENCES

Arruti, A., Fernández-Olmo, I., Irabien, A. (2010).Evaluation of the contribution of local sources to trace metals levels in urban PM2.5 and PM10 in the Cantabria region (Northern Spain), *The Journal of Environmental Management*, 12(7), 1451–1458.

Berezkin, V. G., Dzido, T. H. (2008). The discovery of thin layer chromatograpjy by N. A. Izmailov and M. S. Shraiber, *Journal of Planar Chromatography*, 21, 399–403.

Buhari, O., Atanda, A., Alzharani T., and Faillace, R. (2019). Exposure to heavy metals may increase cardiovascular risk by elevation of serum lipid levels and hemoglobin A1c: a national health and nutrition examination survey based study. *Journal of the American College of Cardiology*, 73.

Chen, J. H., Pelka K., and Hacohen, N. (2017). Heavy metal enlightens tumor immunity. *Cell*, 169, 567–569.

Fergusson, J.E. Eds. (1990). *The Heavy Elements Chemistry, Environmental Impact and Health Effects*, Oxford: Pergamon Press.

Jumde, M. H., & Gurnule, W. B. (2015a). Separation and identification of heavy metal ions by thin layer chromatography on silica gel-G. *Journal of Chemical and Pharmaceutical Research*, 7(4), 889–895.

Jumde, M. H., & Gurnule, W. B. (2015b. Separation and identification of toxic heavy metal ions by thin layer chromatography. *International Journal of Engineering, Science and Advanced Technology Special*, (1), 52–56.

Jumde, M. H., & Gurnule, W. B. (2016). Thin layer chromatographic separation of Cr (VI), Cr (III), Ni (II), Co (II), Cu (II), Fe (III), Zn (II) and Mo (VI) toxic metal ions. *Research Journal of Pharmaceutical, Biological and Chemical Sciences*, 7(6), 368–381.

Pacyna, J. M. (2010). Monitoring and assessment of metal contaminants in the air, *Toxicology of Metals*, Boca Raton, FL: CRC, In: Chang LW, Magos L, Suzuli T, editors. Press, 28.

Shen, Y., Nie, C.,Y. Wei, Z. Zheng, Z.-L. Xu and Xiang, P. (2022). FRET-based innovative assays for precise detection of the residual heavy metals in food and agriculture-related matrices, *Coordination Chemistry Reviews*, 469, 214676.

Stevenson, F. J. (1994). *Humus Chemistry: Genesis, Composition, Reactions*. New York: John Wiley & Sons.

Sträter, E., Westbeld, A., Klemm, O. (2010). Pollution in coastal fog at Alto Patache, Northern Chile. *Environmental Science and Pollution Research*, ahead of print.

Whitaker, K. D., Ross, E. A., Klaunig, J. E., Singh, T., Veggeberg, R., Alpaugh K., and Goldstein, L. J. (2022). Assessing association of heavy metal levels in urine and sera with neurotoxicity in breast cancer(BC) patients undergoing adjuvant/neoadjuvant chemotherapy, *Journal of Clinical Oncology*, 40, e12510–e12510.

Zhao, Y., Gao, J., Wang, Z., Dai H., and Wang, Y. (2021). Responses of bacterial communities and resistance genes on microplastics to antibiotics and heavy metals in sewage environment. *Journal of Hazardous Materials*, 402, 123550.

Chapter 16

An EPQ inventory model with fuzzy deterioration rate and finite production rate

S.P. Singh

NOTATIONS OF THE MODEL

The following notations have been used to develop the model.

1. K = Production Quantity per unit time.
2. d = Demand rate per unit time.
3. θ = deterioration rate of the on-hand inventory per unit time.
4. $\tilde{\theta}$ = fuzzy deterioration rate of the on-hand inventory per unit time.
5. ϕ = deterioration fraction of production rate per unit time.
6. C_1 = constant inventory holding cost per unit time.
7. r = constant purchase cost of raw materials per unit item.
8. b = constant setup cost per order.
9. p = constant selling price of the production per unit item.
10. t_1 = time at which production is stopped.
11. t_2 = time at which inventory level reaches zero.
12. T = the whole period for the plan.
13. q_1 = inventory level at any time t, where $0 \le t \le t_1$
14. q_2 = inventory level at any time t, where $0 \le t \le t_2$
15. q = actual production quantity received per cycle.
16. $K(1-\phi)$ = actual production rate per unit time.

16.1 ASSUMPTIONS OF THE MODEL

The following assumptions are made in developing the model:

1. Replenishment is instantaneous with constant lead time.
2. Shortages are not allowed.
3. The membership function of the fuzzy deterioration rate $\tilde{\theta}$

DOI: 10.1201/9781003642152-16

$$\mu_{\tilde{\theta}}(\theta) = \begin{cases} \dfrac{\theta - \theta_1}{\theta_0 - \theta_1} & \text{for } \theta_1 \le \theta < \theta_2 \\ \dfrac{\theta_2 - \theta_1}{\theta_2 - \theta_0} & \text{for } \theta_0 \le \theta < \theta_2 \\ 0 & \text{elsewhere} \end{cases}$$

where

θ_1, θ_0, and θ_2 are positive numbers.

16.2 DEVELOPMENT OF THE CRISP MODEL

At time t=0, the production starts at the beginning of each cycle and continues up to time $t = t_1$. The actual production rate becomes less than the original production and reduce to faulty machines since some quantities deteriorate at the time of production.

The inventory accumulated during the production period t_1 after meeting demand during the period and deterioration reaches the zero level at time $t = t_2$. Then the cycle repeats itself for the total planning period T .

The following differential equation gives the instantaneous states of q_1 and q_2 over the cycles t_1 and t_2:

$$\frac{dq_1}{dt} + \theta q_1 = K(1-\phi) - d \qquad ;\ 0 \le t \le t_1 \tag{16.1}$$

$$\frac{dq_2}{dt} + \theta q_2 = -d \qquad ;\ t_1 \le t \le t_2 \tag{16.2}$$

With the conditions

$$q_1(0) = 0;\ q_1(t_1) = 0;\ q_1(t_1) = q_2(t_1) \text{ and } q_2(t_2) = 0 \tag{16.3}$$

Eq. (16.1) is a linear differential equation, and its integrating factor (IF) is given by

$$\text{I.F.} = e^{\int \theta dt} = e^{\theta t}$$

Its solution is given by

$$e^{\theta t} q_1(t) = \int \left(K(1-\phi) - d\right) e^{\theta t} dt + C$$

$$e^{\theta t}q_1(t)=\frac{\{K(1-\phi)-d\}}{\theta}e^{\theta t}+C$$

$$t=0,\ q_1(t)=0, \text{we have}$$

$$C=\frac{-\{K(1-\phi)-d\}}{\theta} \tag{16.4}$$

$$\therefore e^{\theta t}q_1(t)=\frac{\{K(1-\phi)-d\}}{\theta}\left(e^{\theta t}-1\right)$$

$$\therefore q_1(t)=\frac{\{K(1-\phi)-d\}}{\theta}\left(1-e^{-\theta t}\right) \quad ;0\le t\le t_1$$

From equation (16.2), we get

$$\frac{dq_2}{dt}+\theta q_2=-d$$

Which is also a linear differential equation; it's

$$\text{I.F}=e^{\int\theta dt}=e^{\theta t}$$

$$\therefore e^{\theta t}q_2(t)=-\int de^{\theta t}dt+C$$

$$=-\frac{de^{\theta t}}{\theta}+C$$

$$\textit{When } t=t_2, q_2(t_2)=0 \text{ we get as}$$

$$0=-\frac{de^{\theta t_2}}{\theta}+C \qquad \Rightarrow C=\frac{d}{\theta}e^{\theta t_2}$$

$$\therefore e^{\theta t}q_2(t)=-\frac{de^{\theta t}}{\theta}+\frac{de^{\theta t_2}}{\theta}$$

$$e^{\theta t}q_2(t)=\frac{d}{\theta}\left(e^{\theta t_2}-e^{\theta t}\right)$$

$$\Rightarrow q_2(t)=\frac{d}{\theta}\left\{e^{\theta(t_2-t)}-1\right\} \tag{16.5}$$

From equation (16.4), we have

$$q_1(t_1)=\frac{\{K(1-\phi)-d\}}{\theta}\left(1-e^{-\theta t_1}\right)$$

$$=\frac{\{K(1-\phi)-d\}}{\theta}\left\{1-\left(1-\theta t_1+\frac{\theta^2t_1^2}{2}-\frac{\theta^3t_1^3}{3}\right)\right\} \tag{16.6}$$

$$= \frac{\{K(1-\phi)-d\}}{\theta}\left(\theta t_1 - \frac{\theta^2 t_1^2}{2} + \frac{\theta^3 t_1^3}{3}\right)$$

$$= \{K(1-\phi)-d\}\left(t_1 - \frac{\theta t_1^2}{2} + \frac{\theta^2 t_1^3}{6}\right)$$

From equation (16.5), we have

$$q_2(t_1) = \frac{d}{\theta}\left\{e^{\theta(t_2-t_1)} - 1\right\}$$

$$= \frac{d}{\theta}\left\{1+\theta(\mathrm{t}_2 - t_1) + \frac{\theta^2(t_2-t_1)^2}{2} - 1\right\}$$

$$= \frac{d}{\theta}\left\{\theta \mathrm{t}_2 - \theta t_1 + \frac{\theta^2\left(t_2^2 + t_1^2 - 2t_1t_2\right)}{2}\right\}$$

$$q_1(t_1) = \frac{\{K(1-\phi)-d\}}{\theta}\left(1-e^{-\theta t_1}\right)$$

$$= \frac{\{K(1-\phi)-d\}}{\theta}\left(1-e^{-\theta t_1}\right)$$

$$q_2(t_1) = \frac{d}{\theta}\left\{e^{\theta(t_2-t_1)} - 1\right\} = \frac{d}{\theta}\frac{\left(e^{\theta t_2} - e^{\theta t_1}\right)}{e^{\theta t_1}} \tag{16.7}$$

given $q_1(t_1) = q_2(t_1)$

$$\Rightarrow \frac{\{K(1-\phi)-d\}}{\theta}\left(\frac{e^{\theta t_1}-1}{e^{\theta t_1}}\right) = \frac{d}{\theta}\frac{\left(e^{\theta t_2} - e^{\theta t_1}\right)}{e^{t_1}}$$

$$\Rightarrow \frac{\{K(1-\phi)-d\}}{d}\left(e^{\theta t_1}-1\right) + e^{\theta t_1} = e^{\theta t_2}$$

$$\Rightarrow \left\{\frac{K(1-\phi)-d}{d} + 1\right\}e^{\theta t_1} - \frac{K(1-\phi)-d}{d} = e^{\theta t_2}$$

$$\Rightarrow \frac{K(1-\phi)}{d}e^{\theta t_1} - \frac{K(1-\phi)}{d} + 1 = 1 + \theta t_2$$

$$\Rightarrow \frac{K(1-\phi)}{d}\left(1+\theta t_1 + \frac{\theta^2 t_1{}^2}{2} + \frac{\theta^3 t_1{}^3}{6}\right) - \frac{K(1-\varphi)}{d} = \theta t_2$$

$$\Rightarrow \frac{K(1-\phi)}{d}\theta t_1\left(1 + \frac{\theta t_1}{2} + \frac{\theta^2 t_1{}^2}{6}\right) = \theta t_2$$

$$\Rightarrow \frac{K(1-\phi)}{d} t_1 \left(1+\frac{\theta t_1}{2}+\frac{\theta^2 t_1^{\,2}}{6}\right) = t_2$$

$$\Rightarrow t_2 = \frac{q}{d}\left(1+\frac{\theta t_1}{2}+\frac{\theta^2 t_1^{\,2}}{6}\right) \tag{16.8}$$

where $q = K(1-\Phi)t_1$ is the actual production quantity during the period t_1.

Let HC denote the holding cost per cycle, which is given by

$$\text{HC} = \text{C}_1\left[\int_0^{t_1} q_1(t)\,dt + \int_{t_1}^{t_2} q_2(t)\,dt\right]$$

$$= \text{C}_1\left[\int_0^{t_1} \frac{\{K(1-\phi)-d\}}{\theta}\left(1-e^{-\theta t}\right)dt + \int_{t_1}^{t_2} \frac{d}{\theta}\left\{e^{\theta(t_2-t)}-1\right\}\right]dt$$

$$= \text{C}_1\left[\begin{array}{l}\displaystyle\int_0^{t_1} \frac{\{K(1-\phi)-d\}}{\theta}\left(1-1+\theta t-\frac{\theta^2 t^2}{2}\right)dt \\ \displaystyle+\int_{t_1}^{t_2} \frac{d}{\theta}\left\{1+\theta(t_2-t)+\frac{\theta^2(t_2-t)^2}{2}-1\right\}\end{array}\right]dt$$

$$= \text{C}_1\left[\int_0^{t_1} \{K(1-\phi)-d\}\left(t-\frac{\theta t^2}{2}\right)dt + d\int_{t_1}^{t_2}\left\{(t_2-t)+\theta\frac{(t_2-t)^2}{2}\right\}dt\right]$$

$$= \text{C}_1\left[\{K(1-\phi)-d\}\left.\left|\frac{t^2}{2}-\frac{\theta t^3}{6}\right|\right._0^{t_1} + d\left\{\left.\left|t_2 t-\frac{t^2}{2}\right|\right._{t_1}^{t_2} - \left.\left|\frac{\theta(t_2-t)^3}{6}\right|\right._{t_1}^{t_2}\right\}\right]$$

$$= \text{C}_1\left[\{K(1-\phi)-d\}\left(\frac{t_1^2}{2}-\frac{\theta t_1^3}{6}\right)+d\left\{\left(t_2^2-\frac{t_2^2}{2}\right)-\left(t_2 t_1-\frac{t_1^2}{2}\right)+\frac{\theta(t_2-t_1)^3}{6}\right\}\right]$$

$$= \text{C}_1\left[\{K(1-\phi)-d\}\left(\frac{t_1^2}{2}-\frac{\theta t_1^3}{6}\right)+d\left\{\left(\frac{t_2^2}{2}-t_2 t_1+\frac{t_1^2}{2}\right)+\frac{\theta(t_2-t_1)^3}{6}\right\}\right]$$

$$\text{HC} = \text{C}_1\left[\{K(1-\phi)-d\}\left(\frac{t_1^2}{2}-\frac{\theta t_1^3}{6}\right)+d\left\{\frac{(t_2-t_1)^2}{2}+\frac{\theta(t_2-t_1)^3}{6}\right\}\right] \tag{16.9}$$

Let LS denote the cost for loss of stock due to deterioration in each cycle, which is given by

$$\begin{aligned}\text{LS} &= p\left[\theta\left\{\int_0^{t_1} q_1(t)\,dt + \int_{t_1}^{t_2} q_2(t)\,dt\right\}\right] \\ &= p\left\{K(1-\phi)t_1 - dt_2\right\}\end{aligned} \tag{16.10}$$

Let LP denote the cost for loss of production quantity per cycle due to a faulty machine, and is given by

$$\text{LP} = rK\varphi t_1 \tag{16.11}$$

Also, the number of cycles in the entire planning horizon T is T/t_2. Hence, the total inventory cost, TC, for the whole planning horizon, T, is given by

$$\begin{aligned}TC &= \left[HC + LS + LP + b\right]\frac{T}{t_2} \\ &= \left[\begin{array}{l} C_1\left[\left\{K(1-\phi)-d\right\}\left(\frac{t_1^2}{2}-\frac{\theta t_1^3}{6}\right)+d\left\{\frac{(t_2-t_1)^2}{2}+\frac{\theta(t_2-t_1)^3}{6}\right\}\right] \\ +p\left\{K(1-\phi)t_1 - dt_2\right\} + rK\phi t_1 + b \end{array}\right] \\ &= f_1(q) + \theta f_2(q)\end{aligned} \tag{16.12}$$

where

$$f_1(q) = \frac{C_1T}{2}\left\{1-\frac{d}{K(1-\phi)}\right\}q + \left(b + \frac{rq\phi}{1-\phi}\right)\frac{dT}{q} \tag{16.13}$$

$$f_2(q) = \frac{T}{2}\left\{\left(b + pq + \frac{rq\phi}{1-\phi}\right) + \frac{C_1q^2}{6}\left\{\frac{d}{K^2(1-\theta)^2} - \frac{1}{d}\right\}\right\} \tag{16.14}$$

The model developed above gives the following particular cases:
Case (i): If $\phi \to 0, \quad K \to \infty, \quad \theta \to 0$, we obtain

$$TC = \frac{C_1qT}{2} + \frac{bdT}{q}$$

This equation is the same as the average total cost of a classical EOQ model with constant demand and no deterioration.

Case (ii): when $\theta \to 0$, we get

$$TC = \frac{C_1 q T}{2} + \frac{bdT}{q} + \frac{\theta T}{2}\left(b + pq - \frac{C_1 q}{6}\right)$$

Which is an EOQ model under deterioration and constant demand.

Case (iii): when $\theta \to 0$, we get

$$TC = \frac{C_1 q T}{2}\left(1 - \frac{d}{k}\right) + \frac{bdT}{q} + \frac{\theta T}{2}\left(1 - \frac{d}{k}\right)\left\{b + pq - \left(1 + \frac{d}{k}\right)\frac{C_1 q^2}{6}\right\}$$

This is the classical Economic Production Quality (EPQ) model under deterioration and constant demand.

Case (IV): For $\phi \to 0,\ \ \theta \to 0$, we get

$$f_1(q) = \frac{C_1 T}{2}\left(1 - \frac{d}{k}\right)q + b.\frac{dT}{q}$$

$$TC = \frac{C_1 q T}{2}\left(1 - \frac{d}{k}\right) + b.\frac{dT}{q}$$

16.3 DEVELOPMENT OF A FUZZY MODEL

Earlier authors have assumed that the deterioration rate is a constant fraction of the on-hand inventory level in the development of Economic Order Quantity models. But in the real world, it is not always easy to determine the deterioration rate exactly, but in most cases, it is uncertain in nature. The Economic Order Quantity model has been developed with the assumption that the deterioration rate is a fuzzy number $\tilde{\theta}$ instead of considering the deterioration rate as constant.

Thus, the function of (1.3.12) can be redefined as

$$\min\ TC = f_1(q) + \tilde{\theta}\, f_2(q) \text{ where } q \geq 0 \tag{16.15}$$

where Wavy bar ~ represents the fuzzification of the parameters.

We express the fuzzy deterioration rate $\tilde{\theta}$ as a normal triangular fuzzy number $(\theta_1, \theta_0, \theta_2)$.

Suppose the membership function of the fuzzy deterioration rate $\tilde{\theta}$ is as follows:

$$\mu_{\tilde{\theta}}(\theta) = \begin{cases} \dfrac{\theta - \theta_1}{\theta_0 - \theta_1} & \text{for } \theta_1 \le \theta \le \theta_0 \\ \dfrac{\theta_2 - \theta}{\theta_2 - \theta_0} & \text{for } \theta_0 \le \theta \le \theta_2 \\ 0 & \text{elsewhere} \end{cases} \tag{16.16}$$

where θ_1, θ_0, and θ_2 are positive values and $0 \le \theta_0 < \theta_2$.

The centroid of $\mu_{\tilde{\theta}}(\theta)$ is given by

$$M_0\ (\theta_1,\ \theta_0,\ \theta_2) = (\theta_1 + \theta_0 + \theta_2)/3$$

For any positive numbers C_1, d, b, T, φ, p, r, k and every fixed value of $q > 0$.

Let $G_q(\theta)$ denote the total cost function $TC(q,\theta)$ for q, θ.

$$G_q\ (\theta) = f_1(q) + \theta\ f_{2\neg}(q) = z$$

Then we have

$$\theta = \frac{z - f_1(q)}{f_2(q)} \qquad \text{for } f_2(q) \ne 0 \tag{16.17}$$

$$\Rightarrow z \ge f_1(q) = z_1(q)$$

By extension principle, the membership function for fuzzy cost function $G_q\ (\tilde{\theta})$ is given by

$$\begin{aligned} \mu_{G_q(\tilde{\theta})} &= Sup_{\theta \in G_q^{-1}(z)} \mu_{\tilde{\theta}} && \text{if } G_q^{-1}(z) \ne \varphi \\ &= 0 && \text{if } G_q^{-1}(z) = \varphi \end{aligned} \tag{16.18}$$

Thus, if $z \ge z_1(q)$, then

$$\begin{aligned} \mu_{G_q(\tilde{\theta})}(z) &= \mu_{\tilde{\theta}}(\theta) \\ &= 0 \quad \text{else.} \end{aligned}$$

Now let

$$C_j(q) = f_1(q) + \theta_j f_2(q)$$

$$\text{So} \quad C_j(q) - z_1(q) = \theta_j f_2(q) \ge 0 \quad \Rightarrow z_1(q) \le C_j(q)$$

$$\begin{aligned}&\text{and}\quad C_i(q)-C_j(q)=(\theta_i-\theta_j)f_2(q) \qquad \text{for } j=0,1,2\\&\text{and}\quad 0\le\theta_1<\theta_0<\theta_2\end{aligned} \tag{16.19}$$

Further,

$$\theta_i \ge \theta_j \Leftrightarrow C_i(q)\ge C_j(q) \quad \text{for i}=0,2,\text{j}=1 \quad \text{or} \quad \text{i}=2,\text{j}=0$$

$$\text{and}\quad \theta_i \le \theta_j \Leftrightarrow C_i(q)\ge C_j(q) \quad \text{for}(\text{i}=0,1,\text{j}=2)\text{or}(\text{i}=1,\text{j}=0)$$

$$\text{P}=\int_{-\infty}^{\infty}\mu_{G_q(\tilde{\theta})}(z)\,dz$$

$$R=\int_{-\infty}^{\infty}\mu_{G_q(\tilde{\theta})}(z)\,dz$$

The centroid for $\mu_{G_q(\tilde{\theta})}(z)$ is given by R/P, which is the estimate of the total cost.

We have $\text{z}_1(\text{q})\le \text{C}_1(\text{q})<\text{C}_0(\text{q})<\text{C}_2(\text{q})$.

Using the conditions (16.18) to (16.19).

The membership function of $G_q(\tilde{\theta})$ can be defined as:

$$\begin{aligned}\mu_{G_q(\tilde{\theta})}(z)&=\frac{z-f_1(q)-\theta_1 f_2(q)}{(\theta_0-\theta_1)f_2(q)} \quad \text{for } C_1(q)\le z\le C_0(q)\\&=\frac{f_1(q)+\theta_2 f_2(q)-2}{(\theta_0-\theta_1)f_2(q)} \qquad \text{for } C_0(q)\le z\le C_2(q)\\&=0, \quad \text{else where}\end{aligned} \tag{16.20}$$

Therefore, P and R can be written as:

$$\begin{aligned}P&=\frac{1}{(\theta_0-\theta_1)f_2(q)}\int_{C_1(q)}^{C_0(q)}\left(z-f_1(q)-\theta_1 f_2(q)\right)dz\\&+\frac{1}{(\theta_2-\theta_0)f_2(q)}\int_{C_1(q)}^{C_2(q)}\left\{f_1(q)+\theta_2 f_2(q)-z\right\}dz\\&=\frac{(\theta_2-\theta_1)f_2(q)}{2}\end{aligned} \tag{16.21}$$

$$R = \frac{1}{(\theta_0 - \theta_1) f_2(q)} \int_{C_1(q)}^{C_0(q)} z\{z - f_1(q) - \theta_1 f_2(q)\} dz$$

$$+ \frac{1}{(\theta_2 - \theta_0) f_2(q)} \int_{C_1(q)}^{C_2(q)} \{f_1(q) - \theta_2 f_2(q) - z\} dz \tag{16.22}$$

$$= (\theta_2 - \theta_1) f_2(q) - \{3 f_1(q) + (\theta_0 + \theta_1 + \theta_2) f_2(q)\} / 6$$

After evaluating the integral of R and P and with simple algebra, we obtain the centroid of

$$\mu_{G_q \tilde{\theta}}(z) \text{ as}$$

$$\frac{R}{P} = f_1(q) + \frac{1}{3}(\theta_1 + \theta_0 + \theta_2) f_2(q) \tag{16.23}$$

16.3.1 Optimal solution of the inventory model

Given θ_0, we have to find $(\theta_1^*, \theta_2^*, q^*)$ such that the centroid of the fuzzy total cost is minimal, that is,

$$\min \frac{R(\theta_1, \theta_2, q)}{P(\theta_1, \theta_2, q)} = \frac{R(\theta_1^*, \theta_2^*, q^*)}{P(\theta_1^*, \theta_2^*, q^*)} \tag{16.24}$$

Since the equation (16.23) is a highly nonlinear equation, it is difficult to minimize the cost function analytically. We have used the standard software package LINGO for minimization of the cost function (16.24).

Then, we regard it as the EPQ.

Also,

$$\hat{\theta}_0 = \frac{1}{3}(\theta_1^* + \theta_0^* + \theta_2^*)$$

is the centroid of the normal fuzzy triangular number $(\theta_1^*, \theta_0^*, \theta_2^*)$.

16.4 NUMERICAL ILLUSTRATION OF THE MODEL

The following numerical example has been considered to illustrate the Inventory model.

Let,

$C_1 = 10$
$D = 2$
$b = 500$
$T = 40$
$p = 3$
$r = 1$
$k = 10$
$\theta = 0.003$
$\phi = 0.005$

In appropriate units.

With these parameter values, the solution of the equation (16.24) and using the software LINGO, the numerical results are obtained as:

$$q^* = 15.89, \theta_1^* = 0.001, \theta_2^* = 0.005$$

$$and \ \frac{R}{P} = 5071.059$$

$$\text{Thus } \theta_0^* = 0.003 \quad if \qquad \theta_0 = 0, \theta_1 = 0, \theta_2 = 0$$

$$\text{Then } \ q^* = 15.821$$

$$\text{and } \ \frac{R}{P} = 5056.866$$

16.5 SENSITIVITY ANALYSIS OF THE MODEL

We have examined the sensitivity of each of the decision variables $q^*, \theta_1^*, \theta_2^*$, using the numerical example and the centroid of the fuzzy total cost R/P, to changes in each of the eight parameters e, d, b, p, r, k, θ_0 and φ.

Here, each parameter has been varied from -50% to +50%, keeping other parameters constant. The results obtained are displayed in Table 16.1.

We see from Table 16.1 that all the decision variables are sensitive to changes in C_1. The total cost decreases by 29% for a 50% decrease in C_1 and increases by 22% for a 50% increase in C_1. All the decision variables are sensitive to changes in d.

For a 50% increase, q and total cost increase by 31% and 14%, respectively. On the other hand, when d is reduced by 50%, q and total cost are decreased by 33% and 24%, respectively. For changes in the parameter b, we observe that q and total cost are sensitive, whereas θ_1 and θ_2 are much less sensitive. The decision variable q and total cost increase and decrease by 22% and 29%, respectively, for a 50% increment and decrement in b.

Table 16.1 Effect of changes in parameters on the decision variables

Parameters	Changes (%)	q^*	θ_1^*	θ_2^*	Total cost (R/P)
C	+50	12.964	0.00000	0.00600	6207.063
	+20	14.501	0.00000	0.00590	5553.479
	−20	17.774	0.00097	0.00503	4537.498
	−50	22.507	0.00000		3590.985
D	+50	22.809	0.00156	0.00444	5801.410
	+20	17.859	0.00300	0.00300	5411.154
	−20	13.872	0.00091	0.00508	4650.951
	−50	10.601	0.00028	0.00572	3811.510
B	+50	19.483	0.00043	0.00557	6213.923
	+20	17.415	0.00300	0.00300	5556.267
	−20	14.205	0.00000	0.00599	4534.615
	−50	11.220	0.00300	0.00300	3583.439
P	+50	15.887	0.00108	0.00492	5072.201
	+20	15.889	0.00106	0.00494	5071.516
	−20	15.892	0.00104	0.00493	5070.602
	−50	15.894	0.00101	0.00499	5069.926
R	+50	15.890	0.00105	0.00495	5071.262
	+20	15.890	0.00105	0.00495	5071.140
	−20	15.890	0.00105	0.00495	5070.978
	−50	15.890	0.00105	0.00495	5070.856
K	+50	15.256	0.00300	0.00300	5281.584
	+20	15.564	0.00107	0.00493	5177.367
	−20	16.421	0.00099	0.00500	4907.379
	−50	18.399	0.00037	0.00563	4380.649
	+50	15.925	0.00158	0.00742	5048.119
	+20	15.904	0.00126	0.00594	5073.886
	−20	15.876	0.00240	0.00240	5068.228
	−50	15.856	0.00150	0.00150	5063.975
	+50	15.895	0.00300	0.00300	5069.640
	+20	15.892	0.00105	0.00495	5070.492
	−20	15.888	0.0105	0.00495	5071.624
	−50	15.885	0.00300	0.00300	5072.470

BIBLIOGRAPHY

1. Perumal, V. and Arivarignass, G. (2002): A production model with two rates of productions and backorders. *International Journal of Management System* Vol. 18, 109–119.
2. Nobel, R.D. and Headen, M. (2000): A lost sales production/inventory model with two discrete production models.
3. Ouyang, L.Y. Chen, C.K. and Chang, H.C. (1999): Lead time and ordering cost reductions in continuous review inventory system with partial back orders. *Journal of Operations Research Society*, Vol. 50, 1277–1279.

4. Harris, F.: *Operations and costs* (Factory Management Series), pp. 18–52. A.W.Shaw Co, Chicago (1915).
5. Chare, P., Schrader, G.: A model for exponentially decaying inventories, *Journal of Industrial Engineering*, Vol. 15, pp. 238–243, 1963.
6. Goel, V.P., Aggarwal S.P.: Order level inventory system with power demand pattern for deteriorating items, Proceedings of all India seminar on Operational Research and Decision making, University of Delhi, Delhi, 1981.
7. Covert, R.P., Philip, G.C.: An EOQ model for items with Weibull distribution deterioration, *AIIE Trans.*, Vol. 5, pp. 323–326, 1973.
8. Aggarwal, S.P.: A note on an order – level inventory model for a system with constant rate of deterioration, *Opsearch*, Vol. **15**, pp. 184–187, 1978.
9. Cohen, A.C.: The reflected Weibull distribution, *Tecnometrics*, Vol. **15**. pp. 867–873, 1973.
10. S.P. Singh and V.K. Sehgal "An EOQ model with power demand, generalized Weibull distributed deterioration and shortages." in *International Journal of Ultra Scientist of Physical Sciences*, Vol. 23, No. 1, 2011, pp. 131–138.
11. S.P. Singh and G.C. Panda "An Inventory model forgeneralised Weibull deteriorating items with price dependent demand and permissible delay in payments under inflation." *Log Forum*, Vol. 11(3), 2015, 259–266.
12. S.P. Singh and Ajay Soni "An EOQ inventory model using ramp type demand with deterioration and shortages." *International Journal of Engineering Technologies and Management Research*,Vol. **5** (Iss. 2: SE): 2018, 334–345.
13. S.P. Singh and V.K. Sehgal "An EOQ Inventory Model for Weibull distributed deteriorating Items with Power Demand Pattern and Shortages." *JP Journal of Mathematical Sciences*, Vol. **1**, No 2, 2011, pp. 99–110
14. Datta, T.K., Pal, A.K.: Order level inventory system with power demand pattern for items with variable rate of deterioration, *Indian Journal of Pure and Applied Mathematics*, Vol. **19**, No. 11, pp. 1043–1053, 1988.
15. Deb, M., Chaudhari, K.S.: An EOQ model for items with finite rate of production and variable rate of deterioration, *Opsearch*, Vol. **23**, pp. 175–181, 1986.
16. Pal, A.K., Mandal, B.: An EOQ model for deteriorating inventory with alternating demand rates, *The Korean Journal of Computational and Applied Mathematics*, vol. **4**, No. 2, pp. 397–407, 1997.
17. Dave, U., Patel, U.K.: (t, s_i) Policy inventory model for deteriorating items with time proportional demand, *Journal of Operational Research Society*, Vol. **32**, pp. 137–142, 1981.
18. Shah, Y.K., Jaiswal, M.C.: An order level inventory model for a system with constant rate of deterioration, *Opsearch*, Vol. **14**, pp. 174–184, 1977.
19. Jalan, A.K., Giri, R.R. and Chaudhuri, K.S.: EOQ model for items Weibull distribution deterioration, shortages and trended demand, *International Journal of Systems Science*, Vol. **27**, pp. 851–855, 1996.
20. S.P. Singh and V.K. Sehgal "A Fuzzy-Stochastic Approach to Multi-Objective Inventory Model of deteriorating Items with Purchasing Cost as Weibull distribution." *JP Journal of Mathematical Sciences*, Vol. 3, Issues 1 & 2, 2011, pp. 1–21.

Chapter 17

A periodic review Inventory model with emergency order using Weibull Distribution for demand items under Bayesian approach

S.P. Singh

17.1 INTRODUCTION

There is a provision for placing an order only at discrete time points in a periodic review inventory system. If the time between two consecutive reviews is large in the presence of random demands, there could be heavy shortages, and the profitability of the system could suffer more when the shortage cost is high. However, it may not be feasible/profitable to carry a high level of stock in anticipation of the uncertain demand, especially if the holding cost is significant. To meet the excess demand in such a situation, one or more emergency orders can be placed before the end of the review period. Taking into account the nature of the demand and the supply source, the timing and size of the emergency order can be determined optimally.

Therefore, for the optimization of the model, an accurate estimation of demand is important. There are quite a few articles dealing with emergency orders, either in the case of continuous or periodic review or even a single-period model.

Rosenshine, M., and Obee, D. (1976) have analyzed a standing order inventory system with emergency orders. Kodama (1995) has developed a probabilistic single-period inventory model with partial returns and additional orders. Zhang (1996) has developed ordering policies for an inventory system with three supply modes. Chiang and Gutierrez (1996) have developed a periodic review inventory system with two supply modes. Chiang and Gutierrez (1998) have obtained an optimal control policy for a periodic review inventory system with emergency orders. Johansen and Kok (1999) have developed a two-supplier inventory model. Johansen and Thorstenson (1998) have developed an inventory model with Poisson demands and emergency orders.

Lau and Lau (1998) have developed a decision model for single-period products with two ordering opportunities. Mohebbi and Posner (1999) have considered a lost-sales continuous review inventory system with emergency ordering. The main intention of placing emergency orders in all these research papers was to reduce the lead time, that is, the delay in meeting the demand. Whereas in this chapter, the purpose is to satisfy the demand as

DOI: 10.1201/9781003642152-17

soon as a shortage occurs. Hence, no lead time is assumed, because of which the purchase price and the setup costs for emergency orders could be high. However, these will be balanced against higher shortage costs.

Kamath and Pakkala (2006) presented an inventory model as a mixture of periodic and continuous review models, which makes the model quite different from other researchers. In fact, they discussed two operational policies, which will be referred to as policy I and policy II. The choice between these policies depends on how the purchasing manager functions. One of the policies is very appealing because, quite often, orders are not placed based on the inventory level. This may be because the managers do not receive information about inventory levels at various points. Those who decide not to monitor the inventory levels may incur additional costs as a result.

The purchasing manager can have two policies, which are as follows:

Policy I: A periodic review policy is considered with regular orders placed according to the (s, S) policy at each review point. Further, emergency orders are placed if required when the inventory level reaches a reorder level; that is, an (s, Q) in continuous time within the review period is considered.

Policy II: At the beginning of each review period, as in policy I, an (s, S) policy is adopted. The timing and the size of the emergency orders to be placed within the review period are also decided at the beginning of the review period according to a (t, Q) policy.

These two policies differ according to when the manager decides to place emergency orders. In policy I, the decision regarding when to place emergency orders is made. Whereas in policy II, the decision is taken at the beginning of the review period.

Policy I is economical when the purchasing manager gets information and when the inventory level depletes, whereas policy II is used when the purchasing manager does not get information regarding inventory levels continuously.

Policy I may be a feasible option when there is a special low buying price agreement with the supplier. If the continuous monitoring incurs significant cost, then policy II may be used.

The mean demand is considered to be a random variable, so that fluctuating demand is considered in the model. The Bayesian approach has been used for updating the resulting demand distribution. It has been found that the Bayesian approach is appropriate for products such as those mentioned in the beginning. Especially when policy II is used, the time between the placement of consecutive emergency orders depends on the estimate of the demand. Even in the case of policy II, the solution depends on the demand parameters. Hence, it would be of interest to know whether the Bayesian updating of the distribution of the demand improves the accuracy of the solution to the policies just discussed.

17.2 NOTATIONS OF THE INVENTORY MODEL

1. N = Number of review periods.
2. D_i = Demand in the ith successive review periods $i = 1, 2, \ldots\ldots N$
3. $f_i(.)$ = Density function for the ith demand.
4. s = Optimum inventory policy for $i = 1, 2, \ldots\ldots N$.
5. C_1 = The holding cost per unit in any period.
6. C_r = The purchase cost per unit.
7. C_e = The emergency purchase cost per unit.
8. C_{3r} = Set up cost per order.
9. C_{3e} = Emergency setup cost per order.
10. p = Selling price per unit.
11. C_2 = Backordered at cost per unit.
12. s_1 = Regular reorder level.
13. s_2 = Emergency reorder level.
14. S_1 = Regular order level.
15. Q = The emergency order quantity.
16. μ = The discount factor.
17. x = Inventory level at the beginning of the review period with $x < s_1$.
18. k = Number of emergency orders.
19. $U_i(S_1, Q, s_2)$ = The function of S_1, Q, and s_2 for the ith review period.
20. $V_k(S_1, Q, s_2)$ = The expected single-period profit
21. λ = Mean of the demand distribution.
22. $\lambda_1 = (C_1 + C_2)/2$
23. $\lambda_2 = \lambda_1 + p(1 - \mu)$
24. $\lambda_3 = \mu p - C_2/2$
25. $P_{N+1}(.)$ = Salvage or penalty costs

17.2.1 Model development and analysis of the inventory model

For policy I

$$P_i(x) = \begin{Bmatrix} U_i(S_i, Q_i, s_2) + C_r x - C_{3r} \\ \qquad\qquad \text{if } x < s_1 \\ U_i(x, Q, s_2) + C_r x \\ \qquad\qquad \text{if } x \geq s_1 \end{Bmatrix} \tag{17.1}$$

$$U_i\left(S_1,Q,s_2\right)=-C_r s_1+\sum_{k=0}^{\infty}V_k\left(S_1,Q,s_2\right)P\left[K=k\right] \tag{17.2}$$

$$\begin{aligned}V_k\left(S_1,Q,s_2\right)&=\left\{\lambda_2-\mu\left(\sigma_1-\sigma_2\right)\right\}\int_0^Z uf\left(u\right)du-\left(\mu\sigma_2+\lambda_1+\lambda_2\right)Z\ F\left(Z\right)\\&+\left(\lambda_3-\mu\sigma_2\right)\lambda+\left(\lambda_1+\lambda_2-C_1+\mu\sigma_2\right)Z-\lambda_1 Z^2\int_Z^{\infty}\frac{f\left(u\right)}{u}du\\&-k\left[C_{3e}+C_eQ+\left\{C_2 s_2{}^2-C_1\left(S_1+Z-\left(Q+s_2\right)^2\right)\right\}\int_0^{\infty}\frac{f\left(u\right)}{2u}du\right]\\&\text{for }\ k=\ 0,1,2.......\end{aligned} \tag{17.3}$$

Next, we want to obtain the probability of placing k emergency orders.

Suppose that an emergency order is placed at t_1; then for the given demand d over the review period, the residual demand $d(1-t_1)$ should be met by the emergency orders. If this has to be met by exactly k emergency orders, then the condition will be

$$(k-1)Q<d(1-t_1)\le kQ$$

In terms of S_1 and s_2, t_1 can be written as $(S_1-s_2)/d$.

Hence, the condition reduces to

$$S_1-s_2+(k-1)Q<d\le S_1-s_2+kQ$$

Thus, the probability of placing k emergency orders is given by

$$\begin{aligned}&P\left[K=k\right]=F\left(S_1-s_2+kQ\right)-F\{(S_1-s_2+\left(k-1\right)Q\\&\text{for }\ k=1,2,.....\\&=F\left(S_1-s_2\right)\qquad\quad \text{for }\ k=0\end{aligned}$$

17.2.2 Analysis of policy II

The solution for policy II can be obtained by using the solutions obtained for policy I. The decision is taken at the beginning of each review period regarding when to place emergency orders for policy II. Thus, the time points at which these emergency orders are placed do not depend on demand variations within the review period.

This is then maximized with respect to the decision variables S_1, Q, and t_1. Once the optimal values of t_1 and Q are fixed based on the expected demand λ, the number of emergency orders can be estimated as

$$\hat{k} = \left[\frac{\lambda(1-t_1)}{Q}\right] + 1$$

The interval between any two consecutive emergency orders is given by

$$\hat{t}_2 = \frac{(1-t_1)}{\hat{k}}$$

Thus, the solutions for policy I and policy II are approximately equivalent, and they differ only operationally. Therefore, the numerical study for policy II is not performed.

17.2.3 The inventory model without emergency orders

The dynamic programming formulation of the model without emergency order is given by

$$p_i[x] = \{k_1(s) + C_{1r}x - C_{3r}\}$$

where S' and s' are the optimal orders and reorder levels if no emergency orders are placed.

$$K_i(y) = -C_r y + N_0(y)$$

with

$$N_0(y) = \int_0^y \{pu - C_1(2y-u)/2\}\, f(u)\, du + \mu \int_0^\infty P_{i+1}(y-u)\, f(u)\, du + \int_y^\infty \left[\{py + \mu p(u-y)\} - \frac{\{C_1 y^2 + C_2(u-y)^2\}}{2u}\right] f(u)\, du \qquad (17.4)$$

17.2.4 Bayesian approach with updating of the demand

The value of the mean demand with the density function given by

$$f(d_i\alpha)=\frac{d_i^{k-1}e^{-d_i/\alpha}}{\Gamma(k)\alpha^k};\ d_i>0,\ k>0,\ \alpha>0 \tag{17.5}$$

The mean demand and variance of the demand are given by

$$E(D_i)=\int_0^\infty x.\frac{x^{k-1}e^{-x/\alpha}}{\Gamma(k)\alpha^k}\,dx$$

$$=\frac{1}{\Gamma(k)\alpha^k}.\frac{k\Gamma(k)}{\frac{1}{\alpha^{k+1}}}$$

$$E(D_i)=\alpha k$$

$$E\left(D_i^2\right)=\frac{1}{\Gamma(k)\alpha^k}\int_0^\infty x^{(k+2)-1}e^{-x/\alpha}\,dx$$

$$=\frac{1}{\Gamma(k)\alpha^k}.\frac{k(k+1)\Gamma(k)}{\frac{1}{\alpha^{k+2}}}$$

$$E\left(D_i^2\right)=\alpha^2k(k+1)$$

now

$$\begin{aligned}\text{Var}(\text{D}_\text{i})&=E\left(D_i^2\right)-\left[E(D_i)\right]^2\\&=k(k+1)\alpha^2-k^2\alpha^2\\&=k\alpha^2\end{aligned}$$

So that the coefficient of variation is given by

$$C.V=\frac{\sigma}{\bar{D}_i}=\frac{\sqrt{\alpha^2k}}{\alpha k}=k^{-\frac{1}{2}} \tag{17.6}$$

$$g(\theta d_i)=\frac{\left(\prod_{i=1}^{n}f(d_i\theta)\right)g(\theta)}{\int\left\{\prod_{i=1}^{n}f(d_i\theta)\right\}g(\theta)\,d\theta}$$

The marginal density function of demand is given by

$$f_{D_i}(d_i) = \int_0^n \left(\prod_{i=1}^n f(d_i\theta) \right) g(\theta)\, d\theta$$

$$= \frac{\Gamma(nk+a)}{\Gamma(a)\left[\Gamma(k)\right]^n} \frac{b^a}{\left(b + \sum_{i=1}^n d_i\right)^{nk+a}} \left(\prod_{i=1}^n d_i \right)^{k-1}$$

$$f(d_i\ \alpha) = \frac{k}{\alpha} d_i^{k-1} e^{-\frac{d_i^k}{\alpha}};\ d_i > 0,\ k > 0,\ \alpha > 0 \tag{17.7}$$

The mean demand and variance of the demand are

$$E(D_i) = \alpha^{1/k} \Gamma\left(\frac{1}{k} + 1\right)$$

and

$$E(D_i^2) = \int_0^\infty x^2 f(x\ \alpha)\, dx$$

$$= \alpha^{2k} \Gamma\left(\frac{2}{k} + 1\right) \tag{17.8}$$

Now, since

$$\begin{aligned} V(D_i) &= E(D_i^2) - \left[E(D_i^2)\right]^2 \\ &= \alpha^{2/k} \left\{ \Gamma\left(\frac{2}{k}+1\right) - \left[\Gamma\left(\frac{1}{k}+1\right)^2\right] \right\} \end{aligned} \tag{17.9}$$

So the coefficient of variation is given as

$$\begin{aligned} C.V = \frac{S.D}{\bar{D}_i} &= \frac{\sqrt{\alpha^{2/k} \left\{ \Gamma\left(\frac{2}{k}+1\right) - \left[\Gamma\left(\frac{1}{k}+1\right)^2\right] \right\}}}{\alpha^{1/k} \Gamma\left(\frac{1}{k}+1\right)} \\ &= \left\{ \Gamma\left(\frac{2}{k}+1\right) - \left[\Gamma\left(\frac{1}{k}+1\right)^2\right] \right\} \end{aligned} \tag{17.10}$$

Suppose α follows the conjugate prior distribution of the Weibull distribution, which is the inverse Gamma distribution; that is, α^{-1} follows the Gamma distribution with known parameters a and b^{-1}; that is,

$$h(\alpha;a,b)=\frac{b^{a}}{\Gamma(a)}\left(\frac{1}{\alpha}\right)^{a+1}e^{-b/\alpha};\ a,b>0,\theta>0 \tag{17.11}$$

that is, $\alpha \sim \mathrm{IG}(a,b)$

that is, $\alpha^{-1} \sim \mathrm{G}(a,b^{-1})$

We know that a, b are shape and scale parameters of the Gamma distribution $\mathrm{G}(a,b)$, respectively.

The mean and variance are

$$E(\alpha;a,b)=\frac{b^{a}}{a-1},\ a>1$$

$$Var(\alpha;a,b)=\frac{b}{(a-1)^{2}(a-2)},\ a>2$$

It is observed that if a < 1, the mean does not exist, and if 1 < a ≤ 2, the variance does not exist.

Then, its posterior distribution is given by n and past demand distributions $D_1, D_2, \ldots D_n$ is given by

$$h(\alpha/d_i)=\frac{\left(\prod_{i=1}^{n}f(d_i/\alpha)\right)h(\alpha)}{\int\left\{\prod_{i=1}^{n}f(d_i/\alpha)\right\}h(\alpha)\ d\alpha}$$

Now

$$\prod_{i=1}^{n}f(d_i/\alpha)=\prod_{i=1}^{n}\left(\frac{k}{\alpha}\right)d_i^{k-1}\ e^{-d_i^k/\alpha}$$

$$=\left(\frac{k}{\alpha}\right)^{n}\left(\prod_{i=1}^{n}d_i\right)^{k-1}e^{-\sum_{i=1}^{n}d_i^k/\alpha}$$

17.3 NUMERICAL STUDY OF THE INVENTORY MODEL

In the present study, we have considered the demand to be a Weibull distribution with a fixed coefficient of variation. The significance of accurate estimation of the demand is investigated through a comparison between Bayesian and non-Bayesian approaches. The following are the values for the various costs and parameters:

μ = discount factor = 1/1.1
$C_r = 1$
$C_{3r} = 5$
$C_1 = 0.30$
$C_2 = 5$
$\sigma_1 = 0.1$
$\sigma_2 = 3$
$N = 2$

17.3.1 Comparison of emergency orders and non-emergency orders

The percentage gain in profit by using emergency orders is given by

$$\nu = \frac{\eta_1 - \eta_2}{\eta_1} \times 100\%$$

where η_1 is the profit for the model with emergency orders.

η_2 is the profit for the model without emergency orders.

The values taken by ν for various values of C_e and C_{3e} are given in Table 17.1.

It is observed that the percentage gain in profit decreases with C_e as well as C_{3e}. This behavior is expected because placing emergency orders becomes more and more expensive as the related cost increases.

The first entry of Table 17.1 corresponds to the case where the regular and emergency ordering costs coincide. Thus, maximum profit can be earned if a supplier agrees to supply additional orders when required.

Further, if either C_e or C_{3e} is quite high, then no emergency orders are placed, so that the model solution reduces to the solution of the model without emergency orders. The percentage gain in profit given for various values of the holding cost and shortage costs is given in Table 17.2. The fixed and variable emergency ordering costs were fixed at $C_{3e} = 20$ and $C_e = 1.5$.

We observe from Table 17.2 that the gain in profit decreases with the holding cost, which is expected because the items ordered through the emergency channel incur additional holding costs, and these costs increase as the holding cost increases. Further, if the shortage is made up immediately through emergency orders, the gain in profit increases as the shortage cost increases.

Table 17.1 The percentage gain in profit by using the option of placing emergency orders for various values of C_{3e} and C_e

C_{3e} X C_e	*5*	*7.5*	*10*	*12.5*	*15*	*17.5*	*20*	*22.5*	*25*
1.00	15.822	15.690	15.563	15.440	15.321	15.205	15.091	14.981	14.872
1.10	13.647	13.484	13.331	13.184	13.034	12.895	12.765	12.631	12.512
1.20	11.668	11.483	11.312	10.993	10.883	10.617	10.435	10.258	10.085
1.25	10.734	10.510	10.303	10.100	9.909	9.724	9.368	9.207	8.966
1.30	9.884	9.637	9.433	9.236	9.054	8.856	8.820	8.678	8.540
1.40	8.201	7.964	7.754	7.554	7.361	7.178	7.001	6.828	6.750
1.50	6.606	6.365	6.147	6.150	5.946	5.759	5.57	5.401	5.232

Table 17.2 The percent gain in profit by using the option of placing emergency orders

C_1 X C_2	*0.05*	*0.10*	*0.15*	*0.20*	*0.25*	*0.30*
4	3.379	2.473	1.744	1.619	1.566	1.507
5	7.785	6.491	5.694	5.611	5.589	5.578
6	11.302	10.452	9.718	8.951	8.552	7.941
7	19.926	18.525	16.361	15.310	14.134	13.465
8	25.492	24.371	22.209	21.473	20.129	19.262
9	33.105	32.051	30.416	29.256	28.157	27.431
10	41.270	39.406	38.115	37.396	37.252	37.114

BIBLIOGRAPHY

Datta, T.K., Pal, A.K.: Order level inventory system with power demand pattern for items with variable rate of deterioration, *Indian Journal of Pure and Applied Mathematics*, Vol. 19, No. 11, pp. 1043–1053, 1988.

Dave, U., Patel, U.K.: (t, s_i) Policy inventory model for deteriorating items with time proportional demand, *Journal of Operational Research Society*, Vol. 32, pp. 137–142, 1981.

Deb, M., Chaudhari, K.S.: An EOQ model for items with finite rate of production and variable rate of deterioration, *Opsearch*, Vol. 23, pp. 175–181, 1986.

Donaldson, W.A.: Inventory replenishment policy for a linear trend in demand – An analytical solution, *Operational Research Quarterly*, Vol. 28, pp. 663–670, 1977.

Dye C.Y. (2002). A deteriorating inventory model with stock Dependent demand and partial Backlogging under condition of permissible delays in payments, *Opsearch*, 39, No. 3 & 4, 189–201.

Ghosh, S.K., Chaudhari, K.S.: An order level inventory model for a deteriorating item with Weibull distribution deterioration, time quadratic demand and shortages, *Advanced Modeling and Optimization*, Vol. 6, pp. 21–35, 2004.

Jalan, A.K., Giri, R.R. and Chaudhuri, K.S.: EOQ model for items Weibull distribution deterioration, shortages and trended demand, *International Journal of Systems Science*, Vol. **27**, pp. 851–855, 1996.

Mondal, M. and Maiti, M. (1997). Inventory Model for damageable items with stock dependent demand and shortages. *Opsearch*, **34**, 155–166.

Pal, A.K., Mandal, B.: An EOQ model for deteriorating inventory with alternating demand rates, *The Korean Journal of Computational and Applied Mathematics*, Vol. **4**, No. 2, pp. 397–407, 1997.

Singh, S.P. and Sehgal, V.K.: "An EOQ model with power demand, generalized Weibull distributed deterioration and shortages." in *International Journal of Ultra Scientist of Physical Sciences*, Vol. **23**, No. 1, 2011 a, pp. 131–138.

Silver, E.A.: A simple inventory replenishment decision rule for a linear trend in demand, *Journal of the Operational Research Society*, Vol. **30**, pp. 71–75, 1979.

Singh, S.P., Panda, G.C. "An Inventory model forgeneralised Weibull deteriorating items with price dependent demand and permissible delay in payments under inflation." *Log Forum*, Vol. **11**(3), 2015, 259–266.

Singh, S.P. and Sehgal, V.K. "A Fuzzy-stochastic approach to multi-objective inventory model of deteriorating items with purchasing cost as Weibull distribution." in *JP Journal of Mathematical Sciences*, Vol. **3**, Nos. 1 & 2, 2011b, pp. 1–21.

Shah, Y.K., Jaiswal, M.C.: An order level inventory model for a system with constant rate of deterioration, *Opsearch*, Vol. **14**, pp. 174–184, 1977.

Chapter 18

Finding an optimal solution and optimal cost percentage to a road green freight transportation problem

Bhaskar Chatterjee, Bhawna Agrawal, Gaurav Sharma, Barkha Sharma, and Arti Shrivastava

18.1 INTRODUCTION

The transportation problem presents a significant challenge that businesses, organizations, and firms must address. Gaspard Monge, a French mathematician and economist, is recognized for his contributions to the formulation of this issue in 1871. However, it was not until 1920 that Sarbjit (2012) conducted a formal analysis of the transportation problem. This dilemma impacts various fields, including business, industry, communication networks, genetics, and the scheduling and allocation of transportation. Transportation models primarily aim to optimize the distribution of goods from multiple sources, such as factories or plants, to various destinations, like warehouses. The goal of each transportation problem is to meet the demands of the destination while adhering to capacity limitations, all while minimizing production costs. It is crucial to understand that transferring goods from any origin to any destination requires cost minimization to improve profitability.

18.2 METHODOLOGY

The objective is to identify an optimal solution for the transportation problem while adhering to quality constraints. The necessary information for this study will be sourced from the internet, libraries, mathematical literature, and journals. Data will be specifically collected from the cold drinks company. The transportation of cold drink products for various depots and a retail outlet presents a classic transportation problem. This issue can be addressed using a transportation model that incorporates intermediate destinations between the source and final destination, meaning that goods are transported from manufacturing plants to distribution centers or warehouses and ultimately to retail stores. The model will include pure supply nodes with associated demands, as well as pure demand nodes and transportation nodes. It is assumed that there is unit transportation from supply

DOI: 10.1201/9781003642152-18

nodes i to transshipment nodes k, and a unit transportation cost from transportation nodes k to demand nodes j. The cold drinks transportation problem will be formulated as a linear programming model of the transportation type, represented in tableau form, and subsequently solved by using a mathematical application.

18.3 MATHEMATICAL METHODS

The following outlines recommended mathematical techniques for developing the mathematical model:

Step 1: Formulate the problem statement and express it in matrix format. The formulation of the transportation problem is analogous to that of the linear programming problem. In this context, the total transportation cost acts as the objective function, while the supply and demand at each source and destination represent the constraints.

Step 2: Derive an initial basic feasible solution:

1. The Northwest Corner Rule is one of three methods to obtain an initial solution.
2. The Row Minimum Method
3. The Least Cost Method
4. The Vogel's Approximation Method

Any initial solution derived from these methods must meet the following criteria:

5. The solution must be feasible.
6. The count of positive allocations must equal m + n–1, where m denotes the number of rows and n signifies the number of columns.

Step 3: Assess the initial solution for optimality: Utilize the method to evaluate the optimality of the solution obtained in Step 2. If the current solution is optimal, the process concludes; if not, identify a new, improved solution.

Step 4: Revise the solution: Continue with Step 3 until an optimal solution is achieved.

18.3.1 North-West corner

The North-West Corner method is one of the primary techniques employed to determine and ensure the first viable solution to a transportation problem. Its widespread use can be attributed to its simplicity. The method derives its name from the fact that it selects units starting from the upper left corner. Conventional understanding dictates that a solution to the transportation problem can only be achieved when supply and demand are balanced. In cases where demand surpasses supply, we introduce dummy variables.

This fictitious origin accounts for the difference between total supply and total demand. It is important to note that this artificial origin incurs no cost.

18.3.2 Algorithm of North-West corner

Step 1: Allocate as much as possible to the selected cell, then subtract the allocated amount to adjust the respective supply and demand figures.

Step 2: Indicate that no further allocations can be made in any row or column with zero supply or demand by crossing it out. If both the row and column sum to zero simultaneously, cross out the row (or column).

Step 3: If there is one row or column that remains uncrossed, cease operations; otherwise, move to the cell directly below a crossed-out row or to the right of a recently crossed-out column. Then, return to Step 1.

18.3.3 Least Cost Method (LCM)

By concentrating on the most cost-effective routes, the Least Cost Method (LCM) identifies a more advantageous initial solution. This strategy prioritizes the lowest unit costs, allowing for the assignment of as many calls as possible at the outset.

18.3.4 Algorithm of Least Cost Method (LCM)

Step 1: Identify the cell in the transportation tableau with the lowest unit cost, allocate the maximum feasible amount to it, and eliminate any rows or columns where supply or demand is fully satisfied. If both a row and a column can be crossed out simultaneously, only one should be removed. In cases where the lowest unit cost is not unique, select the cell that allows for the largest allocation.

Step 2: After adjusting supply and demand for all remaining rows and columns, continue the process with the next lowest unit cost from the available options in the transportation tableau. Allocate the maximum possible to this cell and remove the corresponding row or column where either supply or demand has been depleted.

Step 3: Continue this process until the demand at various locations and the total supply from different sources are fully satisfied.

18.3.5 Vogel Approximation Method (VAM)

Each allocation is determined by considering the opportunity cost that would arise if allocations in specific cells with the lowest unit transportation costs were overlooked. The Vogel Approximation Method (VAM) is a heuristic technique that is preferred over the other two methods previously mentioned. This strategy aims to minimize penalty costs, offering an initial solution that is more aligned with the optimal outcome.

18.3.6 Algorithm of VAM

Step 1: Compute the difference between the two lowest distribution costs for each row and column.

Step 2: Identify the row or column with the largest difference and highlight the corresponding value. In the event of a tie, choose the row or column that permits the greatest movement of units.

Step 3: Allocate the maximum feasible amount within the constraints of the selected row or column to the cell with the lowest cost.

Step 4: Eliminate any row or column that has been satisfied by the allocation made in Step 3.

Step 5: Continue repeating Steps 1 to 4, excluding the crossed-out rows and columns, until all allocations are completed. This method is designed to minimize penalty costs and provides an initial solution that is closer to the optimal result.

18.3.7 Stepping Stone Method (SSM)

The SSM is a method used to derive an optimal solution by imposing a limit on the unallocated cells in the initial basic solution, thereby minimizing transportation costs. In cases where two cells have the same allocation (a tie), priority is given to the cell with the larger allocation. This method utilizes unoccupied cells that have a negative allocation value during adjustments.

18.3.8 Algorithm of SSM

Step 1: Initiate the process by establishing the first basic solution from the three proposed methods. Ensure that the final tableau generated from the first method meets the requirement of $m + n - 1$ filled cells, where m represents the number of rows and n the number of columns.

Step 2: Identify which cells are filled and which are vacant, then compute the net change for the unoccupied cells.

Step 3: If the net change for an unoccupied cell is negative, allocate a minimum unit to that cell; if not, no action is required.

Step 4: Select the unoccupied cell with the highest allocation or randomly choose one if there is a tie in negative values.

Step 5: If the unoccupied cells yield a positive result, update your optimality and finalize your findings.

Step 6: If the results are negative, return to Step 2; if they are positive, conclude the process.

18.4 PROBLEM AND RESULT DISCUSSIONS

Problem: A total of six recipients in India have ordered loading units in the amounts of 18, 29, 15, 34, 26, 21, 36, and 21. Concurrently, six suppliers from Mumbai, Delhi, Kolkata, Chennai, Bhopal, and Hyderabad have available loading units totaling 24, 31, 19, 49, 40, and 37. This situation is illustrated through a transportation network model. The following table details the distances in kilometers between different cities as a numerical example.

	Mumbai	*Delhi*	*Kolkata*	*Chennai*	*Bhopal*	*Hyderabad*	*DEMAND*
Ahmadabad	523	972	2063	1820	580	1180	25
Jaipur	1170	296	1657	2052	580	1425	36
Indore	583	787	1616	1440	190	802	22
Ranchi	1665	1215	402	1735	1052	1266	41
Bangalore	1020	2127	1874	324	1442	574	33
Patna	1736	1042	583	2053	1000	1465	43
Supply	30	25	25	44	45	31	

First, it is essential to create a table that consists of 'm' columns, corresponding to the number of suppliers, and 'n' rows, which represent the number of recipients, taking into account the constraints of supply and demand. This transportation problem can be mathematically framed as a linear programming challenge. The main objective is to lower the overall transportation costs between the suppliers and the recipients.

$$F_{obj}(X,C) = \sum_{i=1}^{m}\sum_{j=1}^{n} C_{ij}\, x_{ij}$$

Subject to

$$x_{11} + x_{12} + x_{13} + x_{14} + x_{15} + x_{16} = 25$$

$$x_{21} + x_{22} + x_{23} + x_{24} + x_{25} + x_{26} = 36$$

$$x_{31} + x_{32} + x_{33} + x_{34} + x_{35} + x_{36} = 22$$

$$x_{41} + x_{42} + x_{43} + x_{44} + x_{45} + x_{46} = 41$$

$$x_{51} + x_{52} + x_{53} + x_{54} + x_{55} + x_{56} = 33$$

$$x_{61} + x_{62} + x_{63} + x_{64} + x_{65} + x_{66} = 43$$

$$x_{11} + x_{21} + x_{31} + x_{41} + x_{51} + x_{61} = 24$$

$$x_{12} + x_{22} + x_{32} + x_{42} + x_{52} + x_{62} = 31$$

$$x_{13} + x_{23} + x_{33} + x_{43} + x_{53} + x_{63} = 19$$

$$x_{14} + x_{24} + x_{34} + x_{44} + x_{54} + x_{64} = 49$$

$$x_{15} + x_{25} + x_{35} + x_{45} + x_{55} + x_{65} = 40$$

$$x_{16} + x_{26} + x_{36} + x_{46} + x_{56} + x_{66} = 37$$

$$\forall x_{ij} \geq 0 \text{i} = 1,2,3,\dots.\text{m and j} = 1,2,3\dots\text{n}$$

The process of completing the table using the North-West corner method begins with the first cell located in the upper left corner, which is associated with specific supply and demand figures. The subsequent step involves selecting the smaller of the two values and recording it in the corresponding cell of the upper left corner. This recorded quantity must then be deducted from both the supply and demand totals. In this instance, the supply and demand for the initial cell are set at 24 and 18, respectively. Following this, the recorded amount is subtracted from both figures. The software will simultaneously identify which supply or demand value has reached zero. If the demand value becomes zero, all remaining cells in that row should be filled with zeros. Conversely, if the supply value is zero, zeros must be entered in all remaining cells of the corresponding column. By executing this procedure step by step, a feasible solution is illustrated in the subsequent table. Non-base elements are defined as zero elements, while all non-zero elements are classified as base elements. A solution is considered degenerate when the number of base elements equals $m + n - 1$, which in this scenario calculates to $8 + 6 - 1 = 13$.

Subsequent iterations and a feasible solution for the North-West corner method.

	Mumbai	*Delhi*	*Kolkata*	*Chennai*	*Bhopal*	*Hyderabad*	*DEMAND*
Ahmadabad	**523(25)**	**0**	**0**	**0**	**0**	**0**	**25**
Jaipur	**1170(5)**	**296(25)**	**1657(6)**	**0**	**0**	**0**	**36**
Indore	**0**	**0**	**1616(19)**	**1440(3)**	**0**	**0**	**22**
Ranchi	**0**	**0**	**0**	**1735(41)**	**0**	**0**	**41**
Bangalore	**0**	**0**	**0**	**0**	**1442(33)**	**0**	**33**
Patna	**0**	**0**	**0**	**0**	**1000(12)**	**1465(31)**	**43**
Supply	**30**	**25**	**25**	**44**	**45**	**31**	

The transportation cost of this solution is computed by substituting all allocations.

$$\begin{aligned} T.P. &= 25 \times 523 + 5 \times 1170 + 25 \times 296 + 6 \times 1657 + 19 \times 1616 + 3 \times 1440 \\ &\quad + 41 \times 1735 + 33 \times 1442 + 12 \times 1000 + 6 \times 1000 + 31 \times 1465 \\ &= Rs\,247427 \end{aligned}$$

18.4.1 Row minimum method

By choosing components from the cost matrix, the RMM finds the lowest cost Cij in the first row. The chosen C_{ij} element's value is equal to x_{ij}, which is where the base matrix $X = [x_{ij}]$ is constructed from. In order to determine the first viable solutions, the matrix must be supplemented with components that correspond to the arcs in the following rows that have the lowest unit transportation costs. Comparing transportation costs with the corresponding supply and demand figures, starting with the first row, is the procedure for finishing the results table using the row minimum approach. The first row's lowest value, 545, represents a supply of 37 and a demand of 18. The next stage was to reduce supply and demand by 18 so that the demand was zero. As a result, the remaining cells in the first row were filled with zeros.

Feasible solution of the row minimum method.

	Mumbai	*Delhi*	*Kolkata*	*Chennai*	*Bhopal*	*Hyderabad*	*DEMAND*
Ahmadabad	**545(25)**	**0**	**0**	**0**	**0**	**0**	**25**
Jaipur	**0**	**296(25)**	**0**	**0**	**580(11)**	**0**	**36**
Indore	**0**	**0**	**0**	**0**	**190(22)**	**0**	**22**
Ranchi	**0**	**0**	**402(25)**	**0**	**1052(12)**	**1266(4)**	**41**
Bangalore	**0**	**0**	**0**	**324(33)**	**0**	**0**	**33**
Patna	**1736(5)**	**0**	**0**	**2053(11)**	**0**	**1465(27)**	**43**
Supply	**30**	**25**	**25**	**44**	**45**	**31**	

The transportation cost associated with this solution is calculated by replacing all allocations with their respective values.

$$
\begin{aligned}
T.P. &= 25 \times 545 + 25 \times 296 + 11 \times 580 + 22 \times 190 + 25 \times 402 + 12 \\
&\quad \times 1052 + 4 \times 1266 + 33 \times 324 + 5 \times 1736 + 11 \times 2053 + 27 \times 1465 \\
&= RS\,140283
\end{aligned}
$$

18.4.2 Least cost matrix method

By adding the lowest unit costs in line with the non-decreasing sequence of transport unit costs, this method improves the table. To guarantee that every row and column has at least one zero value, the cost matrix needs to be modified. All elements in the appropriate row (or column) should have their minimal value deducted from them. The next step is to determine which of the zero elements in the adjusted cost matrix has the lowest cost, C_{ij}. The base matrix $X = [x_{ij}]$ is produced from the C_{ij} element, which is equivalent to the xij. It is crucial to add more entries to the X matrix that represent the lowest unit transportation costs in order to develop a fundamentally viable solution. In the next phase, supply and demand are both cut by 15, which leaves the demand value at zero. As a result, the remaining cells in the third row must be filled with zeros.

Feasible solutions of the Least Cost Matrix Method.

	Mumbai	Delhi	Kolkata	Chennai	Bhopal	Hyderabad	DEMAND
Ahmadabad	**523(25)**	**0**	**0**	**0**	**0**	**0**	**25**
Jaipur	**0**	**296(25)**	**0**	**0**	**580(11)**	**0**	**36**
Indore	**0**	**0**	**0**	**0**	**190(22)**	**0**	**22**
Ranchi	**0**	**0**	**402(25)**	**0**	**0**	**1266(16)**	**41**
Bangalore	**0**	**0**	**0**	**424(33)**	**0**	**0**	**33**
Patna	**1736(5)**	**0**	**0**	**2053(11)**	**1000(12)**	**1465(15)**	**43**
Supply	**24**	**31**	**19**	**49**	**40**	**37**	

The transportation expenses for this solution are assessed by integrating all allocations into the calculation.

$$\begin{aligned} T.P. &= 25\times 523+25\times 296+11\times 580+22\times 190+25\times 402+16 \\ &\quad \times 1266+33\times 424+5\times 1736+11\times 2053+12\times 1000+15\times 1465 \\ &= Rs137275 \end{aligned}$$

18.4.3 Vogel's approximation method

A minimal solution can be determined using the VAM technique, which analyzes the transportation cost matrix. To derive the optimal solution through the VAM approach, the GNU Octave software is employed to compute the ratio, reflecting the difference between the lowest and second-lowest cost options in each row and column. The next step involves identifying the cell with the maximum difference values, represented as $r_k = \max\{r_i\}$ and $s_i = \max\{s_j\}$. If rk is less than sj, the lowest cost in the corresponding column should be highlighted; conversely, if r_k is greater than s_j, the lowest cost in the specified row should be indicated.

Feasible solutions of the VAM Method.

	Mumbai	Delhi	Kolkata	Chennai	Bhopal	Hyderabad	DEMAND
Ahmadabad	**523(25)**	**0**	**0**	**0**	**0**	**0**	**25**
Jaipur	**0**	**296(25)**	**0**	**0**	**580(11)**	**0**	**36**
Indore	**583(5)**	**0**	**0**	**0**	**0**	**802(17)**	**22**
Ranchi	**0**	**0**	**402(25)**	**1735(11)**	**0**	**1266(5)**	**41**
Bangalore	**0**	**0**	**0**	**324(33)**	**0**	**0**	**33**
Patna	**0**	**0**	**0**	**0**	**1000(34)**	**1465(9)**	**43**
Supply	**30**	**25**	**25**	**44**	**45**	**31**	

Overall transportation expenses utilizing the VAM approach.

$$\begin{aligned} T.P. &= 25\times 523+25\times 296+11\times 580+5\times 583+17\times 802+25 \\ &\quad \times 402+11\times 1735+5\times 1266+33\times 324+34\times 1000+9\times 1465 \\ &= Rs136746 \end{aligned}$$

Comparison of results by using the GNU Octave Software

Method	*Northwest Corner method*	*Row minimum method*	*Least cost matrix method*	*VAM*
Total TP cost	RS 247427	RS 140283	RS 137275	RS 136746
Solution Degenerated	Yes	Yes	Yes	Yes

Optimality test by using the Stepping Stone Method (SSM):
Allocation table, 1st Iteration (optimality test).

	Mumbai **(D_1)**	*Delhi* **(D_2)**	*Kolkata* **(D_3)**	*Chennai* **(D_4)**	*Bhopal* **(D_5)**	*Hyderabad* **(D_6)**	*DEMAND*
Ahmadabad **(S_1)**	523**(25)**	972	2063	1820	580	1180**(1)**	25
Jaipur **(S_2)**	1170	296**(25)**	1657	2052	580**(11)**	1425	36
Indore **(S_3)**	583**(5)**	787	1616	1440	190	802**(17)**	22
Ranchi **(S_4)**	1665.	1215	402**(25)**	1735**(11)**	1052	1266**(5)**	41
Bangalore **(S_5)**	1020	2127	1874	324 **(33)**	1442	574	33
Patna **(S_6)**	1736	1042	583	2053	1000**(34)**	1465**(9)**	43
Supply	24	31	19	49	40	37	

1. Establish a closed loop for cells that are not in use.

Unoccupied cells	*Closed path*	*Net cost difference*
S_1D_2	**$S_1D_2 \rightarrow S_1D_1 \rightarrow S_3D_1 \rightarrow S_3D_6 \rightarrow S_6D_6 \rightarrow S_2D_5 \rightarrow S_2D_2$**	**972–523+583–802+1465–1000+580–296=979**
S_1D_3	**$S_1D_3 \rightarrow S_1D_1 \rightarrow S_3D_1 \rightarrow S_3D_6 \rightarrow S_4D_6 \rightarrow S_4D_3$**	**2063–523+583–802+1266–402=2185**
S_1D_4	**$S_1D_4 \rightarrow S_1D_1 \rightarrow S_3D_1 \rightarrow S_3D_6 \rightarrow S_4D_6 \rightarrow S_4D_4$**	**1820–523+583–802+1266–1735=609**
S_1D_5	**$S_1D_5 \rightarrow S_1D_1 \rightarrow S_3D_1 \rightarrow S_3D_6 \rightarrow S_6D_6 \rightarrow S_6D_5$**	**580–523+583–802+1465–1000=303**
S_1D_6	**$S_1D_6 \rightarrow S_1D_1 \rightarrow S_3D_1 \rightarrow S_3D_6$**	**1180–523+583–802=438**
S_2D_1	**$S_2D_1 \rightarrow S_2D_5 \rightarrow S_6D_5 \rightarrow S_6D_6 \rightarrow S_3D_6 \rightarrow S_3D_1$**	**1170–580+1000–1465+802–583=344**
S_2D_3	**$S_2D_3 \rightarrow S_2D_5 \rightarrow S_6D_5 \rightarrow S_6D_6 \rightarrow S_4D_6 \rightarrow S_4D_3$**	**1657–580+1000–1465+1266–402=1476**
S_2D_4	**$S_2D_4 \rightarrow S_2D_5 \rightarrow S_6D_5 \rightarrow S_6D_6 \rightarrow S_4D_6 \rightarrow S_4D_4$**	**2052–580+1000–1465+1266–1735=538**
S_2D_6	**$S_2D_6 \rightarrow S_2D_5 \rightarrow S_6D_5 \rightarrow S_6D_6$**	**1425–580+1000–1465=380**
S_3D_2	**$S_3D_2 \rightarrow S_3D_6 \rightarrow S_6D_6 \rightarrow S_6D_5 \rightarrow S_2D_5 \rightarrow S_2D_2$**	**787–802+1465–1000+580–296=734**

(Continued)

Unoccupied cells	*Closed path*	*Net cost difference*
S_3D_3	**$S_3D_3 \rightarrow S_3D_6 \rightarrow S_4D_6 \rightarrow S_4D_3$**	**1616–802+1266+402=1678**
S_3D_4	**$S_3D_4 \rightarrow S_3D_6 \rightarrow S_4D_6 \rightarrow S_4D_4$**	**1440–802+1266–1735=169**
S_3D_5	**$S_3D_5 \rightarrow S_3D_6 \rightarrow S_6D_6 \rightarrow S_6D_5$**	**190–802+1465–1000=–147**
S_4D_1	**$S_4D_1 \rightarrow S_4D_6 \rightarrow S_3D_6 \rightarrow S_3D_1$**	**1665–1266+802–583=618**
S_4D_2	**$S_4D_2 \rightarrow S_4D_6 \rightarrow S_6D_6 \rightarrow S_6D_5 \rightarrow S_2D_5 \rightarrow S_2D_2$**	**1215–1266+1465–1000+580–296=698**
S_4D_5	**$S_4D_5 \rightarrow S_4D_6 \rightarrow S_6D_6 \rightarrow S_6D_5$**	**1052–1266+1465–1000=251**
S_5D_1	**$S_5D_1 \rightarrow S_5D_4 \rightarrow S_4D_4 \rightarrow S_4D_6 \rightarrow S_3D_6 \rightarrow S_3D_1$**	**1020–324+1735–1266+802–583=1384**
S_5D_2	**$S_5D_2 \rightarrow S_5D_4 \rightarrow S_4D_4 \rightarrow S_4D_6 \rightarrow S_6D_6 \rightarrow S_6D_5 \rightarrow S_2D_5 \rightarrow S_2D_2$**	**2127–324+1735–1266+1465–1000+580–296=3021**
S_5D_3	**$S_5D_3 \rightarrow S_5D_4 \rightarrow S_4D_4 \rightarrow S_4D_3$**	**1874–324+1735–402=2883**
S_5D_5	**$S_5D_5 \rightarrow S_5D_4 \rightarrow S_4D_4 \rightarrow S_4D_6 \rightarrow S_6D_6 \rightarrow S_6D_5$**	**1442–324+1735–1266+1465–1000=2052**
S_5D_6	**$S_5D_6 \rightarrow S_5D_4 \rightarrow S_4D_4 \rightarrow S_4D_6$**	**574–324+1735–1266=719**
S_6D_1	**$S_6D_1 \rightarrow S_6D_6 \rightarrow S_3D_6 \rightarrow S_3D_1$**	**1736–1465+802–583=490**
S_6D_2	**$S_6D_2 \rightarrow S_6D_5 \rightarrow S_2D_5 \rightarrow S_2D_2$**	**1042–1000+580–296=326**
S_6D_3	**$S_6D_3 \rightarrow S_6D_6 \rightarrow S_4D_6 \rightarrow S_4D_3$**	**583–1465+1266–402=–18**
S_6D_4	**$S_6D_4 \rightarrow S_6D_6 \rightarrow S_4D_6 \rightarrow S_4D_4$**	**2053–1465+1266–1735=119**

2. The highest negative cost in the unoccupied cell is S_3D_5, which has a value of –147. Next, examine the closed path that follows $S_3D_5 \rightarrow S_3D_6 \rightarrow S_6D_6 \rightarrow S_6D_5$. This closed path includes both positive and negative allocations for the currently unoccupied cell S_3D_5.

	Mumbai **(D_1)**	*Delhi* **(D_2)**	*Kolkata* **(D_3)**	*Chennai* **(D_4)**	*Bhopal* **(D_5)**	*Hyderabad* **(D6)**	*DEMAND*
Ahmadabad **(S_1)**	523**(25)**	972[979]	2063[2185]	1820[609]	580[303]	1180[438]	25
Jaipur **(S_2)**	1170[344]	296**(25)**	1657[1476]	2052[538]	580**(11)**	1425[380]	36
Indore **(S_3)**	583**(5)**	787[734]	1616[1678]	1440[169]	190[147](+)	802**(17)**(–)	22
Ranchi **(S_4)**	1665[618]	1215[698]	402**(25)**	1735**(11)**	1052[251]	1266**(5)**	41
Bangalore **(S_5)**	1020[1384]	2127[3021]	1874[2883]	324 **(33)**	1442[2052]	574 [719]	33
Patna **(S_6)**	1736[490]	1042[326]	583[–18]	2053[119]	1000**(34)**(-)	1465**(9)**(+)	43
Supply	30	25	25	44	45	31	

3. Minimum allocated value for all negative positions (–) on closed path = 17.

And now subtract 17 from all (–) and add to all (+).

	Mumbai **(D_1)**	*Delhi* **(D_2)**	*Kolkata* **(D_3)**	*Chennai* **(D_4)**	*Bhopal* **(D_5)**	*Hyderabad* **(D_6)**	*DEMAND*
Ahmadabad **(S_1)**	523(**25**)	972	2063	1820	580	1180	25
Jaipur **(S_2)**	1170	296(**26**)	1657	2052	580(**11**)	1425	36
Indore **(S_3)**	583(**5**)	787	1616	1440	190(**17**)	802	22
Ranchi **(S_4)**	1665.	1215	402(**25**)	1735(**11**)	1052	1266(**5**)	41
Bangalore **(S_5)**	1020	2127	1874	324 (**33**)	1442	574	33
Patna **(S_6)**	1736	1042	583	2053	1000(**17**)	1465(**26**)	43
Supply	30	25	25	44	45	31	

Now repeat the process until the optimal solution arises.
2nd Iteration (optimality test)

18.5 ESTABLISH A CLOSED LOOP FOR CELLS THAT ARE NOT IN USE

Unoccupied cells	*Path (Closed)*	*Net cost*
S_1D_2	$S_1D_2 \rightarrow S_1D_1 \rightarrow S_3D_1 \rightarrow S_3D_5 \rightarrow S_2D_5 \rightarrow S_2D_2$	**972-523+583-190+580-296=1126**
S_1D_3	$S_1D_3 \rightarrow S_1D_1 \rightarrow S_3D_1 \rightarrow S_3D_5 \rightarrow S_6D_5 \rightarrow S_6D_6 \rightarrow S_4D_6 \rightarrow S_4D_3$	**2063-253+583-190+1000-1465+1266-402=2332**
S_1D_4	$S_1D_4 \rightarrow S_1D_1 \rightarrow S_3D_1 \rightarrow S_3D_5 \rightarrow S_6D_5 \rightarrow S_6D_6 \rightarrow S_4D_6 \rightarrow S_4D_4$	**1820-523+583-190+1000-1465+1266-1735=756**
S_1D_5	$S_1D_5 \rightarrow S_1D_1 \rightarrow S_3D_1 \rightarrow S_3D_5$	**580-523+583-190=450**
S_1D_6	$S_1D_6 \rightarrow S_1D_1 \rightarrow S_3D_1 \rightarrow S_3D_5 \rightarrow S_6D_5 \rightarrow S_6D_6$	**1180-523+580-190+1000-1465=585**
S_2D_1	$S_2D_1 \rightarrow S_2D_5 \rightarrow S_3D_5 \rightarrow S_3D_1$	**1170-580+190-583=197**
S_2D_3	$S_2D_3 \rightarrow S_2D_5 \rightarrow S_6D_5 \rightarrow S_6D_6 \rightarrow S_4D_6 \rightarrow S_4D_3$	**1657-580+1000-1465+1266-402=1476**
S_2D_4	$S_2D_4 \rightarrow S_2D_5 \rightarrow S_6D_5 \rightarrow S_6D_6 \rightarrow S_4D_6 \rightarrow S_4D_4$	**2052-580+1000-1465+1266-1735=538**
S_2D_6	$S_2D_6 \rightarrow S_2D_5 \rightarrow S_6D_5 \rightarrow S_6D_6$	**1425-580+1000-1465=380**
S_3D_2	$S_3D_2 \rightarrow S_3D_5 \rightarrow S_2D_5 \rightarrow S_2D_2$	**787-190+580-296=881**
S_3D_3	$S_3D_3 \rightarrow S_3D_5 \rightarrow S_6D_5 \rightarrow S_6D_6 \rightarrow S_4D_6 \rightarrow S_4D_3$	**1616-190+1000-1465+1266-402=1825**

Unoccupied cells	*Path (Closed)*	*Net cost*
S_3D_4	$S_3D_4 \rightarrow S_3D_5 \rightarrow S_6D_5 \rightarrow S_6D_{36} \rightarrow S_4D_6 \rightarrow S_4D_4$	1440-190+1000-1465+1266-1735=316
S_3D_6	$S_3D_6 \rightarrow S_3D_5 \rightarrow S_6D_5 \rightarrow S_6D_6$	802-190+1000-1465=147
S_4D_1	$S_4D_1 \rightarrow S_4D_6 \rightarrow S_6D_6 \rightarrow S_6D_5 \rightarrow S_3D_5 \rightarrow S_3D_1$	1665-1266+1465-1000+190-583=471
S_4D_2	$S_4D_2 \rightarrow S_4D_6 \rightarrow S_6D_6 \rightarrow S_6D_5 \rightarrow S_2D_5 \rightarrow S_2D_2$	1215-1266+1465-1000+190-583=1237
S_4D_5	$S_4D_5 \rightarrow S_4D_6 \rightarrow S_6D_6 \rightarrow S_6D_5$	1052-1266+1465-1000=251
S_5D_1	$S_5D_1 \rightarrow S_5D_4 \rightarrow S_4D_4 \rightarrow S_4D_6 \rightarrow S_6D_6 \rightarrow S_6D_5 \rightarrow S_3D_5 \rightarrow S_3D_1$	1020-324+1735-1266+1465-1000+190-583=1237
S_5D_2	$S_5D_2 \rightarrow S_5D_4 \rightarrow S_4D_4 \rightarrow S_4D_6 \rightarrow S_6D_6 \rightarrow S_6D_5 \rightarrow S_2D_5 \rightarrow S_2D_2$	2127-324+1735-1266+1465-1000+580-296=3021
S_5D_3	$S_5D_3 \rightarrow S_5D_4 \rightarrow S_4D_4 \rightarrow S_4D_3$	1874-324+1735-402=2883
S_5D_5	$S_5D_5 \rightarrow S_5D_4 \rightarrow S_4D_4 \rightarrow S_4D_6 \rightarrow S_6D_6 \rightarrow S_6D_5$	1442-324+1735-1266+1465-1000=2052
S_5D_6	$S_5D_6 \rightarrow S_5D_4 \rightarrow S_4D_4 \rightarrow S_4D_6$	574-324+1735-1266=719
S_6D_1	$S_6D_1 \rightarrow S_6D_5 \rightarrow S_3D_5 \rightarrow S_3D_1$	1736-1000+190-583=343
S_6D_2	$S_6D_2 \rightarrow S_6D_5 \rightarrow S_2D_5 \rightarrow S_2D_2$	1042-1000+580-296=326
S_6D_3	$S_6D_3 \rightarrow S_6D_6 \rightarrow S_4D_6 \rightarrow S_4D_3$	583-1465+1266-402=-18
S_6D_4	$S_6D_4 \rightarrow S_6D_6 \rightarrow S_4D_6 \rightarrow S_4D_4$	2053-1465+1266-1735=119

2. The most significant negative cost in the unoccupied cell is found at, that is, cell S_6D_3= −18. And draw a closed path from S_1D_5; then the closed path is $S_6D_3 \rightarrow S_6D_6 \rightarrow S_4D_6 \rightarrow S_4D_3$. This closed path features both positive and negative allocations related to the unoccupied cells S_6D_3.

	Mumbai **(D_1)**	*Delhi* **(D_2)**	*Kolkata* **(D_3)**	*Chennai* **(D_4)**	*Bhopal* **(D_5)**	*Hyderabad* **(D_6)**	*DEMAND*
Ahmadabad **(S_1)**	523**(25)**	972[1126]	2063[2332]	1820[756]	580[450]	1180[585]	25
Jaipur **(S_2)**	1170[197]	296**(25)**	1657[1476]	2052[538]	580**(11)**	1425[380]	36
Indore **(S_3)**	583**(5)**	787[881]	1616[1825]	1440[316]	190**(17)**	802[147]	22
Ranchi**(S_4)**	1665[471]	1215[698]	402**(25)**(–)	1735**(11)**	1052[251]	1266**(5)**(+)	41
Bangalore **(S_5)**	1020[1237]	2127[3021]	1874[2883]	324 **(33)**	1442[2052]	574 [719]	33
Patna **(S_6)**	1736[343]	1042[326]	583[-18](+)	2053[119]	1000**(17)**	1465**(26)**(c)	43
Supply	30	25	25	44	45	31	

Minimum allocated value among all negative positions (-) on closed path =25.

And now subtract 25 from all (-) and add it to all (+).

	Mumbai (D_1)	Delhi (D_2)	Kolkata (D_3)	Chennai (D_4)	Bhopal (D_5)	Hyderabad (D_6)	DEMAND
Ahmadabad (S_1)	523(**25**)	972	2063	1820	580	1180	25
Jaipur (S_2)	1170	296(**25**)	1657	2052	580(**11**)	1425	36
Indore (S_3)	583(**5**)	787	1616	1440	190(**17**)	802	22
Ranchi (S_4)	1665.	1215	402	1735(**11**)	1052	1266(**30**)	41
Bangalore (S_5)	1020	2127	1874	324 (**33**)	1442	574	33
Patna (S_6)	1736	1042	583(**25**)	2053	1000(**17**)	1465(**1**)	43
Supply	30	25	25	44	45	31	

Repeat the above step until the optimal solution is obtained.

3rd Iteration (optimality test)

Create a closed loop for unoccupied cells.

Unoccupied cells	*Path (closed)*	*Net cost*
S_1D_2	$S_1D_2 \rightarrow S_1D_1 \rightarrow S_3D_1 \rightarrow S_3D_5 \rightarrow S_2D_5 \rightarrow S_2D_2$	972-523+583-190+580-296=1126
S_1D_3	$S_1D_3 \rightarrow S_1D_1 \rightarrow S_3D_1 \rightarrow S_3D_5 \rightarrow S_6D_5 \rightarrow S_6D_3$	2063-253+583-190+1000-583 =2350
S_1D_4	$S_1D_4 \rightarrow S_1D_1 \rightarrow S_3D_1 \rightarrow S_3D_5 \rightarrow S_6D_5 \rightarrow S_6D_6 \rightarrow S_4D_6 \rightarrow S_4D_4$	1820-523+583-190+1000-1465+1266-1735=756
S_1D_5	$S_1D_5 \rightarrow S_1D_1 \rightarrow S_3D_1 \rightarrow S_3D_5$	580-523+583-190=450
S_1D_6	$S_1D_6 \rightarrow S_1D_1 \rightarrow S_3D_1 \rightarrow S_3D_5 \rightarrow S_6D_5 \rightarrow S_6D_6$	1180-523+580-190+1000-1465=585
S_2D_1	$S_2D_1 \rightarrow S_2D_5 \rightarrow S_3D_5 \rightarrow S_3D_1$	1170-580+190-583=197
S_2D_3	$S_2D_3 \rightarrow S_2D_5 \rightarrow S_6D_5 \rightarrow S_6D_3$	1657-580+1000-583=1494
S_2D_4	$S_2D_4 \rightarrow S_2D_5 \rightarrow S_6D_5 \rightarrow S_6D_6 \rightarrow S_4D_6 \rightarrow S_4D_4$	2052-580+1000-1465+1266-1735=538
S_2D_6	$S_2D_6 \rightarrow S_2D_5 \rightarrow S_6D_5 \rightarrow S_6D_6$	1425-580+1000-1465=380
S_3D_2	$S_3D_2 \rightarrow S_3D_5 \rightarrow S_2D_5 \rightarrow S_2D_2$	787-190+580-296=881
S_3D_3	$S_3D_3 \rightarrow S_3D_5 \rightarrow S_6D_5 \rightarrow S_6D_3$	1616-190+1000-583=1843
S_3D_4	$S_3D_4 \rightarrow S_3D_5 \rightarrow S_6D_5 \rightarrow S_6D_6 \rightarrow S_4D_6 \rightarrow S_4D_4$	1440-190+1000-1465+1266-1735=316
S_3D_6	$S_3D_6 \rightarrow S_3D_5 \rightarrow S_6D_5 \rightarrow S_6D_6$	802-190+1000-1465=147
S_4D_1	$S_4D_1 \rightarrow S_4D_6 \rightarrow S_6D_6 \rightarrow S_6D_5 \rightarrow S_3D_5 \rightarrow S_3D_1$	1665-1266+1465-1000+190-583=471
S_4D_2	$S_4D_2 \rightarrow S_4D_6 \rightarrow S_6D_6 \rightarrow S_6D_5 \rightarrow S_2D_5 \rightarrow S_2D_2$	1215-1266+1465-1000+580-296=698
S_4D_3	$S_4D_3 \rightarrow S_4D_6 \rightarrow S_6D_6 \rightarrow S_6D_3$	402-1266+1465-583=18
S_4D_5	$S_4D_5 \rightarrow S_4D_6 \rightarrow S_6D_6 \rightarrow S_6D_5$	1052-1266+1465-1000=251

Unoccupied cells	*Path (closed)*	*Net cost*
S_5D_1	$S_5D_1 \rightarrow S_5D_4 \rightarrow S_4D_4 \rightarrow S_4D_6 \rightarrow S_6D_6 \rightarrow S_6D_5 \rightarrow S_2D_5 \rightarrow S_2D_2$	**1020-324+1735-1266+1465-1000+190-583=1237**
S_5D_2	$S_5D_2 \rightarrow S_5D_4 \rightarrow S_4D_4 \rightarrow S_4D_6 \rightarrow S_6D_6 \rightarrow S_6D_5 \rightarrow S_2D_5 \rightarrow S_2D_2$	**2127-324+1735-1266+1465-1000+580-296=3021**
S_5D_3	$S_5D_3 \rightarrow S_5D_4 \rightarrow S_4D_4 \rightarrow S_4D_6 \rightarrow S_6D_6 \rightarrow S_6D_3$	**1874-324+1735-1266+1465-583=2901**
S_5D_5	$S_5D_5 \rightarrow S_5D_4 \rightarrow S_4D_4 \rightarrow S_4D_6 \rightarrow S_6D_6 \rightarrow S_6D_5$	**1442-324+1735-1266+1465-1000=2052**
S_5D_6	$S_5D_6 \rightarrow S_5D_4 \rightarrow S_4D_4 \rightarrow S_4D_6$	**574-324+1735-1266=719**
S_6D_1	$S_6D_1 \rightarrow S_6D_5 \rightarrow S_3D_5 \rightarrow S_3D_1$	**1736-1000+190-583=343**
S_6D_2	$S_6D_2 \rightarrow S_6D_5 \rightarrow S_2D_5 \rightarrow S_2D_2$	**1042-1000+580-296=326**
S_6D_4	$S_6D_4 \rightarrow S_6D_6 \rightarrow S_4D_6 \rightarrow S_4D_4$	**2053-1465+1266-1735=119**

Since all net costs ≥ 0, the final optimal solution is arrived.

	Mumbai **(D_1)**	*Delhi* **(D_2)**	*Kolkata* **(D_3)**	*Chennai* **(D_4)**	*Bhopal* **(D_5)**	*Hyderabad* **(D_6)**	*DEMAND*
Ahmadabad **(S_1)**	523(**25**)	972	2063	1820	580	1180	25
Jaipur **(S_2)**	1170	296(**25**)	1657	2052	580(**11**)	1425	36
Indore **(S_3)**	583(**5**)	787	1616	1440	190(**17**)	802	22
Ranchi **(S_4)**	1665.	1215	402	1735(**11**)	1052	1266(**30**)	41
Bangalore **(S_5)**	1020	2127	1874	324 (**33**)	1442	574	33
Patna **(S_6)**	1736	1042	583(**25**)	2053	1000(**17**)	1465(**1**)	43
Supply	30	25	25	44	45	31	

The optimality solution is

$$= 523\times25+296\times25+580\times11+583\times5+190\times17+1735\times11$$
$$+1266\times30+324\times33+583\times25+1000\times17+1465\times1 = Rs133797$$

It is important to highlight that various methods yield distinct objective function values when evaluating the initial values of the basic solutions for the specified transportation problem. The analysis reveals that the North-West corner method resulted in the highest transportation cost, whereas the matrix method produced a lower cost compared to the frequently utilized RMM. Notably, the VAM approach achieved the lowest value for the objective function. The subsequent discussion outlines the most cost-effective solution for the supply network concerning the transportation problem.

$$\text{The optimize cost} = \text{Rs}\,133797$$

$$\text{Cost percent} = \frac{(\text{Initial cost} - \text{optimized cost})}{(\text{optimized cost})} \times 100$$

The initial transportation cost is determined by multiplying the demand from each warehouse by the highest unit of supply found in the corresponding column. Here, the optimized cost applies to every option that was evaluated.

$$\text{Therefore, cost percentage in NWC method is} = \frac{\text{RS}\,247427 - 133797}{133797} \times 100 = 84.9\%$$

$$\text{Therefore, cost percentage in Row Minimum method is} = \frac{\text{RS}\,140283 - \text{Rs}133797}{133797} \times 100 = 4.83\%$$

$$\text{Therefore, cost percentage in LCM method is} = \frac{\text{RS}137275 - \text{Rs}133797}{133797} \times 100 = 2.60\%$$

$$\text{Therefore, cost percentage in VAM method is} = \frac{\text{RS}136746 - 133797}{1337974} \times 100 = 2.203\%$$

Each depot provides the initial transportation cost based on the highest unit supply; the NWC approach yields the best possible cost reduction of 84.9%.

18.6 CONCLUSION

A significant area of research within optimization and its associated disciplines is the transportation problem. This study examined the challenges encountered by a cold beverage company in road transportation, conceptualizing them as a linear programming issue. Various methods were employed to resolve this problem. The computational outcomes yielded the optimal values for supply costs, the quantity of cases to be delivered, and the regions for potential expansion, ultimately achieving the minimum overall transportation cost. The research highlights several techniques available for addressing the issue, with the most effective strategies being those that maximize savings on transportation costs.

REFERENCES

Latunde, T., Bamigbola, O. M., & Aderinto, Y. O. (2016). Sensitivity of parameters in an optimal control model of the electric power generating system. *Ilorin Journal of Computer Science and Information Technology (ILJCSIT)*, *1*(1), 54–70.

Pursula, M. & Miittymaki, J. (2001). *Mathematical Methods on Optimization in Transportation Systems*. Kluwer Academic Publishers.

Rekha, V. J. (2013). Optimization techniques for transportation problem of three variables. *International Organization of Scientific Research Journal of Mathematics, IOSR-JM 1*, 46–50.

Salami, A. O. (2014). Application of transportation linear programming algorithms to cost reduction in Nigeria soft drinks industry. *International Journal of Social Behavioral, Educational, Economic, Business and Industrial Engineering, 8(2)*, 416–422.

Sarbjit, S. (2012). Note on transportation problem with new method for resolution of degeneracy. *Universal Journal of Industrial and Business Management*, *3*(1), 26–36.

Shraddha, M. (2017). Solving transportation problem by various method and their comparison. *International Journal of Mathematics Trends and Technology (IJMTT)*, *44*(4), 2231–5373.

Susanta, K. M. (2019). An Optimal Solution for Transportation Problem: Direct Approach, A project report, UGC-HRDC, Sambalpur University, Burla, Odisha, India.

Chapter 19

Enhanced solution of transportation problems with Russell's approximation method under the fuzzy environment

Utpal Majee, Bhawna Agrawal, Gaurav Sharma, Neelu Sharma, and Ravendra Prasad Namdeo

19.1 INTRODUCTION

The transportation problem represents one of the earliest instances of linear programming applications. Transportation models are extensively utilized in logistics and supply chain management to minimize costs. Efficient algorithms have been created to address the transportation problem when the cost coefficients, as well as supply and demand quantities, are precisely known. However, the presence of randomness and uncertainty in real-world scenarios is unavoidable due to unforeseen circumstances. There are instances where the cost coefficients and the supply and demand quantities in a transportation problem may be ambiguous due to uncontrollable factors. To quantitatively address imprecise information in decision-making, Bellman et al. [2] introduced the concept of fuzziness. Following this foundational work, numerous authors have expanded on this idea; for instance, Bellman and Zadeh [3] proposed decision-making strategies within a fuzzy environment. Subsequent contributions from researchers such as Shiang-Tai Liu and Chiang Kao [16], Chanas et al. [5], and Pandian et al. [14] have led to various methods for solving fuzzy transportation problems. Chanas and Kuchta [4] introduced the notion of optimal solutions for transportation problems with fuzzy coefficients represented as fuzzy numbers. Additionally, Chanas, Kolodziejczyk, and Machaj [5] developed a fuzzy linear programming model to tackle the transportation problem. Liu and Kao [11] presented a method for solving fuzzy transportation issues based on the extension principle, while Lin introduced a genetic algorithm for addressing transportation problems with fuzzy objective functions. Srinivasan [18–23] has also outlined new methodologies for resolving fuzzy transportation challenges.

Nagoor Gani and Abdul Razak [13] developed a fuzzy solution for a two-stage transportation problem aimed at minimizing costs, where both supplies and demands are represented as trapezoidal fuzzy numbers. A. Nagoor Gani, Edward Samuel, and Anuradha [7] employed Arshamkhan's Algorithm

DOI: 10.1201/9781003642152-19

to address a fuzzy transportation issue. Additionally, Pandian and Natarajan [14] introduced a fuzzy zero-point method to determine an optimal fuzzy solution for a fuzzy transportation problem characterized by trapezoidal fuzzy numbers for all parameters.

19.2 SOME IMPORTANT DEFINITION

19.2.1 Fuzzy transportation problem

A fuzzy transportation problem is a transportation problem in which the transportation costs, supply, and demand quantities are fuzzy quantities. Most of the existing techniques provide only crisp solutions for the fuzzy transportation problem.

In a fuzzy transportation problem, all parameters are fuzzy numbers; fuzzy numbers may be normal or abnormal, triangular, or trapezoidal. The mathematical representation of the fuzzy transportation problem is as follows:

$$\text{Minimize } Z = \sum_{i=1}^{m} \sum_{j=1}^{n} \tilde{c}_{ij} . x_{ij} \; r = 1, 2, 3, \ldots\ldots k \tag{19.1}$$

Subject to

$$\sum_{j=1}^{n} x_{ij} = \tilde{a}_i, \quad i = 1, 2, \ldots, m$$

$$\sum_{i=1}^{m} x_{ij} = \tilde{b}_j, \quad j = 1, 2, \ldots, n$$

$$x_{ij} \geq 0 \quad \forall i, j$$

where $\tilde{c}_{ij}$, $\tilde{a}_i$, and $\tilde{b}_j$ represent the transportation cost, demand of product, and availability of product, respectively, in terms of fuzzy parameters.

19.2.2 Fuzzy set

A fuzzy set A is defined as the set of ordered pairs $(\mathrm{x}, \mu_A(x))$ where x is an element of the universe of discourse U and $\mu_A(x)$ is the membership that is attributed to each $\mathrm{x} \epsilon U$, a real number $\epsilon [0,1]$ describing the degree to which x belongs to the set.

19.2.3 Trapezoidal fuzzy number

A fuzzy number A is a trapezoidal fuzzy number denoted by (a_1, a_2, a_3, a_4), where a_1, a_2, a_3, and a_4 are real numbers, and its membership function $\mu_A(x)$ is given below (Figure 19.1):

$$\mu_A(x) = \begin{cases} 0 & \text{for } x \le a_1 \\ \dfrac{(x - a_1)}{(a_2 - a_1)} & \text{for } a_1 \le x \le a_2 \\ 1 & \text{for } a_2 \le x \le a_3 \\ \dfrac{(a_4 - x)}{(a_4 - a_3)} & \text{for } a_3 \le x \le a_4 \\ 0 & \text{for } x \ge a_4 \end{cases}$$

19.2.4 Pascal triangular graded mean

$$P(a, b.c, d) = \frac{(a + 3b + 3c + d)}{8}$$

19.3 RUSSELL'S APPROXIMATION METHOD

The various steps of this method are as follows:

Step 1. Calculate the quantities $\widetilde{U_i}$, $\widetilde{V_j}$ and $\widetilde{C_{ij}} - \widetilde{U_i} - \widetilde{V_j}$ and nd for all i and j, where

$$\widetilde{U_i} = \max_{1 \le j \le n} \{\widetilde{C_{ij}}\} \text{ for } i = 1, 2, \ldots\ldots\ldots\ldots, m$$
$$\widetilde{V_j} = \max_{1 \le j \le n} \{\widetilde{C_{ij}}\} \text{ for } j = 1, 2, \ldots\ldots\ldots n$$

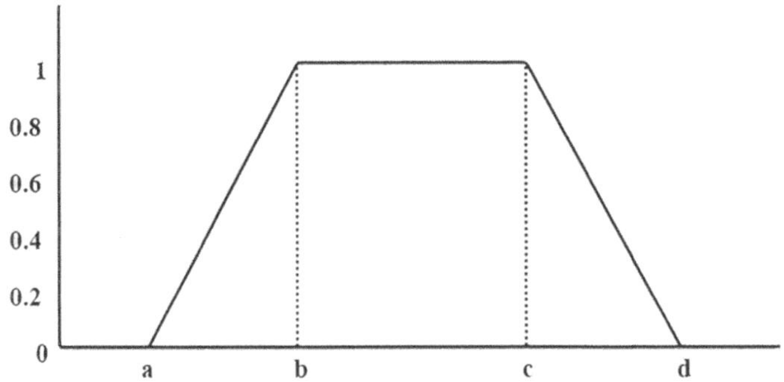

Figure 19.1 Graph of fuzzy trapezoidal number.

Step 2. Select the variables $\widetilde{X_{ij}}$ having the most negative value of $\widetilde{C_{ij}} - \widetilde{U_i} - \widetilde{V_j}$. If there are ties in the value of $\widetilde{C_{ij}} - \widetilde{U_i} - \widetilde{V_j}$, select $\widetilde{X_{ij}}$ having the smallest unit cost $\widetilde{C_{ij}}$. If there are ties again in the value of $\widetilde{C_{ij}}$, select $\widetilde{X_{ij}}$ having the largest amount of remaining source supply or destination demand. Set the activity level of $\widetilde{X_{ij}}$ equal to the smaller value between the source supply $\widetilde{a_i}$ and the destination demand $\widetilde{b_j}$.

Step 3. Subtract $\widetilde{X_{ij}}$ from $\widetilde{a_i}$ or $\widetilde{b_j}$ found in Step 3. Eliminate the row or column from the transportation table that results in a zero supply or destination demand after this subtraction. Stop if all $\widetilde{a_i}$ and $\widetilde{b_j}$ are zero, otherwise go to Step 1.

19.4 NUMERICAL PROBLEM

Balanced transportation problem (Table 19.1):

After the implementation of Pascal Triangular Graded Mean, the above problem changes into a crisp value as follows (Table 19.2):

The optimal solution is given in Table 19.3:

$$\text{The minimum total transportation cost is } (3.38 \times 2.5) + (3.75 \times 3.88) + (4.5 \times 0.49) + (5.50 \times 0.62) + (4.13 \times 6.25) + (0 \times 0.01) + (4.38 \times 3.38) + (2.75 \times 1.63) + (3.5 \times 5) = 91.21$$

Table 19.1 Balanced transportation problem

	D1	*D2*	*D3*	*D4*	*D5*	*Supply*
S1	(1,2,4,8)	(1,3,4,8)	(4,5,8,12)	(1,4,6,8)	(4,5,6,8)	**(3,6,7,9)**
S2	(2,4,7,11)	(2,4,5,7)	(1,4,7,10)	(2,5,7,11)	(2,5,8,9)	**(2,4,6,10)**
S3	(2,3,4,8)	(2,4,7,10)	(2,4,5,6)	(3,4,5,6)	(4,6,9,12)	**(1,3,4,5)**
S4	(2,3,5,6)	(1,2,3,6)	(1,3,6,7)	(1,3,4,6)	(2,5,8,11)	**(4,5,8,10)**
Supply	**(1,2,3,4)**	**(3,5,7,9)**	**(1,3,5,7)**	**(2,4,6,8)**	**(2,3,5,7)**	

Table 19.2 Crisp value of balanced transportation problem

	D1	*D2*	*D3*	*D4*	*D5*	*Supply*
S1	3.38	3.75	6.88	4.88	5.63	**6.38**
S2	5.75	4.50	5.50	6.13	6.25	**5.25**
S3	3.88	5.63	4.38	4.50	7.63	**3.38**
S4	4.00	2.75	4.38	3.50	6.50	**6.63**
Supply	**2.50**	**6.00**	**4.00**	**5.00**	**4.13**	

Table 19.3 Optimal solution of balanced transportation problem

	D1	*D2*	*D3*	*D4*	*D5*	D_{dummy}	*Supply*
S1	3.38 **(2.5)**	3.75 **(3.88)**	6.88	4.88	5.63	0	**6.38**
S2	5.75	4.50 **(0.49)**	5.50 **(0.62)**	6.13	6.25 **(4.13)**	0 **(0.01)**	**5.25**
S3	3.88	5.63	4.38 **(3.38)**	4.50	7.63	0	**3.38**
S4	4.00	2.75 **(1.63)**	4.38	3.50 **(5)**	6.50	0	**6.63**
Supply	**2.50**	**6.00**	**4.00**	**5.00**	**4.13**	0.01	

Unbalanced transportation problem (Table 19.4):

After the implementation of Pascal Triangular Graded Mean, the above problem changes into a crisp value as follows (Table 19.5):

The optimal solution is given in Table 19.6:

$$\text{The minimum total transportation cost is } (5.13\times 0.75)+(3.50\times 5.63) +(4.38\times 5)+(5.50\times 2.13)+(3.75\times 5)+(6.25\times 1)+ (5.63\times 4.5)+(0\times 5.25)=107.50$$

Table 19.4 Unbalanced transportation problem

	D1	*D2*	*D3*	*D4*	*D5*	*Supply*
S1	(4,6,7,9)	(2,4,6,9)	(4,5,7,9)	(1,3,4,6)	(2,3,6,10)	**(3,6,7,9)**
S2	(3,4,6,7)	(1,4,5,7)	(1,4,7,10)	(3,4,6,9)	(4,5,8,11)	**(2,4,6,8)**
S3	(3,4,7,8)	(3,6,9,10)	(4,7,8,10)	(4,7,10,14)	(2,3,4,7)	**(4,6,8,11)**
S4	(3,6,7,8)	(4,7,10,14)	(4,5,6,8)	(4,7,9,11)	(2,5,8,10)	**(2,5,6,9)**
Supply	**(4,7,10,12)**	**(2,5,7,8)**	**(2,3,6,7)**	**(2,4,7,10)**	**(3,4,6,7)**	

Table 19.5 Crisp value of unbalanced transportation problem

	D1	*D2*	*D3*	*D4*	*D5*	*Supply*
S1	6.50	5.13	6.13	3.50	4.88	**6.38**
S2	5.00	4.38	5.50	5.25	6.75	**5.00**
S3	5.50	7.25	7.38	8.63	3.75	**7.13**
S4	6.25	8.63	5.63	7.88	6.38	**5.50**
Supply	**8.38**	**5.75**	**4.50**	**5.63**	**5.00**	

Table 19.6 Optimal solution of unbalanced transportation problem

	D1	*D2*	*D3*	*D4*	*D5*	*Supply*
S1	6.50	5.13 **(0.75)**	6.13	3.50 **(5.63)**	4.88	**6.38**
S2	5.00 **(d)**	4.38 **(5)**	5.50	5.25	6.75	**5.00**
S3	5.50 **(2.13)**	7.25	7.38	8.63	3.75 **(5)**	**7.13**
S4	6.25 **(1)**	8.63	5.63 **(4.5)**	7.88	6.38	**5.50**
S_{dummy}	0 **(5.25)**	0	0	0	0	**5.25**
Supply	**8.38**	**5.75**	**4.50**	**5.63**	**5.00**	

19.5 CONCLUSION

It is crucial to acknowledge that our approach involves applying the Pascal Triangular Graded Mean to the balanced and unbalanced transportation problem. We proposed Fuzzy Russell's method to find the optimal solution using the Pascal Triangular Graded Mean method with trapezoidal fuzzy numbers. This method can be used for all kinds of fuzzy numbers. This method is very easy to apply and can be utilized for the fuzzy transportation problem. Our method proves to be equally effective in determining the optimal solution for both balanced and unbalanced transportation problems.

BIBLIOGRAPHY

[1] Arsham, H. and A. B. Kahn, A simplex type algorithm for general transportation problems: An alternative to stepping-stone, *Journal of Operational Research Society*, 40 (1989), 581–590.

[2] Basirzadeh, H., An approach for solving fuzzy transportation problem, *Applied Mathematics Sciences*, 5(2011) 1549–1566.

[3] Bellman, R.E. and L.A. Zadeh. Decision Making in a Fuzzy Environment, *Management Science*, 17 (1970), 141–164.

[4] Chanas, S., & Kuchta, D. (1996). A concept of the optimal solution of the transportation problem with fuzzy cost coefficients. *Fuzzy sets and Systems*, *82*(3), 299–305.

[5] Chanas, S., W. Kolodziejczyk and A. Machaj, A fuzzy approach to the transportation problem, *Fuzzy Sets and Systems*, 13 (1984), 211–221.

[6] De, P.K. and Yadav, B, Approach to defuzzify the triangular fuzzy number in transportation problem. *International Journal of Cognitive Computing in Engineering*, 8 (2010), 64–67.

[7] Edward Samuel and A. Nagoor Gani, Simplex type algorithm for solving fuzzy transportation problem, *Tamsui Oxford Journal of Information and Mathematical Sciences*, 27 (1) (2011), 89–98.

[8] Gass, On solving the transportation problem, *Journal of Operational Research Society*, 41 (1990), 291–297.

[9] Kaur, A and Kumar A, A new method for solving fuzzy transportation problems using ranking function. *Applied Mathematical Modelling*, 35 (2011), 5652–5661.

[10] Lious, T.S. and Wang, M. J., Ranking fuzzy numbers with integral value, *Fuzzy Sets and Systems*, 50 (3) (1992), 247–255.

[11] Liu, S.T., Kao, C., Solving Fuzzy transportation problem based on extension principle, *European Journal of Operations Research*, 153 (2004), 661–674.

[12] Manimaran. S and Ananthanarayanan, M, A study on comparison between fuzzy assignment problems using triangular fuzzy numbers with average method, *IJST*, 5 (4) (2012), 2610–2613.

[13] Nagoor Gani, and K. A. Razak, Two stage fuzzy transportation problem, *Journal of Physical Sciences*, 10 (2006), 63–69.

[14] Pandian, P. and Natrajan, G., An optimal more-for-less solution to fuzzy transportation problems with mixed constraints. *Applied Mathematics Sciences*, 4 (2010), 1405–1415.

[15] Pandian, P and Nagarajan, G, A new algorithm for finding a fuzzy optimal solution for fuzzy transportation problem, *Applied Mathematics Sciences*, 4 (2) (2010), 79–90.

[16] Shiang-Tai Liu and Chiang Kao, Solving fuzzy transportation problems based onextension principle *Journal of Physical Science*, 10 (2006), 63–69.

[17] Kumar, S. K., Lal, I. B., & Verma, S. P. (2011). An alternative method for obtaining initial feasible solution to a transportation problem and test for optimality. *International Journal of Computer Science and Communication*, 2(2), 455–457.

[18] Srinivasan, R. and Muruganandam. S, A new algorithm for solving fuzzy transportation problem with trapezoidal fuzzy numbers, *International Journal of Recent Trends in Engineering and Research*, 2 (3) (2016), 428–437.

[19] Srinivasan, R., Modified method for solving fully fuzzy transportation problem, *Global Journal of Research Analysis*, 5 (4) (2016), 177–179.

[20] Srinivasan, R. and Muruganandam, S, A new approach for solving unbalanced fuzzy transportation problem, *Asian Journal of Research in Social Sciences and Humanities*, 6 (5) (2016), pp. 673–680.

[21] Srinivasan, R. and Muruganandam, S, Optimal solution for multi-objective two stage fuzzy transportation problem, *Asian Journal of Research in Social Sciences and Humanities*, 6 (5) (2016), pp. 744–752.

[22] Srinivasan, R and Muruganandam. S, A Method of Solution to Intuitionistic Fuzzy Transportation Problem, *Asian Journal of Research in Social Sciences and Humanities*, 6 (5) (2016), pp. 753–761.

[23] Srinivasan, R, Muruganandam, S and Vijayan, V, A New Algorithm for Solving Fuzzy Transportation Problem with Triangular Fuzzy Number, *Asian Journal of Information Technology*, 15 (18) (2016), pp. 3501–3505.

[24] Tanaka, H. (1984). A formulation of fuzzy linear programming problem based on comparison of fuzzy numbers. *Control and cybernetics*, *13*, 185–194.

[25] Yager, R. R., A procedure for ordering fuzzy subsets of the unit interval, *Information Sciences*, 24 (1981), 143–161.

[26] Zadeh, L. A., Fuzzy sets, *Information Control*, 8 (1965), 338–353.

[27] Zimmermann, H. J., Fuzzy programming and linear programming with several objective functions, *Fuzzy Sets and Systems*, 1 (1978), 45–55.

[28] Zimmermann, H. J., *Fuzzy Set Theory and Its Applications*, Kluwer Academic, Norwell, MA, 1991.

Chapter 20

A study of digitalization on impulsive buying behavior of consumers with special reference to Madhya Pradesh

Amrita Soni, Ajay Soni, and S.P. Singh

20.1 INTRODUCTION

The era of consumer behavior has undergone a special change due to the rapid advancement of digital technology. The emergence of e-commerce platforms, social media, and mobile applications has revolutionized the way consumers interact with products and make purchasing decisions. Among the various facets of consumer behavior, impulsive buying has garnered particular attention due to its spontaneous and often unplanned nature. Impulsive buying refers to the tendency of consumers to make sudden and unplanned purchases, driven by emotions, urges, or impulses rather than rational decision-making. With the proliferation of digital channels, the dynamics of impulsive buying have experienced a paradigm shift. The convenience, accessibility, and personalized experiences offered by digital platforms have not only facilitated impulsive buying but have also reshaped their underlying drivers and triggers. MP, the largest, represents a diverse consumer market characterized by varying socioeconomic profiles and cultural influences. Understanding the role of digitalization in shaping impulsive buying behavior in this context is crucial for businesses and policymakers alike. The main objective of the study is to delve into the intricate relationship between digitalization and impulsive buying behavior among consumers in Madhya Pradesh. By examining the impact of factors such as online shopping platforms, social media marketing, mobile applications, and digital payment systems, the research seeks to unravel the mechanisms through which digitalization influences impulsive buying tendencies. Furthermore, the study will explore the moderating effects of personal variables, including age, gender, income, and educational background, in the relationship between digitalization and impulsive buying behavior. By identifying key demographic segments that are most susceptible to impulsive buying in the digital era, businesses can tailor their marketing strategies and product offerings more effectively. Model of impulsive buying behavior of Hawkins Stern and digitalization model: Hawkins Stern Model and digitalization models can be combined to understand impulsive buying behavior in the context of digitalization. Let's break down each model and then

 DOI: 10.1201/9781003642152-20

explore how they can be integrated: Hawkins Stern Model: The Hawkins Stern Impulse Buying Model is a psychological model that suggests several factors influencing impulsive buying behavior. It highlights the following four main components:

Environmental Stimuli: These are outside cues, inclusive of product presentations, advertisements, or promotions, that trigger impulsive buying.

Internal State: This refers to the psychological state of the purchaser, consisting of emotions, mood, and arousal degrees, which could influence impulsive behavior.

Response: This is the impulsive shopping behavior itself, which can vary from spontaneous purchases to unplanned spending.

Consumption Situation: The context in which the shopping behavior happens, including elements like time pressure or social influences.

Digitalization Model: Digitalization Model: Digitalization has transformed customer behavior with the aid of introducing new platforms, channels, and modes of interaction. A digitalization version might remember elements, including online shopping platforms: The layout, capabilities, and personal experience of e-commerce websites or apps.

Social Media Influence: How social media platforms impact patron preferences and purchasing choices via targeted advertising and marketing, influencers, and consumer-generated content.

Mobile Technology: The role of smartphones and cell apps in facilitating impulsive purchases whenever and anywhere.

Data Analytics: How consumer data collection and analysis enable customized marketing techniques that inspire impulsive shopping. Combining these fashions, researchers should recommend a complete framework for expertise-impulsive buying conduct in the digital age. For example, environmental stimuli in virtual environments could consist of targeted commercials on social media or personalized suggestions on e-trade websites.

20.2 LITERATURE REVIEW

1. Patel, K., & Jain, R. (2018). The Role of Social Media in Impulsive Buying Behavior: Evidence from Madhya Pradesh. Patel and Jain delve into the influence of social media platforms on impulsive buying behavior. Their study highlights the significant role of social media in triggering impulsive purchases and provides empirical evidence supporting the connection between social media usage and impulsive buying tendencies among consumers in the region.

2. Khan, A., & Siddiqui, N. (2017). Digitalization and Consumer Behavior: A Study of E-commerce Adoption in Madhya Pradesh. Journal of Marketing Research, Khan and Siddiqui's research investigates the relationship between digitalization and consumer behavior, specifically focusing on the adoption of e-commerce platforms in Madhya Pradesh. Their study elucidates how the growing prevalence of e-commerce channels influences impulsive buying behavior among consumers in the region, shedding light on the dynamics of digitalization-driven consumer decisions.
3. Mishra, S., & Tiwari, A. (2016). Impulsive Buying Behavior in the Era of Digitalization: A Case Study of Madhya Pradesh. Mishra and Tiwari conduct a case study focusing on impulsive buying behavior in the era of digitalization, with Madhya Pradesh as the primary context. Their research examines the impact of digital marketing strategies, such as online promotions and discounts, on consumer decision-making processes, providing insights into the mechanisms through which digitalization influences impulsive purchases in the region.

Identified Gap: While existing research is helpful into the relationship between digitalization and impulsive buying behavior in Madhya Pradesh, there is a noticeable gap in the literature concerning the role of mobile technology and apps in facilitating impulsive purchases among consumers in the region. Given the widespread adoption of smartphones and mobile apps, particularly in urban and semi-urban areas of Madhya Pradesh, understanding how mobile technology influences impulsive buying behavior could provide valuable insights for marketers and policymakers. Future research could focus on investigating the impact of mobile apps, push notifications, and location-based services on impulsive buying tendencies, thereby addressing this gap in the literature.

Research objective:

1. To assess the influence of digitalization on impulsive buying behavior.
2. To evaluate whether age, gender, income, and updated information on the digital platform have any impact on the impulsive buying behavior of consumers.
3. To study different digital platforms and their effect on the impulse buying role of consumers.
4. To identify and evaluate what attracts consumers to a particular digital platform which may trigger their impulsive buying behavior.
5. To study the various factors affecting impulse buying among consumers due to digital platforms.
6. To study the linkage among the factors of digitalization and its influential phenomenon on impulsive buying behavior.

Hypothesis:

1. Are consumers using digital platforms for shopping acting more impulsively?
2. Does the presence of different digital platforms show any effect on impulse buying role of consumer?
3. Is a digital platform attractive enough to appeal to and trigger impulsive buying behavior among consumers?
4. Are there any factors affecting impulse buying among consumers due to the digital platform?

20.3 RESEARCH METHODOLOGY

This section presents the research methodology used to demean the study. The research methods have need of the research design, study population, sample design, sample frame, sample technique and sample number, information collection methods, and research procedures.

Research is mainly designed to gather data to answer questions and find solutions to problems. This research work identifies the buying behavior of consumers in a digital mode and the factors that influence the online purchasing mode. The research work can also identify which group is more prone to impulsive buying behavior in Madhya Pradesh.

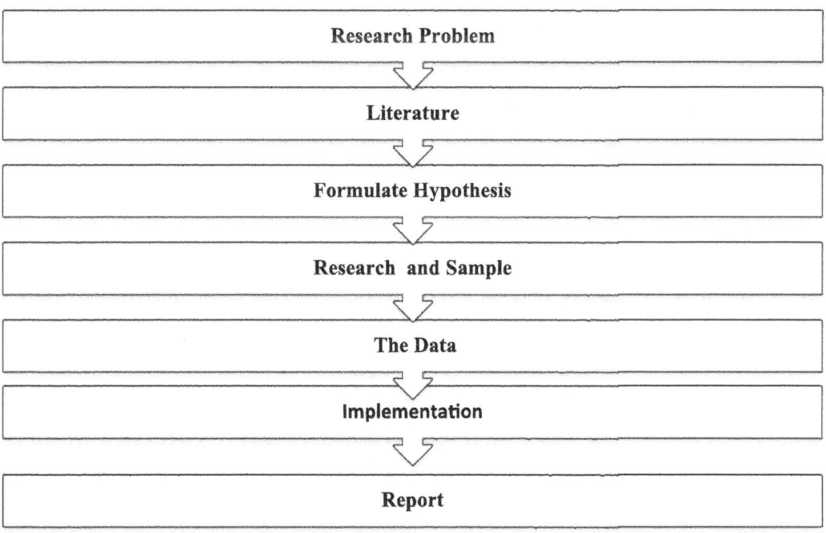

20.3.1 Research design and research problem

These plans explain how facts need to be accrued and analyzed. Research design is the basis for the direction of the study of the problem after formulating speculation. The study designs are recognized via distinct names, including study outline, blueprint, and so forth. It includes surveys and fact-finding enquiries of different kinds.

20.3.2 Classification of research

Research work is classified in three ways:

1. **On the basis of intent**
2. **On the basis of the nature of the data**
3. **On the basis of the methods of data**

This research work applied descriptive research, exploratory research, and survey research; the first two are on the basis of intent, and the last one is on the basis of methods of data.

It collects data that is used to answer a huge range of what, when, and how questions related to a specific population or companies. The intention of all descriptive research is to discover the historical past, details, and existing patterns within the trouble to completely understand it.

Exploratory research: Exploratory research is a study technique used to investigate research issues that are not fully described or understood. It is critical for researchers to, without a doubt, recognize their study's trouble before trying to solve it, which allows them to determine whether or not the topic is worth investigating. This helps make certain that time and assets aren't spent seeking to have a look at research issues that aren't relevant, legitimate, or viable to investigate.

Survey research: Survey research is the method of gathering facts from a predefined group (e.g., clients or ability customers) with the primary purpose of uncovering insights about your products, services, or brand overall. As a quantitative fact collection method, survey research can provide you with a goldmine of statistics that can reveal vital enterprise and product selections. But survey research needs careful planning and execution to get the affects you want.

Research Problem: This study aims to explore the effect of digitalization on impulsive buying conduct among consumers in Madhya Pradesh. With the appearance of virtual platforms, online purchasing has turned out to be more and more prevalent, leading to significant modifications in customer purchasing styles. However, there may be constrained research on how these virtual improvements specifically have an impact on impulsive shopping behavior in this area. The examiner will look at the volume of digitalization, the effectiveness of virtual advertising strategies, and the mental and situational elements contributing to impulsive purchases. Additionally, it's going to examine variations in impulsive buying throughout extraordinary demographic segments. The findings will offer insights for groups and marketers to expand effective and moral digital techniques tailor-made to the clients of Madhya Pradesh.

20.3.3 Sample population

The people for study are the customers of different age groups and different income levels of the six top cities of Madhya Pradesh, which was formed on November 1, 1956, comprising the 17 districts of Mahakoshal, the districts of Bhopal, eight districts of Vindhya Pradesh, and 16 districts of Madhya Bharat. At the time of its formation, the nation had 43 districts. Subsequently, two big districts had been bifurcated. Sixteen extra districts were formed in the year 1998. The Chhattisgarh region, comprising 16 districts, was separated to form a new nation as consistent with the provisions of the Madhya Pradesh Reorganization Act, 2000, and reorganized State of Madhya Pradesh came into existence on November 1, 2000. Again, three new districts, that is, Burhanpur, Ashok Nagar, and Anooppur, were formed in 2003.

Table 20.1 shows the city-wise population of Madhya Pradesh in 2023.

Table 20.1 City-wise population of Madhya Pradesh in 2023

S.NO	*City name*	*Population-2023*
1	Betul	120,937
2	Bhind	231,253
3	Bhopal	2,104,634
4	Burhanpur	2,104,634
5	Chhattarpur	156,206
6	Chhindwara	161,856
7	Damoh	146,418
8	Datia	117,372
9	Dewas	338,889
10	Guna	211,766
11	Gwalior	1,251,481
12	Hoshangabad	138,093
13	Indore	2,298,766
14	Jabalpur	1,235,386
15	Khandwa	234,944
16	Khargone	124,594
17	Mandsaur	165,807
18	Morena	234,644
19	Murwara	259,692
20	Nagda	117,086

(Continued)

Table 20.1 (Continued) City-wise population of Madhya Pradesh in 2023

S.NO	*City name*	*Population-2023*
21	Neemuch	149,922
22	Pithampur	147,704
23	Ratlam	310,055
24	Rewa	275,809
25	Sagar	319,866
26	Satna	327,972
27	Sehore	127,467
28	Seoni	119,782
29	Shivpuri	210,645
30	Singaurli	257,789
31	Ujjain	603,008
32	Vidisha	182,525

Table 20.2 shows the whole population of Madhya Pradesh in 2023.

Table 20.2 Whole population of Madhya Pradesh State in 2023 with Male and Female bifurcation

S.NO	*Estimated population-2023*	*Total population*
1	Estimated population of Madhya Pradesh in 2023	85,002,417
2	Estimated population of males in 2023	44,019,895
3	Estimated population of females in 2023	40,982,522

20.3.4 Sample Design

The selected topmost cities on the basis of population size and digitalization market purpose are Bhopal, Gwalior, Indore, Jabalpur, Ujjain, and Mandsaur.

20.3.5 Sample Techniques

Under this study, the population size is large, and it is divided into a number of districts and cities. For the time being, constant to prove that this research work applied non-probability sampling. The techniques applied under the study were convenience and purposive sampling methods. Non-possibility sampling is described as a sampling approach in which the researcher selects samples based totally on the subjective judgment of the researcher instead of random selection.

Convenience sampling is a non-chance sampling technique wherein samples are decided on from the populace simply because they are with no trouble to be had to the researcher. Researchers pick those samples just because they may be easy to recruit, without considering whether the sample accurately represents the entire population.

Purposive sampling refers to a set of non-probability sampling techniques in which units are selected due to the fact that they have traits which you need for your pattern. In other words, gadgets are selected "on reason" in purposive sampling. Also called judgmental sampling, this sampling approach relies on the researcher's judgment while figuring out and choosing the individuals, instances, or occasions, which can provide excellent data to achieve the research goals.

Sample size: The manner of choosing a number of people for a study, in the sort of way that the individuals were selected, constitutes the big group from which they were selected. The purpose of sampling is to form a consultant group, which enables the researcher to collect information while confronted with constraints of time, budget, and strength.

Selected cities for data gathering purposes are shown in Table 20.3:

Table 20.3 List of selected cities in Madhya Pradesh for Study

S.NO.	*City Name*	*Population*
1	Bhopal	2,104,634
2	Gwalior	1,251,481
3	Ujjain	603,008
4	Indore	2,298,766
5	Jabalpur	1,235,386
6	Mandsaur	165,807
Total Population		7,659,082

There are four things required for sample size:

1	Population size	7659082
2	Margin of error	.05
3	Degree of variability	50
4	Confidence level	1.96

Yamane's 1967 formula of sample size:

$$n = N / 1 + N(e)^2$$

$$n = 7659082 / 1 + 7659082 * (0.05)^2$$

$$n = 7659082 / 19147.705$$

$$n = 400 \text{ sample size}$$

Tools Used for Analysis: Evaluation equipment is described as a chain of charts, maps, and diagrams designed to collect and interpret statistical tools, that is, percentage analysis. The research work was proved or disproved by conducting a chi-square test.

Data Collection Method: The procedure of gathering and studying correct facts from numerous sources to locate answers to analyze troubles, traits, and probabilities, and so forth, to evaluate possible effects is known as data collection. Knowledge is strength, information is know-how, and facts are information in digitized shape, at least as described in research work. The study followed each nominal and ordinal scale to a degree range of factors, establishing the effect on purchaser behavior and an interval scale in figuring out the relationship between digital advertising and consumer behavior.

20.3.6 Primary Data Collection

Initially, the statistics series includes the gathering of real information at once from the supply. This approach permits researchers to attain firsthand statistics, especially tailored to their research targets. There are numerous strategies for primary records collection, together with:

Surveys and Questionnaires: Researchers design structured questionnaires or surveys to acquire data from people or companies. These may be carried out through face-to-face interviews, telephone calls, mail, or online systems.

Secondary Data Collection: Secondary data collection involves using existing facts accrued through another person for a cause other than the original intent. Researchers examine and interpret these statistics to extract applicable facts. Secondary data can be obtained from diverse sources, which include:

Published Sources: Researchers refer to books, academic journals, magazines, newspapers, government reviews, and other published sources that incorporate applicable records.

Online Databases: Numerous online databases provide the right of entry to an extensive range of secondary data, including studies, articles, statistical records, financial data, and social surveys.

20.4 NEED FOR THE STUDY

The present goals are to offer a broad overview of impulsive shopping through literature assessment, locate stimuli that trigger impulse shopping at some point in online shopping, and examine the effect of respondents' demographic and psychographic elements on impulsive shopping behavior. This research wants to have a look at examining the idea of virtual advertising and marketing and patron shopping behavior in the digital generation, to research diverse gear and channels of social media that without delay affect customer behavior, to become aware of the categories of merchandise that customers buy on digital media structures, and to analyze the impact of virtual marketing on the patron decision-making process. Nowadays, the internet has emerged as the main source of statistics, and it is used widely in everyday life. As the internet is an effective supply, it is utilized by corporations for competitive advantage. The clients' opportunity to make impulsive purchases is expected to grow in addition to the convenience of online buying. There are a number of things which have an impact on the online impulsive buying behavior of the purchasers. To examine the factors affecting the online buying behavior of the customers, factor evaluation was used to show the results. Personal behavior, scenario at the time of purchase, merchandising, and ethical issues have an impact on the net impulsive buying of the clients. The consequences of examining are beneficial for the web firms that want to grow the online impulse buying behavior of the purchasers.

Digital marketing and the era allow marketers to connect with their clients and permit clients to receive statistics anytime and anywhere, whether they are at work, in a store, or browsing the web. It also helps marketers to supply their offerings to their customers by way of customizing the offers based on their clients' mood. Digital advertising messages can act as outside cues that help purchasers to understand a need, triggering a selection-making process that could result in a purchase.

20.5 DATA ANALYSIS AND INTERPRETATION

20.5.1 Summary of data analysis and interpretation

20.5.1.1 Demographical profile

Variables	*Options*	*No. of respondent*	*%age*
Gender	male	175	43.75%
	Female	225	56.25%
	Total	400	100%

Variable	*Options*	*No. of respondent*	*%age*
Age group	Below 25 years	120	30%
	26-35 years	140	35%
	36-45 years	100	25%
	Above 45 years	40	10%
	Total	400	100%

Variable	*Options*	*No. of respondent*	*%age*
Education-qualification	School level	30	7.5%
	Diploma	50	12.5%
	UG	170	42.5%
	PG	150	37.5%
	Total	400	100%

Variable	*Options*	*No. of respondent*	*%age*
Occupation	Students	135	33.75%
	Private employee	140	35%
	Government employee	70	17.5%
	business	55	13.75%
	Total	400	100%

Marital status	*No. of respondent*	*%age*
single	220	55%
married	180	45%
Total	400	100%

Variable	*Options*	*No. of respondent*	*%age*
Monthly Income	Below 15000	45	11.25%
	15000-25000	85	21.25%
	26000-35000	120	30%
	Above 35000	150	37.5%
	Total	400	100%

The demographic profile of the 400 respondents provides insights into the customer base in Madhya Pradesh regarding impulsive buying behavior motivated by digitalization:

Gender: The pattern consists of slightly more females (56.25%) than males (43.75%).

Age Group: The largest age group is 26–35 years (35%), followed by the ones below 25 years (30%). Respondents aged 36–45 years constitute 25% and only 10% are above 45 years.

Educational Qualification: Most respondents have an undergraduate (42.5%) or postgraduate (37.5%) qualification. Those with a diploma make up 12.5%, and the best 7.5% have college-level education.

Occupation: Private employees (35%) and students (33.75%) form the majority of the sample, followed by members of authorities' personnel (17.5%) and commercial enterprise proprietors (13.75%).

Marital Status: A better proportion of respondents are single (55%) as compared to married (45%).

Monthly Income: The profit distribution indicates that 37.5% earn above 35,000 INR, 30% earn between 26,000 and 35,000 INR, 21.25% earn between 15,000 and 25,000 INR, and 11.25% earn below 15,000 INR.

This demographic breakdown is well-known and shows a predominantly younger, educated, and numerous client base, with a substantial proportion of college students and personal employees, indicating a capability for varied impulsive shopping behaviors inspired with the aid of digitalization (Table 20.4).

Table 20.4 Digital media and consumer behaviors

Options	*Strongly agree*	*Agree*	*Neutral*	*Disagree*	*Strongly disagree*	*Frequency*
Facebook	150	110	40	60	40	400
%age	37.5%	27.5%	10%	15%	10%	100%
Twitter	60	40	110	150	40	400
%age	15%	10%	27.5%	37.5%	10%	100%
YouTube	110	90	70	100	30	400
%age	27.5%	22.5%	17.5%	25%	7.5%	100%
Instagram	90	120	80	70	40	400
%age	22.5%	30%	20%	17.5%	10%	100%
LinkedIn	60	40	70	130	100	400
%age	15%	10%	17.5%	32.5%	25%	100%
Skype	50	40	80	100	130	400
%age	12.5%	10%	20%	25%	32.5%	100%
Google	140	120	50	40	50	400
%age	35%	30%	12.5%	10%	12.5%	100%
Website/blog	120	90	70	80	40	400
%age	30%	22.5%	17.5%	20%	10%	100%
Multimedia advertising	140	130	40	60	30	400
%age	35%	32.5%	10%	15%	7.5%	100%
Emails	40	60	100	70	130	400
%age	10%	15%	25%	17.5%	32.5%	100%
SMS	30	50	120	150	50	400
%age	7.5%	12.5%	30%	37.5%	12.5%	100%

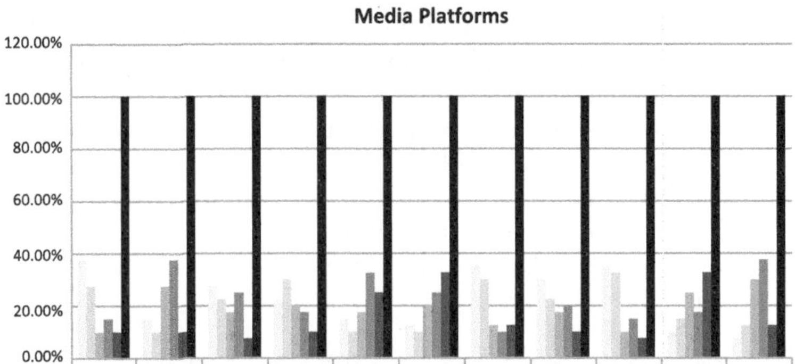

	Facebook	Twitter	Youtube	Instagram	Linkedin	Skype	Google	Website/blog	Multi Media adverti sing	Emails	SMS
Strongly agree	37.50%	15%	27.50%	22.50%	15%	12.50%	35%	30%	35%	10%	7.50%
Agree	27.50%	10%	22.50%	30%	10%	10%	30%	22.50%	32.50%	15%	12.50%
Neutral	10%	27.50%	17.50%	20%	17.50%	20%	12.50%	17.50%	10%	25%	30%
disagree	15%	37.50%	25%	17.50%	32.50%	25%	10%	20%	15%	17.50%	37.50%
Strongly disagree	10%	10%	7.50%	10%	25%	32.50%	12.50%	10%	7.50%	32.50%	12.50%
frequency	100%	100%	100%	100%	100%	100%	100%	100%	100%	100%	100%

Analysis: Digital media platforms that influence consumer behavior include Facebook, with 27.50% to 37.50% of respondents indicating agreement and strong agreement on the scale. A total of 10% to 15% respondents indicated agree to strongly agree for Twitter. A total of 22.50% to 27.50% respondents indicated agree to strongly agree for YouTube. A total of 22.50% to 30 % respondents indicated strongly agree to agree for Instagram. A total of 10% to 15% respondents indicated strongly agree to agree for LinkedIn. A total of 10% to 12.50% respondents indicated agree to strongly agree for Skype. A total of 30% to 35% respondents indicated agree to strongly agree for Google. A total of 22.50% to 30% respondents indicated agree to strongly agree for Websitelink. A total of 32.50% to 35% respondents indicated agree to strongly agree for multimedia advertising. A total of 15% to 10% respondents indicated agree to strongly agree for emails, and 12% to 7.50 % respondents indicated agree to strongly agree for SMS. Neutral results came from respondents regarding digital media like 10% for Facebook, 27.50% for Twitter, 17.50% for YouTube, 20% for Instagram, 17.50% for LinkedIn, 20% for Skype, 12.50% for Google, 17.50% for Websitelink, 10% for multimedia advertising, 25% for emails, and 30% for SMS. The rest of the respondents either indicated disagree or strongly disagree (Table 20.5).

Table 20.5 Merchandise purchase on virtual media

Product categories	*SA*	*A*	*N*	*D*	*SD*	*Frequency*
Convenience Products	160	70	50	40	80	400
%age	40%	17.5%	12.5%	10%	20%	100%
Shopping Products	150	110	30	70	40	400
%age	37.5%	27.5%	7.5%	17.5%	10%	100%
Specialty Products	120	90	70	80	40	400
%age	30%	22.5%	17.5%	20%	10%	100%
Unsought products	40	60	100	70	130	400
%age	10%	15%	25%	17.5%	32.5%	100%

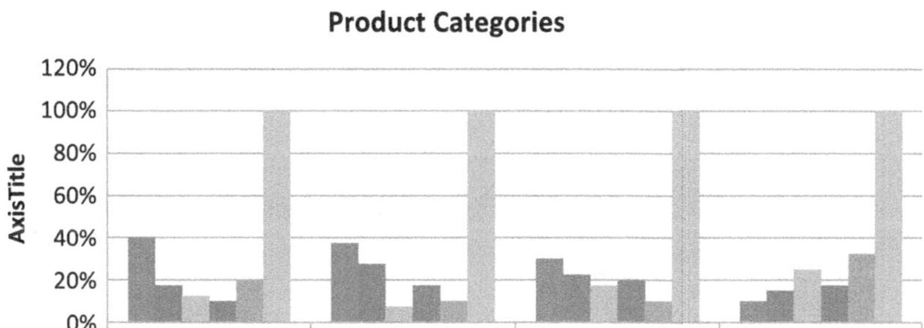

	Convenience Products	Shopping Products	Specialty Products	Unsought Products
SA	40%	37.50%	30%	10%
A	17.50%	27.50%	22.50%	15%
N	12.50%	7.50%	17.50%	25%
D	10%	17.50%	20%	17.50%
SD	20%	10%	10%	32.50%
frequency	100%	100%	100%	100%

Analysis: Convenience products indicated 17.50% to 40% respondents agree to strongly agreed, 12.50% respondents were neutral, and the rest of the respondents disagree to strongly disagree. For shopping products, respondents indicated 27.50% to 37.50% agree to strongly agree, 7.50% respondents are neutral, and the rest of the respondents disagreed to strongly disagreed. For specialty products, respondents indicated 22.50% to 30% agree to strongly agree, 17.50% respondents were neutral, and the rest of the respondents disagree to strongly disagree. For unsought products, respondents indicated 15% to 10% agree to strongly agree, 25% respondents were neutral, and the rest of the respondents disagree to strongly disagree (Table 20.6).

Table 20.6 Products and digital buying

Options	*SA*	*A*	*N*	*D*	*SD*	*Frequency*
Clothes and footwear	170	80	50	60	40	400
%age	42.5%	20%	12.5%	15%	10%	100%
Foods	90	120	80	40	70	400
%age	22.5%	30%	20%	10%	17.5%	100%
Home decor	140	60	70	80	50	400
%age	35%	15%	17.5%	20%	12.5%	100%
Cosmetics and Jewelry	120	80	70	90	40	400
%age	30%	20%	17.5%	22.5%	10%	100%
Medicine	110	90	80	90	30	400
%age	27.5%	22.5%	20%	22.5%	7.5%	100%
Electronics items	130	90	60	80	40	400
%age	32.5%	17.5%	15%	20%	10%	100%
Utensils	90	110	60	40	100	400
%age	22.5%	27.5%	15%	10%	25%	100%

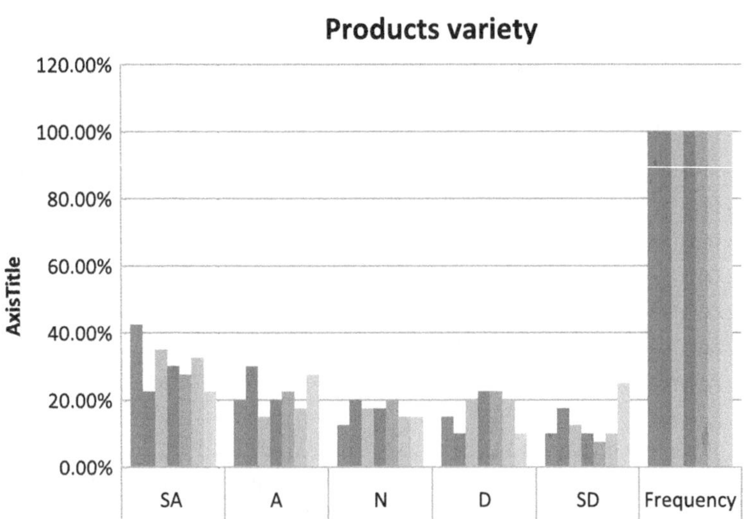

	SA	A	N	D	SD	Frequency
Clothsandfootwear	42.50%	20%	12.50%	15%	10%	100%
foods	22.50%	30%	20%	10%	17.50%	100%
Homedecor	35%	15%	17.50%	20%	12.50%	100%
Cosmetics & Jewell	30%	20%	17.50%	22.50%	10%	100%
Medicine	27.50%	22.50%	20%	22.50%	7.50%	100%
Electronics items	32.50%	17.50%	15%	20%	10%	100%
utensils	22.50%	27.50%	15%	10%	25%	100%

Analysis: A total of 20% to 42.50% of respondents from agree to strongly agree, and they love to purchase garments and footwear, 22.50% to 30% of respondents were strongly agree to agree purchased food products, home décor respondents 15% to 35% agreed to strongly agree, and 20% to 30% respondents were agree to strongly agree for purchased medicines through digital mode. A total of 20% to 30% respondents agreed to strongly agree for purchased cosmetics and jeweler through digital mode and 22.50% to 27.50% respondents were agree to strongly agree for purchased medicines through digital mode. A total of 17.50% to 32.50% respondents agree to strongly agree on purchasing electronics items through digital mode. A total of 27.50% to 32.50% respondents agree to strongly agree with purchasing utensils through digital mode (Table 20.7a).

Table 20.7a Buying behavior of consumers

Cultural	*SA*	*A*	*N*	*D*	*SD*	*Frequency*
Culture	70	110	90	50	80	400
%age	17.5%	27.5%	22.5%	12.5%	20%	100%
Subculture	90	120	80	60	50	400
%age	22.5%	30%	20%	15%	12.5%	100%
Social class	140	110	60	70	20	400
%age	35%	27.5%	15%	17.5%	5%	100%

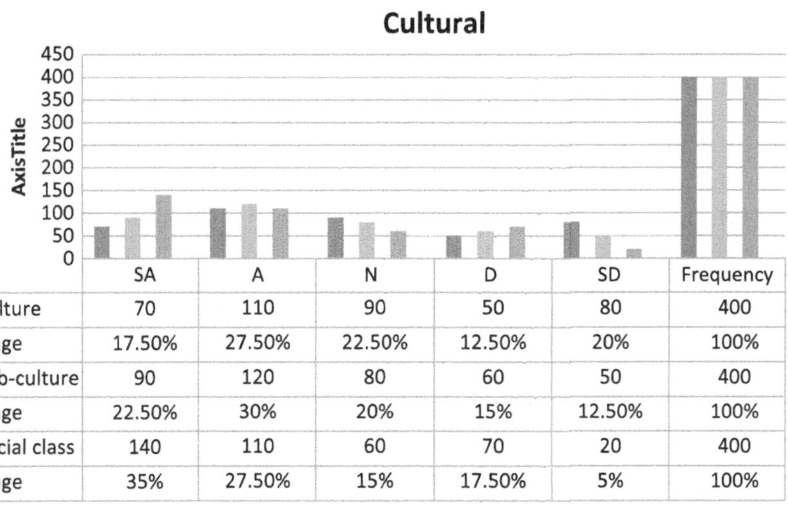

Analysis: From the above study, 27.50% to 17.50 % of respondents agree to strongly agree were influenced by their culture. A total of 30% to 22.50% of respondents agreed to strongly agree were influenced by their subculture.

A total of 27.50% to 35% of respondents agree to strongly agree were influenced by their social class. The neutral respondents were 22.50%. Subculture 20%, social class 15%, and the rest of the respondents strongly disagree to disagree (Table 20.7b).

Table 20.7b Buying behavior of consumers

Social	*SA*	*A*	*N*	*D*	*SD*	*Frequency*
Reference Group	90	80	70	110	50	400
%age	22.5%	20%	17.5%	27.5%	12.5%	100%
Family	110	80	60	70	80	400
%age	27.5%	20%	15%	17.5%	20%	100%
Roles and status	130	90	80	40	60	400
%age	32.5%	22.5%	20%	10%	15%	100%

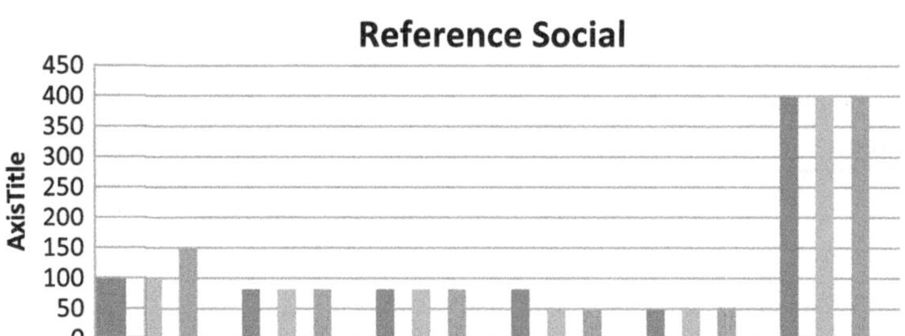

	SA	A	N	D	SD	Frequency
Reference Group	90	80	70	110	50	400
%age	22.50%	20%	17.50%	27.50%	12.50%	100%
Family	110	80	60	70	80	400
%age	27.50%	20%	15%	17.50%	20%	100%
Roles and status	130	90	80	40	60	400
%age	32.50%	22.50%	20	10%	15%	100%

Analysis: A total of 20% to 22.50% of respondents agree to strongly agree from the reference group. A total of 20% to 27.50% of respondents were from agree to strongly agree on family status, and 22.50% to 32.50 % respondents were from agree to strongly agree that consumers are influenced by roles and status. Neutral respondents were from the 17.50% reference group, 25% family, and 20% role and status. The rest of the respondents disagree and strongly disagree.

Personal	*SA*	*A*	*N*	*D*	*SD*	*Frequency*
Age and lifecycle	80	100	90	80	50	400
%age	20%	25%	22.5%	20%	12.5%	100%
Occupation	140	70	80	60	50	400
%age	35%	17.5%	20%	15%	12.5%	100%
Lifestyle and personality	130	100	70	70	30	400
%age	32.5%	25%	17.5%	17.5%	7.5%	100%

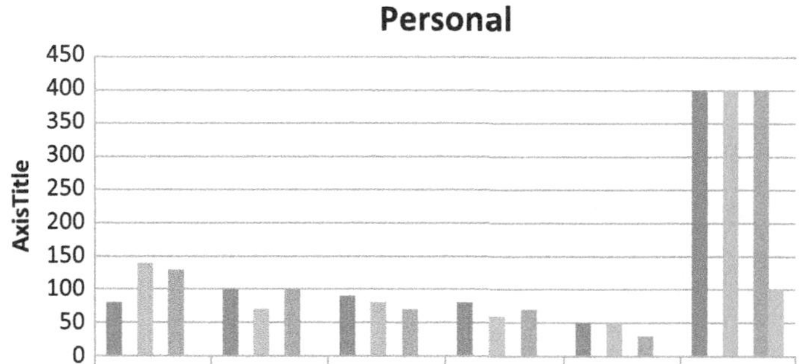

	SA	A	N	D	SD	Frequency
Age & lifecycle	80	100	90	80	50	400
%age	20%	25%	22.50%	20%	12.50%	100%
occupation	140	70	80	60	50	400
%age	35%	17.50%	20%	15%	12.50%	100%
Life-style & personality	130	100	70	70	30	400
%age	32.50%	25%	17.50%	17.50%	7.50%	100

Analysis: From the above study, 25% to 20 % of respondents agreed to strongly agree to influence personal decisions with their age life cycle, and neutral decisions came from 22.50% respondents. A total of 17.50% to 35% of respondents were agree to strongly agree to influence personal with their occupation and neutral decisions came from 20% respondents. A total of 25% to 32.50% of respondents agreed to strongly agree that influence on personal decisions with their lifestyle and personality, and neutral decision came from 17.50% respondents. The rest of the respondents were strongly disagree to disagree.

Psychological	*SA*	*A*	*N*	*D*	*SD*	*Frequency*
Motivation	140	80	70	60	50	400
%age	35%	20%	17.5%	15%	12.5%	100
Perception	90	80	70	110	50	400
%age	22.5%	20%	17.5%	27.5%	12.5%	100%
Learning	150	90	60	80	20	400
%age	37.5%	22.5%	15%	20%	5%	100%
Beliefs and attitude	40	60	100	70	130	400
%age	10%	15%	25%	17.5%	32.5%	100%

Psychological

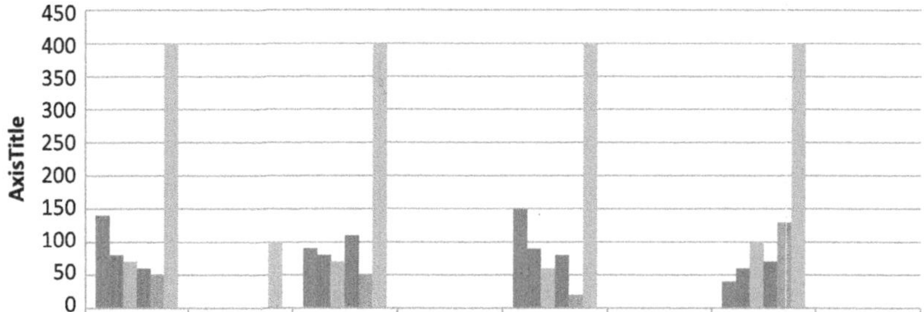

	Motivation	%age	Perception	%age	Learning	%age	Beliefs & attitude	%age
SA	140	35%	90	22.50%	150	37.50%	40	10%
A	80	20%	80	20%	90	22.50%	60	15%
N	70	17.50%	70	17.50%	60	15%	100	25%
D	60	15%	110	27.50%	80	20%	70	17.50%
SD	50	12.50%	50	12.50%	20	5%	130	32.50%
Frequency	400	100	400	100%	400	100%	400	100%

Analysis: A total of 20% to 35% of respondents were motivated to purchase products in digitalize mode, and 17.50% respondents were neutral. A total of 20% to 22.50% of respondents depend on perception for purchasing products in digitalize mode, and 17.50% respondents were neutral. A total of 22.50% to 37.50% of respondents were learning from different modes for purchasing products in digitalize mode, and 15% respondents were neutral. A total of 15% to 10% of respondents had their own beliefs and attitudes regarding purchasing products in digitalize mode, and 25% respondents were neutral. The rest of the respondents were in the mindset of disagree and strongly disagree regarding psychological behavior, showing any impact on purchasing through digitalized mode.

20.6 FINDINGS

It is undeniable that economic digitalization is to blame for shifts in consumer behavior and that digitalization is now a language that is understood by all. Due to the advent of web services that give customers freedom of choice, consumers now have access to product information. This has changed consumers' attitudes, which has led to a change in business models to reach customers as soon as possible. Internet-based shoppers first choose the goods they want, and only then do they contact the business that can meet their demands. This demonstrates how corporations' marketing strategies have altered as a result of digitization. The business focuses on elements like presentation, security, services, after-sales care, and many more to attract and keep customers. The major findings are listed below:

Digital media influence consumers to purchase products. Consumers have strong feelings about Facebook, multimedia advertising, the Google search engine, YouTube, Instagram, and website blogs. Facebook is practical and well-known. Users are familiar with the platform's rules and may have even seen your adverts there. It makes sense that people would also use Facebook to contact customer service. According to a Facebook survey, 45% of Facebook users who shop do so for clothing, while 41% do so for "Health, Household and Personal Care" purchases. The term "multimedia" refers to content that uses a combination of different forms of media, such as text, audio, images, animations, and video. Customers can also get information about products from the Google search engine, and the most important source is YouTube, where customers can go through the reviews of products and can easily evaluate the pros and cons associated with the product before actually buying it.

Mostly, consumers purchase products in a digital mode, like convenience and shopping products. Convenience goods are those that are regularly, instantly, and with the least amount of effort purchased. These include sweets, ice cream, chilly beverages, cigarettes, books, and pharmaceuticals, among others. Convenience stores are the businesses that stock these items. Convenience products frequently lack durability. Shopping is the practice of examining and choosing products or services from merchants with the intention of making a purchase right away. Comparison of products based on their compatibility, quality, price, style, and other factors leads to the selection and purchase. Examples include clothing, furniture, electronics, and appliances, among others. The majority of retail products are robust. Occasionally, or during the festive season, consumers purchase specialty products, and rarely or mostly upper-class families are aware of and purchase unsought products. Special purchasing procedures are required for specialty goods, and the commodities have some unique characteristics. In contrast to purchasing items, purchasers are willing to spend a lot of time and money on them. Examples include rare art collections, antiquities, high-end brands, fashion items, and automobiles. Unsought products are often ones that people are unaware of and are not familiar with.

20.7 CONCLUSION

Consumers can be reached and informed about new items, offers, discounts, or promotions using digital marketing methods like social media and email marketing. Consumers are more likely to buy a specific product if they believe the brand offers something special. Trust in the business and their product is thereby increased. Brands may now establish open communication with customers and give them more information about their products and services. The introduction of digital media has exposed consumers to a wide range of goods and information from all brands. They desire high-quality goods with superior financial returns. Consumer loyalty can be increased through consumer interaction, promotions, and consumer service. With the help of digital media, businesses may promote their goods at the precise moment and location where customers are most likely to make an impulse or spontaneous buy. Brands utilize strategies like online offers, deals, and discounts to persuade customers to purchase their products right away.

By giving customers more control over their buying experiences, digital media has fundamentally changed the way that businesses connect with their customers. Convenience items, shopping items, specialty items, and unsought items are among the product categories that customers purchase through digital platforms. The data a consumer discovers while researching a product has a significant influence on the selection of which product to buy. Because people evaluate brands and companies based on their online presence, it is imperative that they have strong online presences. In light of the fact that a goodwill self is the material posted online, its online presence provides its information more precisely and effectively.

BIBLIOGRAPHY

Bashir, S., Zeeshan, M., & Sabbar, S. (2013).Impact of cultural values and life style on impulse buying behavior: A case study of Pakistan. *International Review of Management and Business Research*, 2(1), 193–200.

Chen, Y. F., & Wang, R.Y. (2016). Are humans rational? Exploring factors influencing impulse buying intention and continuous impulse buying intention. *Journal of Consumer Behaviour*, 15(2), 186–197.

Cho, J., Ching, G. S., & Luong, T. H. (2014). Impulse buying behavior of Vietnamese consumers in supermarket setting. *International Journal of Research Studies in Management*, 3(2), 33–50.

Khan, A., & Siddiqui, N. (2017). Digitalization and consumer behavior: A study of E-commerce adoption in Madhya Pradesh. *Journal of Marketing Research*, 15(4), 220–235.

Mishra, S., & Tiwari, A. (2016). Impulsive buying behavior in the era of digitalization: A case study of Madhya Pradesh. *Journal of Consumer Behavior*, 8(1), 55–70.

Patel, K., & Jain, R. (2018). The role of social media in impulsive buying behavior: Evidence from Madhya Pradesh. *International Journal of Marketing Studies*, 10(3), 78–92.

LIST OF WEBSITES

https://researchguides.ben.edu/

https://extrudesign.com/a-study-on-consumer-behaviour-towards-onlineshopping/

https://www.tandfonline.com/doi/full/10.1080/23311975.2018.1514940

Chapter 21

A Novel Design of Piezometric Energy Harvester for Bicycle

Tejkaran Narolia, Anjali Pathak, and T. Ravi Kiran

21.1 INTRODUCTION

From the past to the present, a bicycle is the best means of transportation for a single man for short distances. In the recent era, bicycles are equipped with various electronic devices such as sensors, speedometers, and other communication devices. Mobile phones and smartwatches are an unavoidable need of humans. The power consumed by these devices is very low and is supplied by the batteries or dynamos. Batteries are not eco-friendly and need to be replaced frequently after discharge. To run the dynamo, more effort has been made by the cyclists to compensate for the friction between the dynamo and the cycle wheel. Also, the dynamo creates the noise during the operation and generates the power intermittently, not continuously.

During the running of the bicycle, various factors such as mechanical jerks and dynamic loads are subjected to it. To increase the lifespan of the bicycle, continuous monitoring is needed (Wang & Jianu, 2010). Wireless sensors can be applied for structural health monitoring and the reliability of bicycles. The use of small power consumption components in electronic devices enables an energy harvester to supply the power.

Energy harvester is a device which can generate a small amount of power through mechanical vibrations. A piezoelectric energy harvester is simple and easy to convert mechanical vibrations into electricity. However, there are other transducers also available, such as electromagnetic and electrostatic types, but they have low efficiency and are rarely applied (L. Xie & Du, 2012). A complete state-of-the-art of design and application of the piezo energy harvesting device has been performed by Khaligh et al. (2010). It was found that the harvester has two main components: one is a transducer, which converts the vibration into electricity, and an electronic circuit works as a bridge between the transducer and the electrical storage device. As per the maximum power theorem, a circuit is required which is able to match the impedance of the piezoelectric patch with the load to obtain maximum power (Roundy et al., n.d.). Satpute et al. (2014) suggested a hydraulic shock absorber based on electromagnetic principles to harvest the power from the vibration of a vehicle. Various components such as mechanical

 DOI: 10.1201/9781003642152-21

amplifications, fluid dampers, and generators have been used. Fang et al. (2013) designed and tested another hydraulic shock absorber with electromagnetic conversion to harness the vibration energy from the vehicle, and results showed that 200 W can be obtained at 10 Hz and 3 mm vibration amplitude. Shi et al. (2015) proposed a suspension system energy-regenerative type to convert the vibration of the suspension. From the experiment, 46.57 W of power was calculated. Tang et al. (2014) suggested a linear electronic transducer to convert the suspension of the vehicle into power. Electrical power of 26–33 was calculated at an RMS of 0.25 m/s of velocity of suspension. A rack-and-pinion mechanism to harvest the power from the vibration of the damper using an electromagnetic transducer was presented by Li et al. (2012). From the transducer, an average power of 19 W was recorded with vehicle speed of 13.33 m/s.

In this study, a new design of piezoelectric transducer has been proposed to convert the bicycle vibration into electricity during the running of the bicycle. Bicycle frame vibrations induced by rolling have been used as a source of energy. An analytical model has been developed for a spring-damper system with dual mass to compute the power. With the help of MATLAB Simulink software, a model for road roughness is created to vibrate the harvester, considering geometrical and process parameters to compute the output of power.

21.2 DESIGN AND MODELING OF HARVESTING DEVICE

The working principle and configuration of the piezo energy harvesting device to harness the energy from the bicycle during cycling are depicted in Figures 21.1(a) and (b). The piezo energy harvester with d_{31} working mode is mounted with rear wheel support on the bicycle as shown in Figure 21.1(a). The main parts of the device are the three piezoelectric cantilevers, the right support to mount the base of the piezoelectric cantilevers, and the left support to swing the ends of the piezoelectric cantilevers, as shown in Figure 21.1(a). Piezoelectric cantilevers are composed of steel plates and piezo patches. Piezoelectric cantilevers are free to vibrate under vibrations induced into the rear wheel support of the bicycle due to the motion of the bicycle on the road. Considering l_b, w_b, t_b, and E_b as the length, width, thickness, and Young's modulus of the beam, then the stiffness of each cantilever beam can be given as $k_1 = k_2 = k_3 = E_b w_b t_b^3 / 4L_b^3$. The total stiffness of the harvester is given as $k_b = 3E_b w_b t_b^3 / 4L_b^3$. The dimensions of the piezoelectric patches are taken as length = lp, width = wp, and thickness = tp. Considering that the piezo energy harvesting device is fitted between the bicycle frame and rear wheel to convert the excitations into electric energy. Further, the bicycle wheel can be modeled as a mass attached to a spring and

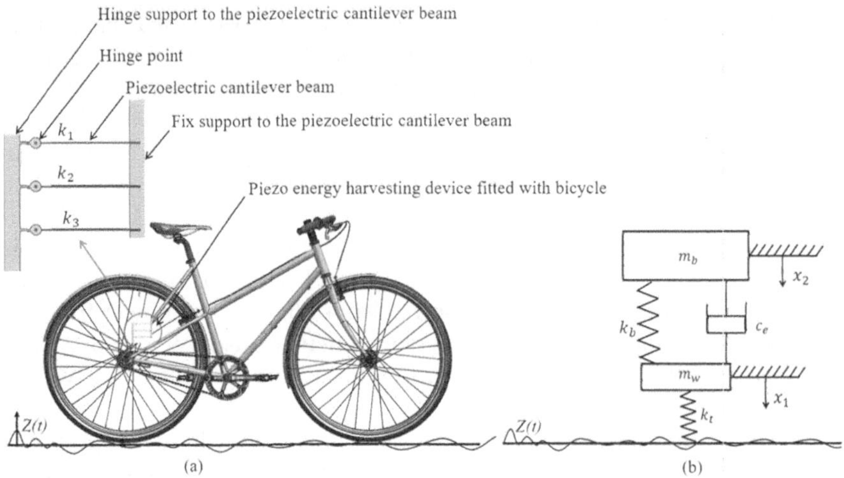

Figure 21.1 (a) Bicycle model with piezo energy harvesting device, (b) Schematic view of the spring-damper system dual-mass.

generate a motion function $z(t)$ due to running on the surface of the road. Analysis of the system is done by assuming the back wheel of the bicycle as an unsprung mass m_w, and the chassis with rider as a sprung mass m_b connected to a spring (k_b) and damper (c_e) in series, as shown in Figure 21.1(b). Hence, the whole configuration (chassis, wheel, bicycle rider, and harvesting device) can be considered a spring-damper system with dual-mass, as shown in Figure 21.1(b). Where k_t is tire stiffness, c_e is an electrical damping due to resistance created by PZT patches during work conversion into electricity (Mitcheson et al., 2008), and can be given as (X. D. Xie & Wang, 2015):

$$c_e = d_{31}^2 k_e^2 / \pi^2 c_v f \tag{21.1}$$

where d_{31} is the piezoelectric coefficient, c_v is the capacitance of the PZT patches, and f is the system's natural frequency. Generated power is calculated by MATLAB simulation of the mathematical model.

A CAD model of the bicycle equipped with a piezoelectric harvester is shown in Figure 21.1(a). The configuration illustrates that the device is mounted between the back-supporting frame, ensuring direct transmission of vibrations from the rear wheel to the harvester. The device primarily consists of a fixed base, which supports three cantilevers, and a moving plate, to which the free ends of the cantilevers are attached via hinge supports. Each cantilever is constructed using a PZT patch bonded to a steel beam, as depicted in Figure 21.2. The dimensions of the cantilever are detailed in the front and top views of Figure 21.2.

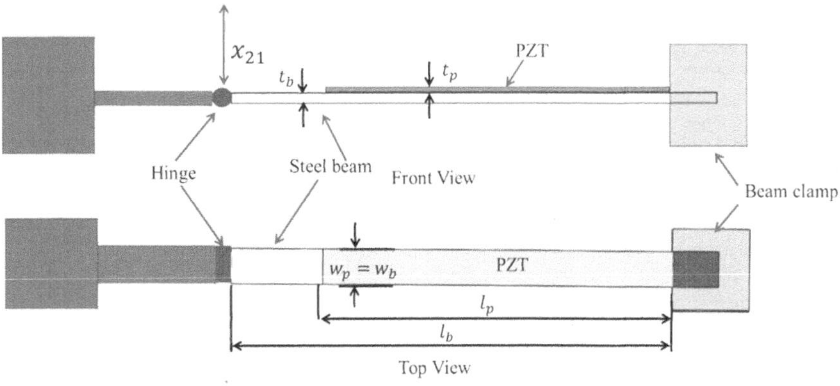

Figure 21.2 Piezoelectric cantilever with hinge.

The governing differential equations, as per Newton's second law of motion, representing the spring-damper system with dual-mass, as shown in Figure 21.1(b), can be given as:

$$\begin{cases} m_w \ddot{x}_1 - c_e(\dot{x}_2 - \dot{x}_1) - k_b(x_2 - x_1) + k_t(x_1 - z(t)) = 0 \\ m_b \ddot{x}_2 + c_e(\dot{x}_2 - \dot{x}_1) + k_b(x_2 - x_1) = 0 \end{cases} \tag{21.2}$$

where x_1 and m_w are the displacement and mass of the rear wheel, while x_2 and m_b are the displacement and combined mass of the bicycle chassis with the rider; at the time of riding of the bicycle the on the road with respect to the equilibrium position, function of the transverse motion, $z(t)$, can be given as (He et al., 2008):

$$\dot{z}(t) + 2\pi n_0 v z(t) = 2\pi n_0 \sqrt{G_q(n_0) v} w(t) \tag{21.3}$$

where

$G_q(n_0)$ = road roughness coefficient in m^3; reference spatial frequency $n_0 = 0.1/m$; bicycle velocity = vm/s; zero-mean white noise signal = $w(t)$. Simulink tool of MATLAB used to model the road roughness traverse function, $z(t)$, as presented in Figure 21.3.

Since the spring-damper with dual-mass system is a vibration system of 2 degrees of freedom, hence the natural frequencies, ν_1, ν_2, and mode shapes $\{\Omega\}_{1,2}$ of the system can be expressed as (Singiresu, 2011):

$$\begin{cases} \nu_1 = \left(\frac{1}{2} \left[\frac{k_t + k_b}{m_w} + \frac{k_b}{m_b} - \sqrt{\left(\frac{k_t + k_b}{m_w} + \frac{k_b}{m_b} \right)^2 - 4\frac{k_t k_b}{m_w m_b}} \right] \right)^{1/2} \\ \nu_2 = \left(\frac{1}{2} \left[\frac{k_t + k_b}{m_w} + \frac{k_b}{m_b} + \sqrt{\left(\frac{k_t + k_b}{m_w} + \frac{k_b}{m_b} \right)^2 - 4\frac{k_t k_b}{m_w m_b}} \right] \right)^{1/2} \end{cases} \tag{21.4}$$

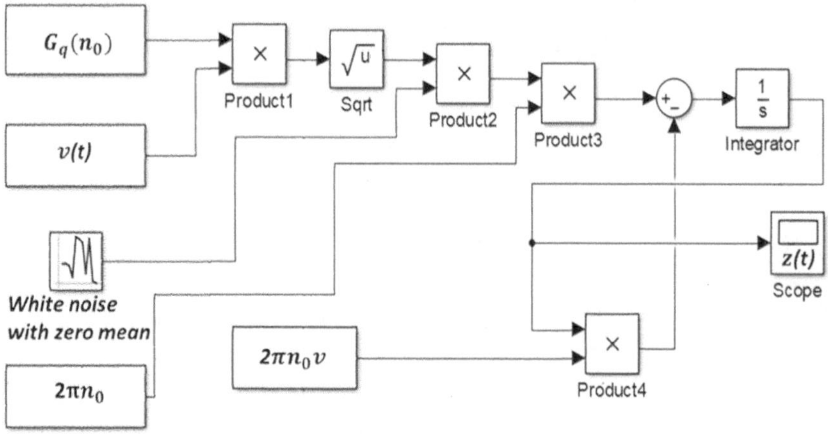

Figure 21.3 Simulink model for producing the transverse function, z(t).

$$\{\Omega\}_{1,2} = \frac{k_b}{(k_t + k_b) - m_w\omega_{1,2}^2} \tag{21.5}$$

Equation (21.2) can be rearranged as:

$$\begin{bmatrix} m_w & 0 \\ 0 & m_b \end{bmatrix}\begin{pmatrix} \ddot{x}_1 \\ \ddot{x}_2 \end{pmatrix} + \begin{bmatrix} c_e & 0 \\ 0 & c_e \end{bmatrix}\begin{pmatrix} \dot{x}_1 \\ \dot{x}_2 \end{pmatrix} + \begin{bmatrix} k_t + k_b & -k_b \\ -k_b & k_b \end{bmatrix}\begin{pmatrix} x_1 \\ x_2 \end{pmatrix} = \begin{pmatrix} k_t z(t) \\ 0 \end{pmatrix} + \begin{bmatrix} 0 & c_e \\ c_e & 0 \end{bmatrix}\begin{pmatrix} \dot{x}_1 \\ \dot{x}_2 \end{pmatrix}, \tag{21.6}$$

or written as:

$$[M]\{\ddot{X}\} + [C]\{\dot{X}\} + [K]\{X\} = \{Q\} + [C']\{\dot{X}\} \tag{21.7}$$

The solution of the above equation can be given with the help of the modal analysis method:

$$\{X\} = [\Omega]\{Y\} = \{\Omega\}_1 Y_1 + \{\Omega\}_2 Y_2, \tag{21.8}$$

where $[\Omega]$ is a mode matrix; $\{Y\}$ generated coordinatesvector.

From Equations (21.7) and (21.8) and multiplying by $\{\Omega\}_1^T$ and $\{\Omega\}_{2,}^T$ we can get,

$$M_1\ddot{Y}_1 + \left(C_1 - C_{11}'\right)\dot{Y}_1 + K_1Y_1 = Q_{11} + C_{12}'\dot{Y}_2 = Q_1; \tag{21.9}$$

$$M_2\ddot{Y}_2 + \left(C_2 - C_{22}'\right)\dot{Y}_2 + K_2Y_2 = Q_{22} + C_{21}'\dot{Y}_1 = Q_2, \tag{21.10}$$

where, $M_1 = \{\Omega\}_1^T[M]\{\Omega\}_1$, $M_2 = \{\Omega\}_2^T[M]\{\Omega\}_2$, $C_1 = \{\Omega\}_1^T[C]\{\Omega\}_1$, $C_1 = \{\Omega\}_2^T[C]\{\Omega\}_2$, $K_1 = \{\Omega\}_1^T[K]\{\Omega\}_1$, $K_2 = \{\Omega\}_2^T[K]\{\Omega\}_2$, $Q_{11} = \{\Omega\}_1^T\{Q\}$, $Q_{22} = \{\Omega\}_2^T\{Q\}$, $C_{11}' = \{\Omega\}_1^T[C']\{\Omega\}_1$, $C_{12}' = \{\Omega\}_1^T[C']\{\Omega\}_2$, $C_{22}' = \{\Omega\}_2^T[C']\{\Omega\}_2$, $C_{21}' = \{\Omega\}_2^T[C']\{\Omega\}_1$.

The generalized coordinates, Y_1 and $Y_{2,}$ are obtained by the method of iteration. Small incremental segments of forces Q_1 and Q_2 are considered in the time domain, such that each and every increment is linear with time to make the solution feasible. Hence, the governing differential equations can be written at each interval of time:

$$M_1\ddot{Y}_1(\tau) + \left(C_1 - C_{11}'\right)\dot{Y}_1(\tau) + K_1Y_1(\tau) = Q_{11}(\tau) + C_{12}'\dot{Y}_2(\tau) = Q_1(\tau) = Q_{1j} + \beta_{1j}\tau, \tag{21.11}$$

$$M_2\ddot{Y}_2(\tau) + \left(C_2 - C_{22}'\right)\dot{Y}_2(\tau) + K_2Y_2(\tau) = Q_{22}(\tau) + C_{21}'\dot{Y}_2(\tau) = Q_2(\tau) = Q_{2j} + \beta_{2j}\tau \tag{21.12}$$

Where, $t_j \le \tau \le t_{j+1}$, $t_{j+1} = t_i + \Delta t$, $Q_{1j} = Q_{11}(t_j) + C_{12}'\dot{Y}_2(t_j)$, $Q_{2j} = Q_{22}(t_j) + C_{21}'\dot{Y}_1(t_j)$, $\beta_{1j} = \left(Q_{1(j+1)} - Q_{1j}\right)/\,\Delta t$, $\beta_{2j} = \left(Q_{2(j+1)} - Q_{2j}\right)/\,\Delta t$, $j = 0,1,2,\ldots\ldots\infty$

The displacements x_1 and x_2 are the wheel mass and the bicycle mass, respectively. If the relative displacements of the bicycle and wheel masses (deflection of the cantilever at the hinge point) is $x_{21} = x_2 - x_1$, then the generated power from each PZT patch in the time interval from *0* to *t* can be found out from Equation (2.16) (Narolia et al., 2023) and Equation (2.25) by Narolia et al. (Lu et al., 2004) and expressed as:

$$P(\tau) = \frac{(2\pi n)^2\, w_p^2 t_p^2 e_{31}^2 x_{21}^2}{2\sqrt{\left(1 + (2\pi n c_p R)^2\right)^2}} R \tag{21.13}$$

If there are six PZT patches mounted on the device, then the total power the device can be given is:

$$P(\tau) = 3\frac{(2\pi n)^2 w_p^2 t_p^2 e_{31}^2 x_{21}^2}{\sqrt{\left(1+(2\pi n c_p R)^2\right)^2}} R \tag{21.14}$$

where electric capacitance of PZT patch $c_p = \frac{\varepsilon_{33} w_p l_p}{t_p}$ and $V = IR$

n = frequency of vibration, e_{31} = piezoelectric constant, $\mathcal{E}_{33}$ = dielectric constant, w_p = width of PZT patch, l_p = length of PZT patch, R = External load, V = voltage across load, and I = Current flowing through the load.

21.3 PARAMETER STUDY AND RESULTS ANALYSIS

To calculate the electrical power harvested from the device simulation, a simulation has been carried out. MATLAB software is applied to execute the simulation considering the various parameters such as road surface, bicycle velocity, length, and thickness of PZT patches. Properties of materials and dimensions of various components of the device are listed in Table 21.1. Parameters of bicycle and road roughness coefficient $G_q(n_0)$ are given in Tables 21.2 and 21.3 (Jerath & Gurav, 2008). With the help of the Simulink model of the MATLAB road roughness for various categories of roads

Table 21.1 Dimensions and material properties of PZT-5H

K_{31}	d_{31} (C/m^2)	E_p (N/m^2)	$\mathcal{E}_{33}$ (F/m)	l_p (mm)	w_p (mm)	t_p (mm)
0.60	11.2	53e9	1.5937×e–8	50-100	25	0.1-0.5

Table 21.2 Parameters of bicycle and beam

m_w (kg)	m_b (kg)	k_t (kN/m)	l_b (mm)	w_b (mm)	t_b (mm)
3	90	90 (Rothhämel, 2023)	150	25	1

Table 21.3 Coefficient of road roughness $G_q(n_0)$

Road class	*Excellent* A	*Good* B	*Average* C	*Poor* D
$G_q(n_0) \times (10^{-6}\,m^3/\text{cycle})$	G1 16	G2 64	G3 256	G4 1024

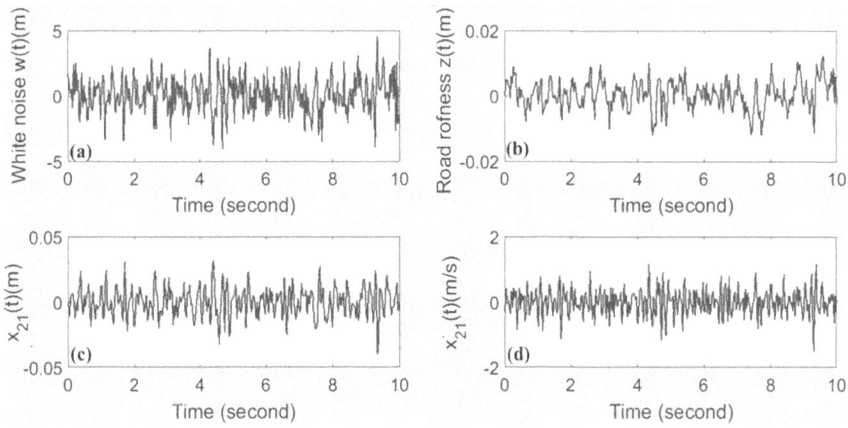

Figure 21.4 (a) Random Gaussian white noise with zero mean, (b) Transverse function for road roughness, (c) Transverse relative displacement of the m_b and m_t, and (d) Transverse relative velocities of the m_b and m_t.

computed by considering random Gaussian white noise with zero-mean $w(t)$; transverse relative displacement $z(t)$; the velocity of the bicycle v; mass with rider m_b; and the mass of the wheel with tire (m_t). Simulink simulation has been executed by taking 10 seconds of time and a 10 m/s velocity of the bicycle for the rough surface road, as given in Figure 21.4(a)–(d).

21.3.1 Effect of road roughness on power output

The effect of the road roughness on the power output for various categories of roads has been depicted in Figure 21.5. From the figure, it was observed that with poor-category roads, more power has been harvested. From Equation (21.13), it is clear that the more the deflection of the cantilever, the more the strain induced, leading to more charge generation and consequently more power generation. Simulation has been conducted for 10 seconds for random Gaussian white noise with zero mean (Figure 21.4(a)), and the bicycle velocity of 10 m/s, length, width, and thickness of PZT patches are selected as 100 mm, 25 mm, and 0.5 mm, respectively. The road roughness coefficients are selected as follows: for poor roads, G4 1024 (n_0 = 1024e-6 m^3/cycle); for average roads, G3 256 (n_0 = 256e-6 m^3/cycle); for good road, G2 64 (n_0 = 64e-6 m^3/cycle); and for excellent roads, G1 16 (n_0 = 16e-6 m^3/cycle). The plot shows the variation in power with external load R. Optimum power is obtained at an external load, which is equal to the impedance of the PZT patches. Maximum power of 237 mW was observed at a 50 kΩ external load when the bicycles ran on the poor road.

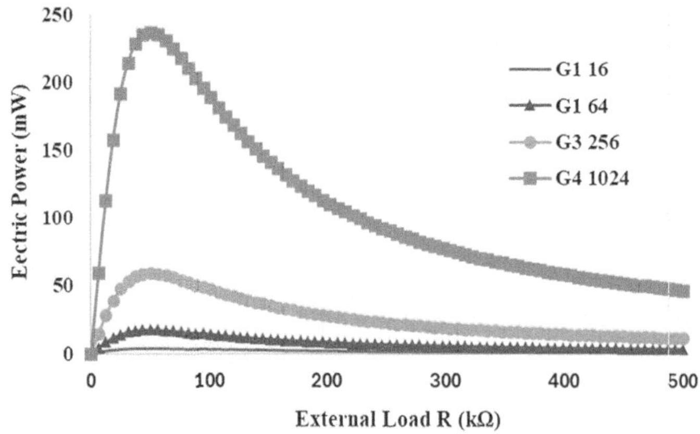

Figure 21.5 Electric power (RMS) with load R for various categories of road.

21.3.2 Effect of the bicycle velocity on power output

Velocity of bicycle and road roughness $(G_q(n_0))$ are the key factors that determine the output electrical power from the harvesting device. As the velocity of the bicycle increases frequency of power generation increases. Various curves in Figure 21.6 show the variation in power with load R. For simulation purposes, bicycle velocities are taken as 1 m/s, 3 m/s, 6 m/s, and 10 m/s, while the dimensions of the PZT patches are selected as $100mm \times 25mm \times 0.5mm$. To obtain the transverse relative displacement of the bicycle and wheel road roughness coefficient, G4 1024 $(n_0 = 1024e - 6\,m^3/cycle)$ has been selected. The higher the value of the velocity, the more power is obtained. This can be interpreted to mean that with higher velocity, the working frequency is high and more power is generated. From Figure 21.6, at minimum velocity (1 m/s), the power obtained is 28 mW, while at maximum velocity (10 m/s), 237 mW is observed.

21.3.3 Effect of the length of the PZT patch

The effect of the length of the piezo patches on the electrical power output is presented in Figure 21.7. It is observed from the figure that maximum power is gained at 75 mm length of the PZT patch. This can be interpreted from Equation (21.13) that power increases with increasing PZT patch length. After 75 mm of PZT, it starts to decrease because of the lower strain in the PZT patch. Simulation is carried out at 10 m/s bicycle velocity and poor road conditions. The width and thickness of the PZT patch are selected as 25 mm and 0.5 mm. The maximum power of 249 mW is obtained from the simulation. Also, external load depends on the length, width, and thickness

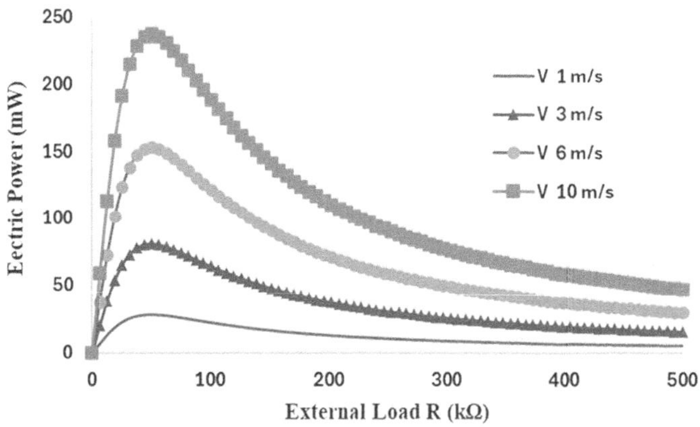

Figure 21.6 Electrical power with external load R at various bicycle velocities.

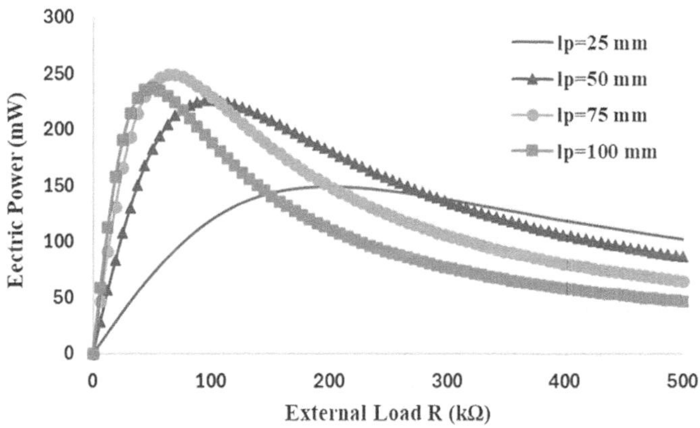

Figure 21.7 Electric power with external load at different thicknesses of PZT.

of PZT patch as $R = \left(\frac{t_p}{l_p w_p \varepsilon_{33} \omega} \right)$. With an increase in the length, the external load is reduced, as depicted in Figure 21.7.

21.3.4 Effect of the thickness of piezo patch on power

Variation in the harvested power with the thickness of PZT patches is given in Figure 21.8. A similar pattern has been depicted by the curves with external load. Geometric parameters such as the length and width of the PZT patches are 100 mm and 25 mm, selected to carry out the simulation.

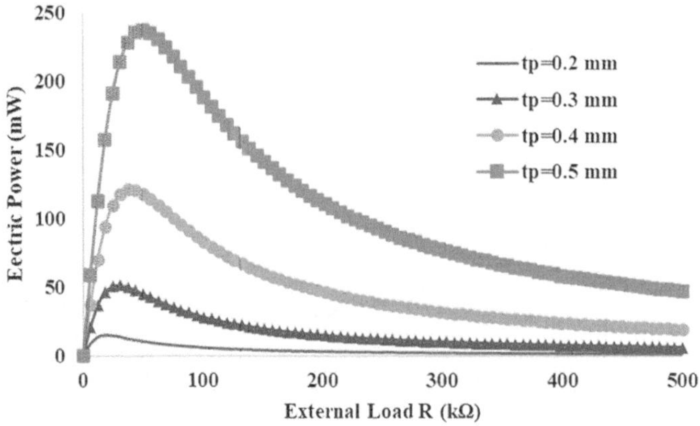

Figure 21.8 RMS of electric power with electric load R at various thicknesses of PZT.

Velocity of the bicycle and road category is taken as 10 m/s and G4 1024 $\left(n_0 = 1024e^{-6}\,m^3\,/\,cycle\right)$. From the simulation data, it was clear that with an increase in the thickness of the PZT patches, power output increases. Equation (21.13) shows that electric power is proportional to the thickness. Minimum power (36 mW) is calculated at 0.2 mm thickness, while maximum power of 237 mW is observed at 0.5 mm thickness of the PZT patch.

21.4 CONCLUSION

This study proposes a novel design for a PZT energy harvesting device utilizing the motion of a bicycle. Simulations were conducted using MATLAB to calculate the electrical power output, considering various road surface conditions, geometrical configurations, and process parameters. An analytical model was developed based on a spring-damper system with dual-mass to estimate power generation. The simulation results indicate that power output increases with higher road roughness, greater bicycle speed, and larger PZT patch dimensions (both length and thickness). The optimal power output of 237 mW was obtained at a bicycle speed of 10 m/s on a poor road surface. The parameters selected for the study included as mass of wheel $m_w = 3\,kg$, mass of bicycle with rider $m_b = 90\,kg$, stiffens of the tire $k_t = 90\,kN/m$, dimensions of the PZT patch 100**mm × 50 mm × 0.5 mm**, dimension of cantilever beam **150 mm × 50 mm × 1 mm**, and road roughness coefficient of $G_q\left(n_0\right) = 11024\,e - 6\,m^3/cycle$. It is anticipated that this energy harvester can be practically mounted on a bicycle to generate power for small electronic devices attached to the bicycle.

ACKNOWLEDGMENT

The authors gratefully acknowledge the Core Research and Innovation Group, Rabindranath Tagore University, Raisen, M.P., for funding and supporting this research under Grant Project ID: CRIG AGU/RNTU/2023/04.

REFERENCES

Fang, Z., Guo, X., Xu, L., & Zhang, H. (2013). Experimental study of damping and energy regeneration characteristics of a hydraulic electromagnetic shock absorber. *Advances in Mechanical Engineering, 2013*. https://doi.org/10.1155/2013/943528

He, L., Qin, G., Zhang, Y., & Chen, L. (2008). Non-stationary random vibration analysis of vehicle with fractional damping. *Proceedings - International Conference on Intelligent Computation Technology and Automation, ICICTA 2008, 2* (December 2014), 150–157. https://doi.org/10.1109/ICICTA.2008.348

Jerath, S., &Gurav, S. (2008). Road surface roughness generation by power spectral density in bridge design. *Proceedings of the 2008 Structures Congress - Structures Congress 2008: Crossing the Borders, 314*(August). https://doi.org/10.1061/41016(314)312

Khaligh, A., Zeng, P., & Zheng, C. (2010). Kinetic energy harvesting using piezoelectric and electromagnetic technologiesstate of the art. *IEEE Transactions on Industrial Electronics, 57*(3), 850–860. https://doi.org/10.1109/TIE.2009.2024652

Li, Z., Zuo, L., Luhrs, G., Lin, L., & Qin, Y. X. (2012). Design, modeling, and road tests of electromagnetic energy-harvesting shock absorbers. *ASME 2012 5th Annual Dynamic Systems and Control Conference Joint with the JSME 2012 11th Motion and Vibration Conference, DSCC 2012-MOVIC 2012, 3*(3), 675–684. https://doi.org/10.1115/DSCC2012-MOVIC2012-8734

Lu, F., Lee, H. P., & Lim, S. P. (2004). Modeling and analysis of micro piezoelectric power generators for micro-electromechanical-systems applications. *Smart Materials and Structures, 13*(1), 57–63. https://doi.org/10.1088/0964-1726/13/1/007

Mitcheson, P. D., Yeatman, E. M., Rao, G. K., Holmes, A. S., & Green, T. C. (2008). Energy harvesting from human and machine motion for wireless electronic devices. *Proceedings of the IEEE, 96*(9), 1457–1486. https://doi.org/10.1109/JPROC.2008.927494

Narolia, T., Mandaloi, G., & Gupta, V. K. (2023). Design and experimental analysis of low wind speed rotary piezoelectric energy harvester. *International Journal of Mechanics and Materials in Design, 19*(4), 793–804. https://doi.org/10.1007/s10999-023-09663-8

Rothhämel, M. (2023, September 22). *Measuring vertical tyre stiffness of bicycle tyres*. https://doi.org/10.59490/650c004def25b2f0766f520b

Roundy, S., Wright, P. K., &Rabaey, J. (n.d.). *A study of low level vibrations as a power source for wireless sensor nodes*. www.elsevier.com/locate/comcom

Satpute, N. V., Singh, S., & Sawant, S. M. (2014). Energy harvesting shock absorber with electromagnetic and fluid damping. *Advances in Mechanical Engineering, 2014*. https://doi.org/10.1155/2014/693592

Shi, D., Chen, L., Wang, R., Jiang, H., & Shen, Y. (2015). Design and experiment study of a semi-active energy-regenerative suspension system. *Smart Materials and Structures*, *24*(1). https://doi.org/10.1088/0964-1726/24/1/015001

Singiresu, S. Rao. (2011). *Mechanical vibrations* (5th ed.). Pearson.

Tang, X., Lin, T., & Zuo, L. (2014). Design and optimization of a tubular linear electromagnetic vibration energy harvester. *IEEE/ASME Transactions on Mechatronics*, *19*(2), 615–622. https://doi.org/10.1109/TMECH.2013.2249666

Wang, W., & Jianu, O. A. (2010). A smart sensing unit for vibration measurement and monitoring. *IEEE/ASME Transactions on Mechatronics*, *15*(1), 70–78. https://doi.org/10.1109/TMECH.2009.2016956

Xie, L., & Du, R. (2012). Harvest human kinetic energy to power portable electronics. *Journal of Mechanical Science and Technology*, *26*(7), 2005–2008. https://doi.org/10.1007/s12206-012-0503-7

Xie, X. D., & Wang, Q. (2015). Energy harvesting from a vehicle suspension system. *Energy*, *86*(July), 385–392. https://doi.org/10.1016/j.energy.2015.04.009

Chapter 22

Designing optimization frameworks for ICT-enabled e-leadership strategies

Amitabh Roy, Akhilesh Tiwari, Sanjay Rastogi, Satendar Singh, and Madhuranjan Vatsa

22.1 INTRODUCTION

Communication takes up a significant percentage of the job of organizational leadership since it is one of the fundamental skills demanded of a leader (Ahmad and Loch, 2020). The efficacy of communication is one of the most important characteristics of leadership in online contexts, in contrast to the many styles of leadership that exist in offline situations (Brammer et al., 2020). Digital communication has become an essential tool for information sharing inside enterprises across all sectors as a result of the rapidly changing technology environment.

The difficulty of addressing the rising ubiquitous dissemination of telecommunications and information technologies, which are fundamentally altering interactions among employees, has recently been addressed in leadership literature. Thus, the support of these technologies is fostering further leadership development. This new phenomenon is known as "e-leadership," and it has been researched to better understand how it differs from conventional leadership.

22.1.1 Realizing the key aspect of leadership

When advanced technologies were combined with COVID-19, a time when representatives of creative organizations were required to communicate, or rather collaborate, via e-mail, text messaging, teleconferencing, real-time collaboration apps, as well as social media, communication devices in the creative industries had become extremely important in workplaces. While it could be practical, this isn't always effective. As a result, it has become vital for leadership research to address how to successfully interact in a digital world and employ digital communication technologies (Atwater et al., 2021). Success in the company and in life are related, and variables affecting business competencies will also impact ICT abilities (Figure 22.1).

Organizational behavior studies often focus on leadership, a traditional subject in the management industry. Leaders are also required to develop their ability to cope with and navigate the difficulties of leading in the digital

DOI: 10.1201/9781003642152-22

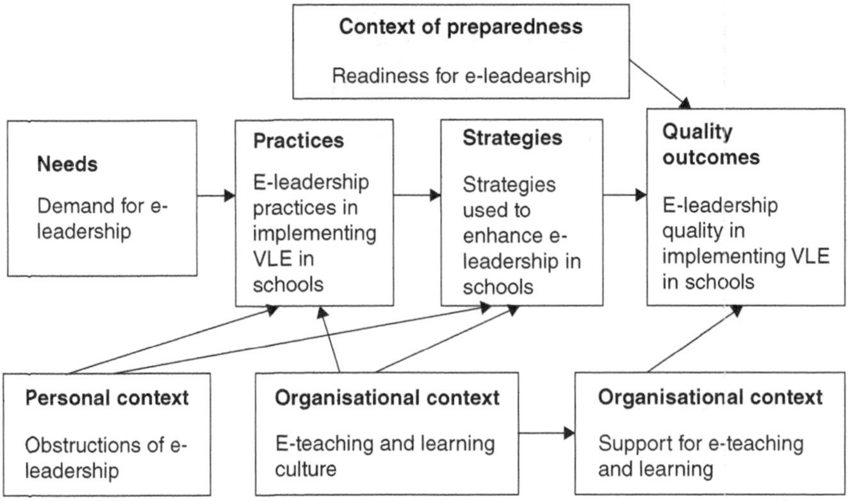

Figure 22.1 Need for e-leadership.

era. Currently, there are two relevant literatures that could serve as the theoretical foundation for creating a unified theoretical framework on e-leadership: the literature on adoption of technology as well as the literature on resource planning for businesses, which can be seen as an important special case of leadership achieving organizational enhancement in literature (Doğru, 2019; Roman et al., 2019; Klein, 2020). Our study aims to investigate whether organizations that use leadership as a crucial "tool" to manage personnel as efficiently as possible are aware of this evolution, regardless of whether their behavior is supportive of the associated needs, as well as the manner in which they are structuring themselves to deal with the challenges and opportunities presented by this new environment.

22.1.2 Literature review

According to a thorough assessment of the literature, the idea of e-leadership is the inevitable result of the e-work environment created by advancements in information technology and their integration into organizational processes. While ICT has a significant influence on enterprises and society, leadership theory has not done a good job of addressing ICT adoption (Gani and Shabir, 2022). There is particularly little research on e-leadership in poor nations.

The term "leadership" may be defined in many different ways (Ishak and Kamil, 2016). Although the word "leadership" is often used to refer to any action taken by people in control, most leadership researchers use the term to refer to how leaders inspire, guide, coordinate, promote, and empower their subordinates to achieve optimum performance (Wang and Wang, 2022;

Eberl and Drews, 2021; Brown and Hartman, 2017; Johansson and Edwards, 2021). There is also a case to be made for the idea that leadership is a two-way influence flow between those in leadership and those who follow for the benefit of shared objectives (April and Dalwai, 2019). Similar to leadership, there are many different definitions of creativity. It has been defined as the propensity to unleash one's capacity in order to provide fresh and useful ideas, options, or anything else physical or intangible that might be used to communicate, amuse, resolve issues, or achieve greatness (Zupanzic et al., 2016). Education, current affairs, fashion, and one's own social and cognitive or emotional relationships are just a few areas where creativity may be seen in action (Anderson et al., 2017). A digital leader has a somewhat non-hierarchical leadership style in contrast to conventional leaders, who exert influence and authority over their followers, who act on their own as sole decision-makers (Akinbode and Al Shuhumi, 2018). They use coaching and mentoring to direct the learning process and develop potential inside the team rather than employing disciplinary power. Eberl and Drews (Eberl and Drews, 2021) assert that e-leadership may be characterized as genuine, transformational, and transactional based on their analysis of the literature on digital leadership. Genuine leadership encourages innovation among staff members and improves group decision-making. Transactional techniques enable incentive systems to promote tech-savviness among staff. A transformational leader motivates people to create via a positive and constructive, and liberating leadership style. The definition of e-leadership roles as well as styles suggests that visionary, technologically aware, collaborative, flexible, and inspiring qualities are expected of an e-leader.

22.2 LEADERSHIP AND E-LEADERSHIP THEORY

In respect of both human and technological development, the effects of the ongoing digital revolution—which gave birth to the modern information era—have been thoroughly researched. Past research has shown that there has been a rise in competitiveness as a consequence of globalization, open economic borders, technological improvements, and digitalization. Some essential skills evolve into dependable survival abilities in the current environment. When individuals are seeking leaders, some organizations evaluate candidates based on company values and cultures. The strategy emphasized the benefits of using the shared experiences' background information to create a cohesive corporate culture in huge organizations.

Leadership and inspiration go hand in hand. The ability of a leader to motivate people to strive toward a shared goal or vision is a critical component of effective leadership and is frequently used to define it. As a consequence of new networking and communication methods, more organizational flexibility, and the growing acceptance of remote work, managers now have to deal with a variety of brand-new problems. Traditional hierarchies, physical presence in the workplace, and rigid organizational structures are all becoming obsolete. They are being replaced by employees' expressed preferences for more autonomous work, lax time management, and overall more self-fulfillment.

Leadership styles have evolved as society has transitioned from an industrial to a post-industrial state. Modern civilization is characterized by rapid and substantial breakthroughs in technology for communication and information. Another word for digitization is the information society, sometimes referred to as the digital society. In contrast to the networked economy, which regulates both information and power via hierarchies to create legal authority, the industrial economy controlled power and knowledge by informal cross-referencing. As a result, today's digital leaders need to possess a broad variety of social and emotional capabilities in addition to possessing strong technical expertise. This new leadership strategy places a stronger emphasis on networking and trust than dominance and control. Instead of imposing outcomes on workers, options are presented to them. Hence, digital leaders should prioritize their roles as networkers, mediators, and facilitators. Even though these developments are beneficial, it may be challenging for the personnel to adjust. It may be difficult to keep employees motivated throughout a digital shift owing to a number of variables, such as their fear of losing their jobs and general resistance to change.

22.3 ROLE OF ICT IN MEDIATING E-LEADERSHIP

The way we operate and create value is fundamentally changing as a result of information technology. Technology is currently acting as a mediator in crucial human connections. A leadership review is now necessary since a new organizational structure, as well as workplace atmosphere, has evolved as a consequence. The new organizational form is the virtual organization, the new working circumstances are the e-environments, as well as the new leadership style is the e-leadership. A defining characteristic of e-leadership is the fresh connection between the leader and the people he leads. Face-to-face contacts are progressively being replaced by encounters mediated by technology. Traditional leaders, including e-leaders, must have the capacity to lead on a global scale. E-leaders stand out due to their skills, attitudes, knowledge, and professional as well as personal experiences.

E-leadership is defined as the efficient use of both electronic and conventional forms of communication. It suggests knowledge of existing ICTs, deliberate adoption of new ICTs for oneself and the organization, the ability to choose from a variety of conventional and electronic means to effectively achieve specific goals, and technical proficiency in utilizing the ICTs chosen. Individual e-leaders make and carry out policies in e-settings, notwithstanding the institutional importance of e-government and e-administration. The first stage in developing a theory and doing empirical research is to define these behaviors. In this research, we use a case study to explore the integrated parts of e-leadership, including e-communication, e-social skills, team-building skills, e-change management, e-technological abilities, and e-trustworthiness, in order to enhance the theory-building process.

In the same way that e-leadership is often characterized as a process that employs ICTs to impact social behavior, e-teaching may also be considered as a social influence mechanism mediated by ICTs, which may result in a change in perspectives, attitudes, reasoning, and behavioral patterns, including performance. That is to say, instructors distinctively act as leaders of students in terms of their right and effective use of authority and influence. Teachers have a variety of powers, including expert authority derived from their experience and supported by technical credentials; referent power to act as role models, instructors, and mentors; coercive power to punish copyright infringement and cheating; reward value of the goal to give the students better marks for outstanding efforts; as well as coercive power to punish students for engaging in illegal behavior. For a number of reasons, educators should use ICTs while serving as e-leaders.

22.4 THE VALUE OF LEADERSHIP

Not much study has been conducted to identify e-leadership as a moderator on these interactions, despite a rise in the importance of e-leadership and present cooperation across organizational networks, suppliers, partners, and customers. E-leadership, which has a well-established moderating function in the literature, is crucial to this research because it plays a crucial role in the coordination of various cooperation partners to link between collaboration and creativity (Darics, 2020; Cardon et al., 2019; Grint, 2020). Due to the growing usage of dispersed teams by enterprises, the leadership position for remote distributed employees is crucial. Owing to COVID-19, tasks and duties have now been completed via digital channels, highlighting the need for virtual collaborations—team members' collective strategies utilized to preserve effective connections with peers in collaboration scenarios, including geographical separation. As a result, cross-cultural leadership is crucial for developing trust, cooperating across organizational networks, creating common mental models, and resolving conflicts—all of which need more work than in a typical team environment.

In many businesses, face-to-face leadership has traditionally been the standard. Because of the COVID-19 epidemic, several businesses have shifted to remote work as per the Swedish Public Health Agency's advice (Giustiniano et al., 2020). Due to the pandemic, this change has intensified, with a 400% rise in the number of people working remotely and a 40% belief that this trend would continue [6]. Some businesses have successfully adapted their whole operations to telework, while others have not been successful since the job demands physical presence. This has caused managers, especially those working in the public sector, to communicate online, while those who do hands-on work continue to be located on-site. According to surveys, when working remotely, executives have found it more difficult to encourage and stay in contact with every person, and some have even felt a stronger need to exert more control over them (Anderson et al., 2017).

While the potential need for a new conception of leadership is acknowledged, little study has been done on what this may entail conceptually. Three key forms of study provide the majority of the evidence and debate around e-leadership: consulting conference papers, survey-based perceptual data, and exploratory and quasi-experimental research.

Many studies have looked at how contacts between the government and citizens through websites, online services, social media, and other means have an impact on governments. Government-owned businesses are known for having poor management. Moreover, poorly implemented leadership affects relationships among employers, workers, and both (Kane et al., 2019). Not only may it lead to inefficiency, but it can also harm workers' morale and dedication.

There is currently a dearth of studies on the proper and efficient use of digital communication tools, despite the growing focus on innovation and leadership style as well as the general agreement that visionary leaders ought to have good "digital media literacy" and communication styles. The extant research focuses on the organizational and task issues brought on by the revolutionary information-transfer capabilities of digital communication platforms. Nevertheless, it misses the small psychological adjustments that take place at the interpersonal level in online communication spaces, which have a big influence on how well creative activities are ultimately performed. Effective leaders develop relationships with their followers based on trust (Figure 22.2).

22.5 TRUSTWORTHINESS

Effective leaders develop relationships with their followers based on trust. This study examines the leadership qualities that boost one's credibility or undermine it. We address three elements that influence a leader's capacity to build credibility: communication, cognitive diversity, and inspiring and involving people (or lack thereof).

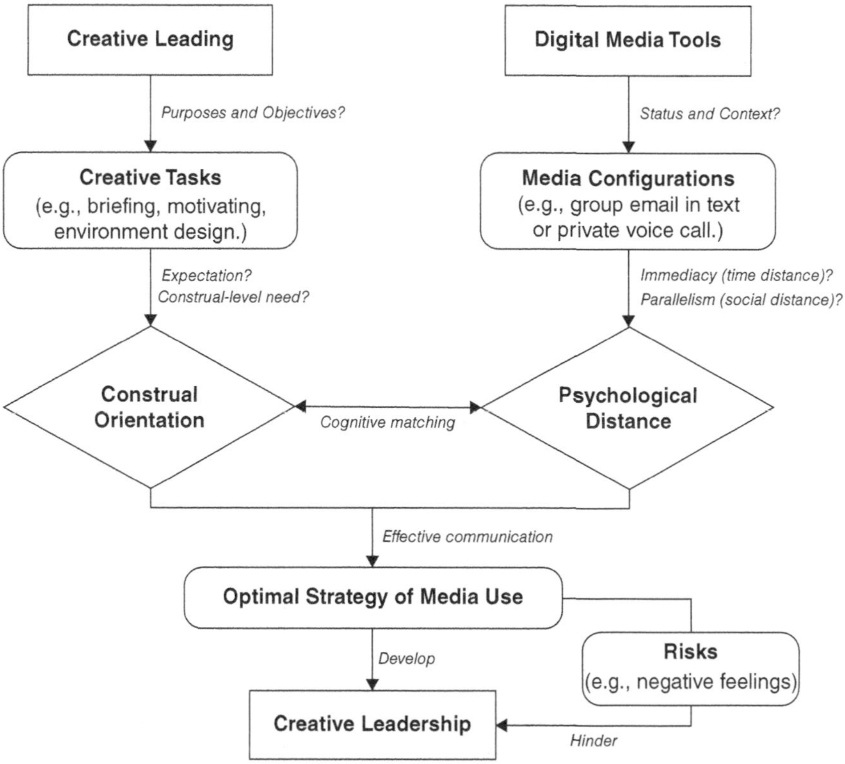

Figure 22.2 Strategic and creative model of leadership.

22.6 LIMITATIONS

There are many restrictions that may be found. We conducted a thorough analysis that was intended to be comprehensive. Due to space restrictions, we only mention research that is directly connected to e-leadership. The only purpose of this essay is to construct an idea and test it in experimental contexts. We note the significance of additional academic papers in the omitted set of articles published, giving attention to the issue of leaders' contact with followers through ICTs, despite past citations asserting a lack of scholarly focus on the subject of e-leadership. As a result, we recommend further study that examines several constructs that are comparable to the concept of "e-leadership" and that are often used as synonyms (e.g., digital leadership).

22.7 DISCUSSION

The widespread nature of ITCs has altered how businesses are structured in recent years. More particularly, they have a growing intensity that permeates the interaction between leaders and followers. Because of this, and as other writers have argued (Akinbode and Al Shuhumi, 2018), leadership is now evolving via the "intermediation" of new ICTs, the existence of which necessitates a shift in the way that leadership is conducted. These technologies now permeate practically every aspect of the workplace, including the internet, intranet, e-mail, instant messaging, virtual meeting systems, collaboration software systems, texts and emails, blogs, file sharing, smart applications, and social media (Grint, 2020; Giustiniano et al., 2020).

While e-leadership is a developing phenomenon in the industrialized world, it has received less attention from poor countries. ICT adoption by SMEs is a typical but limited definition of e-leadership. E-leadership is a concept that refers to more than only how leaders make use of cutting-edge IT (Alsharo et al., 2017). We may "zoom out" in our examination of leadership and current information technology by looking at the connections between these two concepts more broadly. It's odd that there has not been a great deal of exploration of how the tech transformation has affected organizational leadership (Cardon et al., 2019; Grint, 2020; Giustiniano et al., 2020; Kane et al., 2019; Alsharo et al., 2017; Kraft, 2019).

22.8 CONCLUSION

In the past 15 years, e-administration as well as e-governance have been hot subjects, but e-leadership in governmental settings has been almost nonexistent. This is hardly unexpected considering how fragmented and mostly lacking in theory beyond the micro level the liberal research on e-leadership has so far been. This is not to suggest that elements of e-leadership haven't been investigated using different organizational effectiveness theories. They still need to be incorporated and specified in a meaningful manner that takes into consideration the leadership function in actual virtual situations, however. In conclusion, it has been shown that e-leadership is relevant to the actual operations of organizations (as well as its theoretical implications, which have not yet been completely explored, as evidenced by researchers' continued interest in the topic). E-leadership will contribute significantly to our knowledge of leadership since instances of ICT-mediated situations are anticipated to multiply throughout time.

REFERENCES

Ahmad, M.G., & Loch, C. What do the followers want? The core functions of leadership. *The Leadership Quarterly*. 2020;31(2), 1–23.

Akinbode, A.I., & Al Shuhumi, S.R.A. Change management process and leadership styles. *International Journal of Social Science Studies*. 2018;4:609–618.

Alsharo, M., Gregg, D., & Ramirez, R. Virtual team effectiveness: The role of knowledge sharing and trust. *Information & Management*. 2017;54:479–490.

Anderson, H.J., Baur, J.E., Griffith, J.A., & Buckley, M.R. What works for you may not work for (Gen) Me: Limitations of present leadership theories for the new generation. *The Leadership Quarterly* 2017;28:245–260.

April, K., & Dalwai, A. Leadership styles required to lead digital transformation. *Effective Executive* 2019;22:14–45.

Atwater, L., Yu, J., Tawse, A., Fields, L.H., McFarren, J.A., & Nae, E.Y. Relevance of culture in studies of leadership: ignored or dismissed? *Asia Pacific Journal of Management*. 2021 Jun;38:687–708.

Brammer, S., Branicki, L., & Linnenluecke, M.K. COVID-19, societalization, and the future of business in society. *Academy of Management Perspectives*. 2020 Nov;34(4):493–507.

Brown, S.L., & Hartman, A. Feminist Creative Leadership Approaches. In *Encyclopedia of Strategic Leadership and Management 2017* 2017 (pp. 233–243). IGI Global.

Cardon, P.W., Huang, Y., & Power, G. Leadership communication on internal digital platforms, emotional capital, and corporate performance: The case for leader-centric listening. *International Journal of Business Communication*. 2019 Feb;21:1–27.

Darics E. E-leadership or "how to be boss in instant messaging?" The role of nonverbal communication. *International Journal of Business Communication*. 2020 Jan;57(1):3–29.

Doğru, Ç. The Effects of Perceived Organizational Support and Leader-Member Ex-change on Contextual Performance: A Study in the Banking Sector. In *Handbook of Research on Contemporary Approaches in Management and Organizational Strategy* 2019 (pp. 18–35). IGI Global.

Eberl J.K., & Drews P. Digital Leadership–Mountain or molehill? A literature review. In *Innovation Through Information Systems: Volume III: A Collection of Latest Research on Management Issues*; Springer, Cham, 2021:223–237.

Gani, A., & Shabir, S. Confluence of Workplace Spirituality and Leadership. In *Religion and Its Impact on Organizational Behavior*; IGI Global: Hershey, PA, USA, 2022:175–202.

Giustiniano, L., Cunha, M.P., Simpson, A.V., Rego, A., & Clegg, S. Resilient leadership as paradox work: notes from COVID-19. *Management and Organization Review*. 2020 Dec;16(5):971–975.

Grint, K. Leadership, management and command in the time of the Coronavirus. *Leadership*. 2020 Jun;16(3):314–319.

Ishak, A.K.; Kamil, B.A.M. Succession planning at HEIs: Leadership style, career development and knowledge management practices as its predictors. *International Review of Management and Marketing*. 2016;6:214–220.

Johansson, J., & Edwards, M. Exploring caring leadership through a feminist ethic of care: The case of a sporty CEO. *Leadership*. 2021 Jun;17(3):318–335.

Kane, G.C., Phillips, A.N., Copulsky, J., & Andrus, G. How digital leadership is (n't) different. *MIT Sloan Management Review*. 2019;60:34–39.

Klein, M. Leadership characteristics in the era of digital transformation. *Business & Management Studies: An International Journal*. 2020;8:883–902.

Kraft, M.H.G. How to lead with digital media effectively? A literature-based analysis of media in a E-leadership context. *Journal of Economic Development, Environment and People*. 2019;8:42.

Roman, A.V., Van Wart, M., Wang, X., Liu, C., Kim, S., & McCarthy, A. Defining e-leadership as competence in ICT – mediated communications: an exploratory assessment. *Public Administration Review*. 2019 Nov;79(6):853–866.

Wang, Y., & Wang, Y. Developing creative leadership in the use of digital communication tools: A psychological perspective. *Sustainability*. 2022 Sep 20;14(19):11796.

Zupanzic, T., Verbeke, J., Achten, H., Herneoja, A. *Digital leadership complexity, simplicity; eCAADe and Oulu School of Architecture*, University of Oulu, 2016; vol. 1:63–68.

Chapter 23

Models for analyzing the impact of leadership and followership values on organizational outcomes

Sanjay Rastogi, Amitabh Roy, Akhilesh Tiwari, Satendar Singh, and Madhuranjan Vatsa

23.1 INTRODUCTION

Structured organizational groups are composed of two primary groups: leaders and followers. Of the two groups, the success or failure of organizations, such as schools, is dependent on the leader and his or her leadership abilities. Effective leadership and followership in the organizational setting are essential to improving and sustaining academic success (Rukh et al., 2018). With the ever-increasing pressures in education from national, state, and district accountability standards, school organizations should also focus on fostering relationships between the leaders and followers (Mwesigwa et al., 2020). Leadership and followership roles are similar to the Chinese principles of Yin and Yang, representing duality yet harmonious relationships. Organizations should keep in mind that future leaders will come from the pool of individuals currently serving as followers; however, it is equally important to recognize leadership and followership as an undeniable symbiotic relationship between those who lead and those who choose to follow (Rukh et al., 2018; Mwesigwa et al., 2020; Lee et al., 2019).

Leadership is a high organizational force under the combination of the leader's quality, environment, and work system. It is also a key to the realization of the core goal. Leadership patterns are beginning to reflect the transition from industrial to post-industrial society. The new society, variously called the information society, knowledge society, or digital society, is marked by rapid and far-reaching changes in digitalization of information and communication technology (ICT) (Young et al., 2021). In the industrial economy, power and information were filtered through hierarchical structures and formal authority; in the networked economy, power and information are informal and hyperlinked.

The digitization of information is fundamentally changing how we work, how we organize work, and how we create value. In the realm of organizations, a quiet revolution has been taking place around the world. Now significant human interactions are mediated through information technology. Consequently, a new organizational form and a new work environment are shaping up, asking for reconstruction of the concept of leadership. The new

DOI: 10.1201/9781003642152-23

organizational form is called networked or virtual organizations, the new work environment is termed as the e-environment, and the emerging leadership is labeled as e-leadership or electronic leadership (Roman et al., 2019).

23.1.1 What do the leaders do?

Leaders are people who make decisions within organizations and cause others to support and implement these decisions. Leaders utilize rewards, imperative force, and interpretations of culture to obtain the collaboration of others. Leadership in its highest form develops a common understanding with followers and changes their basic ways of thinking (Kim and Beehr, 2020). Leaders engage in distinct forms of behavior, such as

- Taking initiative.
- Evaluating followers' needs, aspirations, and capabilities.
- Maintaining relationships within the organization.
- Founding relationships with other organizations.
- Collecting and disseminating information.
- Providing the resources needed to achieve success.
- Representing members' aspirations and values.

23.1.2 Objectives

There has been limited research investigating the effects of the recognition form of respect between leaders and their followers within the organizational literature. Thus, the goal of this research is to analyze how team performance is influenced by leadership styles in the organizational sector, confronting the perspectives of the leaders with the perception of their followers.

Drawing on leader–member exchange and crossover theory, this paper reviews how leaders' work engagement can spread to followers, highlighting the role of leader–member exchange as an underlying explanatory process. Specifically, we investigate if leaders who are highly engaged in their work have better relationships with their followers, which in turn can explain elevated organizational outcomes.

23.2 LITERATURE REVIEW

The concept of authenticity has been around since Aristotle (Leė et al., 2019). Authentic leadership phenomena have become a point of discussion over the past decade (Lacroix and Pircher Verdorfer, 2017). Authentic leadership is referred to as the behavior that draws from their psychological capacities and the positive ethics of organization that results in more self-awareness, internalized morals, transparency, and balanced information processing of relationships between leaders and employees (Young et al., 2021). There are four components of the authentic leadership proposed in the Authentic Leadership Questionnaire (ALQ) (Young et al., 2021).

Although there have been several studies concerning organizational commitment, little theoretical and empirical attention has been paid to the impact of leadership styles and organizational commitment among academic staff in public universities, as mediated by job satisfaction (Alblooshi et al., 2021). Furthermore, Clarke and Mahadi (2017) had earlier noted that the relationship between leadership styles, job satisfaction, and their influence on the organizational commitment among academic staff in public universities has received minimal attention and thus calls for more research. Thus, this study helps in examining the extent to which leadership styles influence organizational commitment, as mediated by job satisfaction among academic staff in universities, where reports have shown that public universities have witnessed a mass exodus of academics in the recent past to find better alternatives elsewhere.

In order to better understand the process of leadership effectiveness, authors developed a leadership model called the Full Range Leadership (FRL) model (Gutermann et al., 2017). This model proposes that leaders use multiple behaviors to influence their followers. This model is based on the theory of transformational and transactional leadership and incorporates both transformational and transactional leadership behaviors. The FRL model includes three leadership styles: transformational, transactional, and passive/avoidant leadership. There are nine variables in the model, which are represented by these three styles. The transformational leadership style is used to inspire and motivate followers through compelling vision, individual support, and empowerment.

Amiri, Rahima, and Ahmed (Al Amiri et al., 2020) defined job satisfaction simply as how contented an employee is with his or her job. Others have noted that job satisfaction varies in the extent to which they measure feelings about the job (affective) or cognitions about the job (cognitive). Furthermore, it is defined as the degree to which a person is pleased or satisfied with his/her job. More satisfied and happy employees would be more productive than the unsatisfied ones (Bligh, 2017). Job satisfaction creates positive attitudes in employees, boosts their morale, improves their performance, and creates pleasant relationships with coworkers. Employees who are satisfied with their jobs tend to be more creative and innovative, flourish, and bring positive changes to their organizations. This study thus operationalized job satisfaction into two indicators: intrinsic and extrinsic. Intrinsic job satisfaction refers to how employees feel about the nature of the job tasks themselves, whereas extrinsic job satisfaction refers to how employees feel about the aspects of the work situation that are external to the job tasks or work itself.

Leadership styles have continued to be one of the most widely discussed areas by researchers from all over the world (Moss et al., 2020; Schaubroeck et al., 2017). Guo, Zhu, and Zhang (2020) described leadership as a process to influence people in order to get the desired results. They noted that leadership plays a vital role in determining the success or failure of a firm. Gerpott et al. (2020) noted that leaders help to stimulate, motivate, encourage, and recognize their followers in order to get key performance results. Leadership styles, according to this study, have been operationalized as transformational and transactional leadership styles.

23.3 LEADERSHIP AND FOLLOWERSHIP

23.3.1 Leadership style

The concept of leadership is the most discussed and observed, but there is a lack of mutual understanding on what leadership approach or style is deemed to fit in a particular context or culture. In many circumstances, the leader and follower relationship has been viewed as two separate entities. However, there are two resilient, yet connected, active roles that are required and exist on every organizational level. Leadership and followership have been described as two sides of the same coin (Alblooshi et al., 2021). Researchers have referenced the roles of leadership and followership as a two-sided coin due to the symbiotic relationship each role possesses. If we were to physically examine a coin, one side would represent leadership, while the flip side would represent followership. Each side of the coin, when abilities are abundant, will produce outcomes that are advantageous to the organization. Leadership is exceptionally important to a school's or organization's performance; on the other hand, followership must have a role in the performance as well (Al Amiri et al., 2020). Inquisitively, followership receives only a small portion of the spotlight that leadership does.

The role of followers in an organization is significant at all levels. Followers may consist of stakeholders, executives, employees, or individuals who are in support of or trust the cause. It was the work of Kelley that prompted significant dialogue and research about followership (Gerpott et al., 2020). Asrar-ul-Haq and Anwar (2018) emphasized the need to pay attention to followers and that followership is worthy of its own distinct research. According to Li et al. (2017), conversations about leadership should include followership because leaders neither exist nor act in a vacuum without followers. Specifically, leaders have followership ability, and followers have leadership ability (Aldighrir, 2020).

Everyone, whether in the leader or the follower role, will depend on each other, which requires some giving and taking in the relationship. Leaders cannot exist without individuals following, and individuals cannot follow without a leader for guidance. The nature of leadership can perhaps be best understood by turning the coin over and studying followership. Why do people follow leaders? If we can understand this, then we will be a long way down the road to creating those followers and hence becoming an effective leader. People don't just follow anyone. One can't just say "follow me" and expect people to follow out of the goodness of their hearts. One has to give them a good reason for them to follow.

However, the distinction between the activities performed to achieve a leadership position and the activities required to be an effective leader suggests a weakness of human resource systems that are designed to identify and reward leadership potential (Pradhan and Jena, 2019). The talents and skills that may help someone rise in the hierarchy to reach a leadership position may not, later on, facilitate the group coordination effectiveness of the leader. Identifying how leadership contributes to success and what the essential leadership activities are from the followers' perspective is necessary for team and organizational effectiveness.

Individualized consideration, which focuses on followers' development and paying attention to followers' needs, showing empathy, and showing appreciation and support of individual followers' initiatives and viewpoints. Charisma, or idealized influence, has to do with serving as the followers' charismatic role model. Finally, inspirational motivation involves energizing followers by articulating a compelling vision. It is also defined as the leader moving the subordinate beyond immediate self-interest.

23.3.2 Leadership styles and organizational commitment

Various scholars have found a positive relationship between leadership and organizational commitment. Pradhan and Jena (2019) noted that leadership styles that encourage employee commitment are essential for an organization to successfully implement business strategies, achieve their goals, and gain a competitive advantage. Salas-Vallina et al. (2020) suggested that participative and inclusive leadership styles are more positively associated with commitment than task-oriented or structured styles. He noted that employees are highly committed to their organizations, more satisfied with their jobs, and produce higher job performance when they work under the supervision of leaders who adopt consultative or participative leadership behavior.

23.3.3 Followership

The existence of followers is also very important in leadership. First of all, "Every leader has to start as a follower." For example, every soldier cannot become a general leading the army unless he was once a follower. Second, without followers, there is no leader. For every organization or project to succeed, someone must follow the leaders closely, just like someone must lead the project effectively. According to Kelley, 80% of the success of an organization is due to the obedience of followers. So, what is an effective follower? "Followers are subordinates who have less power, authority, and influence than do their superiors, and who therefore usually, but not invariably, fall into line" (Mwesigwa et al., 2020). In an organization, though most people act as followers rather than leaders, there is still limited attention to the followers. This would cause ignorance of followers' importance to the organization's success and their potential in a better organization, just like leaders. In fact, effective followers are crucial to effective leadership; leaders need the strength of followers to achieve project success. And based on research by Alblooshi et al. (2021), which states that sometimes following is more difficult than leading.

To be an effective follower, we need to embrace the following characteristics. First is control of emotions. Followers should not be driven by their emotions, especially when their emotions would affect the work of others. Second, initiative. They will actively complete their tasks according to the leaders' requirements on time or even ahead of time. Third, honesty and trustworthiness. Followers are responsible for other followers, leaders, and their organization (Moss et al., 2020). With this characteristic, there is mutual trust and confidence among the whole organization; that is, leaders can safely hand over the work to the follower, and colleagues can cooperate effectively with the follower. Lastly, diligence. Followers are eager to learn every piece of new knowledge for the tasks assigned by the leader, and in some cases, they will also ask others for advice to better complete the project.

23.4 DISCUSSION

The exercise of leadership requires the consent of subordinates to the leader. In this context, leadership only exists after being recognized by other members of the group. Leaders tend to emerge as different from followers in their ability to initiate and maintain interaction. These leaders serve to increase the participation of members with less ability, accept diverse personalities, and are very tolerant of deviations. In this sense, the relationship between leaders and followers creates a basis for the beginning of the studies on transactional leadership (Aldighrir, 2020). The transactional leader is a negotiator who establishes agreements in exchange for desired results. The transactional leader can both reward satisfactory performance and punish unmet goals.

One of the great challenges of leaders is to direct the potential of human capital toward the achievement of organizational goals and results. As the importance of teams grows in the organizational environment, so does the role of the leader who guides their team (Salas-Vallina et al., 2020). Leaders cannot solve problems alone. The complex world requires a combination of resources and skills from several employees (Hayes et al., 2020). For these reasons, strong emphasis is placed on the promotion of teamwork and strong leadership.

Within a humanistic-based leadership framework, many scholars think that participative leadership style, consultative leadership style, organizational commitment, and job satisfaction are distinct constructs but highly interrelated. For example, leaders who can properly implement participative style (i.e., discussion, empowerment, joint decision-making, and power sharing) and consultative leadership style (i.e., leaders appreciate followers' voices in the establishment of goal settings and task assignments) in planning and administering organizational functions can lead to increased organizational commitment (Mwesigwa et al., 2020; Lee et al., 2019; Young et al., 2021; Roman et al., 2019; Kim and Beehr, 2020).

23.5 CONCLUSIONS

The organizational performance is usually tied to the effectiveness of leadership. This is because the social relations built in the organizational context can influence the perceptions of effectiveness and leadership by the meanings attributed and reconstructed by the group. This study further suggests that the willingness of leaders and followers to properly implement participative and consultative styles may not only lead to higher organizational commitment but also induce subsequent positive attitudinal and behavioral outcomes. When staff are recognized, inspired, and rewarded by the organizational managers, their level of job satisfaction will tend to increase, and organizational commitment will be promoted. We thus note that the style and quality of leadership employed by organization authorities contribute to the satisfaction of its employees and subsequently improve organizational commitment.

REFERENCES

Al Amiri N, Rahima RE, Ahmed G. Leadership styles and organizational knowledge management activities: A systematic review. *Gadjah Mada International Journal of Business*. 2020 Dec;22(3):250–275.

Alblooshi M, Shamsuzzaman M, Haridy S. The relationship between leadership styles and organisational innovation: A systematic literature review and narrative synthesis. *European Journal of Innovation Management*. 2021 May 18;24(2):338–370.

Aldighrir WM. Mediating Effect of Leadership Outcomes on the Relationship between the Leadership Style of Leaders and Followers' Administrative Creativity. *Journal of Education in Black Sea Region*. 2020 May 23;5(2):125–139.

Asrar-ul-Haq M, Anwar S. The many faces of leadership: Proposing research agenda through a review of literature. *Future Business Journal.* 2018 Dec 1;4(2):179–188.

Bligh MC. Leadership and trust. Leadership today: Practices for personal and professional performance. 2017:21–42.

Clarke N, Mahadi N. Mutual recognition respect between leaders and followers: Its relationship to follower job performance and well-being. *Journal of Business Ethics.* 2017 Mar;141(1):163–178.

Gutermann D, Lehmann-Willenbrock N, Boer D, Born M, Voelpel SC. How leaders affect followers' work engagement and performance: Integrating leader–member ex-change and crossover theory. *British Journal of Management.* 2017 Apr;28(2):299–314.

Gerpott FH, Fasbender U, Burmeister A. Respectful leadership and followers' knowledge sharing: A social mindfulness lens. *Human Relations.* 2020 Jun;73(6):789–810.

Guo Y, Zhu Y, Zhang L. Inclusive leadership, leader identification and employee voice behavior: The moderating role of power distance. *Current Psychology.* 2020 Feb 15;41(3):1–10.

Hayes TL, Oltman KA, Kaylor LE, Belgudri A. How leaders can become more commit-ted to diversity management. *Consulting Psychology Journal: Practice and Research.* 2020 Dec;72(4):247.

Kim M, Beehr TA. Empowering leadership: leading people to be present through affective organizational commitment?. *The International Journal of Human Resource Management.* 2020 Sep 7;31(16):2017–2044.

Lacroix M, PircherVerdorfer A. Can servant leaders fuel the leadership fire? The relationship between servant leadership and followers' leadership avoidance. *Administrative Sciences.* 2017 Feb 27;7(1):6.

Lee J, Cho J, Baek Y, Pillai R, Oh SH. Does ethical leadership predict follower outcomes above and beyond the full-range leadership model and authentic leadership?: An organizational commitment perspective. *Asia Pacific Journal of Management.* 2019 Sep 15;36:821–847.

Li SL, Huo Y, Long LR. Chinese traditionality matters: effects of differentiated empowering leadership on followers' trust in leaders and work outcomes. *Journal of Busi-ness Ethics.* 2017 Sep;145:81–93.

Moss SE, Song M, Hannah ST, Wang Z, Sumanth JJ. The duty to improve oneself: How duty orientation mediates the relationship between ethical leadership and followers' feedback-seeking and feedback-avoiding behavior. *Journal of Business Ethics.* 2020 Sep;165:615–631.

Mwesigwa R, Tusiime I, Ssekiziyivu B. Leadership styles, job satisfaction and organizational commitment among academic staff in public universities. *Journal of Management Development.* 2020 Feb 27;39(2):253–268.

Pradhan S, Jena LK. Does meaningful work explains the relationship between trans-formational leadership and innovative work behaviour? *Vikalpa.* 2019 Mar;44(1):30–40.

Roman AV, Van Wart M, Wang X, Liu C, Kim S, McCarthy A. Defining e-leadership as competence in ICT-mediated communications: an exploratory assessment. *Public Administration Review.* 2019 Nov;79(6):853–866.

Rukh L, Shahrukh HM, Iqbal KZ. Effect of authentic leadership on organization commitment: mediating role of job satisfaction. *Journal of Entrepreneurship & Organization Management*. 2018;7(3):247–258.

Salas-Vallina A, Simone C, Fernández-Guerrero R. The human side of leadership: Inspirational leadership effects on follower characteristics and happiness at work (HAW). *Journal of Business Research*. 2020 Feb 1;107:162–171.

Schaubroeck JM, Shen Y, Chong S. A dual-stage moderated mediation model linking authoritarian leadership to follower outcomes. *Journal of Applied Psychology*. 2017 Feb;102(2):203.

Young HR, Glerum DR, Joseph DL, McCord MA. A meta-analysis of transactional leadership and follower performance: Double-edged effects of LMX and empowerment. *Journal of Management*. 2021 May;47(5):1255–1280.

Index